Bernhard Ganter   Gerd Stumme
Rudolf Wille (Eds.)

Formal
Concept Analysis

Foundations and Applications

 Springer

Series Editors

Jaime G. Carbonell, Carnegie Mellon University, Pittsburgh, PA, USA
Jörg Siekmann, University of Saarland, Saarbrücken, Germany

Volume Editors

Bernhard Ganter
TU Dresden, Institut für Algebra
01062 Dresden, Germany
E-mail: ganter@math.tu-dresden.de

Gerd Stumme
Universität Kassel, Fachbereich Mathematik und Informatik
Wilhelmshöher Allee 73, 34121 Kassel, Germany
E-mail: stumme@cs.uni-kassel.de

Rudolf Wille
Technische Universität Darmstadt, Fachbereich Mathematik
Schloßgartenstr. 7, 64289 Darmstadt
E-mail: wille@mathematik.tu-darmstadt.de

Library of Congress Control Number: 2005929194

CR Subject Classification (1998): I.2, G.2.1-2, F.4.1-2, D.2.4, H.3

ISSN   0302-9743
ISBN-10   3-540-27891-5 Springer Berlin Heidelberg New York
ISBN-13   978-3-540-27891-7 Springer Berlin Heidelberg New York

Springer is a part of Springer Science+Business Media

springeronline.com

© Springer-Verlag Berlin Heidelberg 2005
Printed in Germany

Typesetting: Camera-ready by author, data conversion by Olgun Computergrafik
Printed on acid-free paper   SPIN: 11528784   06/3142   5 4 3 2 1 0

Lecture Notes in Artificial Intelligence 3626

Edited by J. G. Carbonell and J. Siekmann

Subseries of Lecture Notes in Computer Science

Preface

Formal Concept Analysis has been developed as a field of applied mathematics based on a mathematization of concept and concept hierarchy. It thereby allows us to mathematically represent, analyze and construct conceptual structures. That has been proven useful in a wide range of application areas such as medicine and psychology, sociology and linguistics, archaeology and anthropology, biology and chemistry, civil and electrical engineering, information and library sciences, information technology and software engineering, computer science and even mathematics itself.

More than 25 years of research have built up a rich mathematical theory and many application methods and procedures which are presented in more than 500 scientific publications. The basics of the mathematical theory were represented in the monograph "Formale Begriffsanalyse: Mathematische Grundlagen" (Springer 1996) and its English translation "Formal Concept Analysis: Mathematical Foundations" (Springer, 1999). Applications with the focus on conceptual knowledge processing are broadly discussed in the volumes "Begriffliche Wissensverarbeitung: Grundfragen und Aufgaben" (B.I.-Wissenschaftsverlag 1994) and "Begriffliche Wissenverarbeitung: Methoden und Anwendungen" (Springer, 2000). Applications of Formal Concept Analysis in text retrieval and mining were recently published by C. Carpineto and G. Romano in their book "Concept Data Analysis: Theory and Applications" (Wiley 2004). From the manifold developments of software for formal concept analysis applications we only mention the open source project ToscanaJ which is creating a large, flexible framework for conceptual knowledge processing and is documented in http://www.tockit.org and http://sourceforg.net/project/toscanaj.

A new field of research needs scientific communication and discourse which is stimulated best by scientific conferences. For the successful development of formal concept analysis such conferences have been above all the annual conferences of the German Classification Society in the 1980s and early 1990s, and since 1995 the International Conferences on Conceptual Structures (Springer LNAI 954, 1115, 1257, 1453, 1640, 1867, 2120, 2393, 2746, 3127). Since 2003, an International Conference on Formal Concept Analysis has been taking place every year: 2003 in Darmstadt, Germany, 2004 in Sydney, Australia (Springer LNAI 2961), 2005 Lens, France (Springer LNAI 3403) and 2006 in Dresden, Germany. Furthermore, in 2005, there will be already the 3rd International Workshop on Concept Lattices and Applications in the Czech Republic.

This volume is the outcome of a project inspired by the 1st International Conference on Formal Concept Analysis in Darmstadt. The idea was to use the expertise of the participating experts to elaborate a comprehensive presentation of the state of the art of formal concept analysis and its applications. Of course, it is clear that such a presentation could not completely cover all current developments in detail. Therefore the goal of this volume is rather to convey essential information which gives readers an orientation and enough knowledge to use formal concept analysis for projects of interest. In any case, this volume should inspire further research and applications, even in directions completely different from the represented content.

The first part of this volume treats foundational themes of formal concept analysis. (1) R. Wille in his contribution shows the surprisingly rich correspondences between the multifarious aspects of concepts in the human mind and the structural properties and relationships of formal concepts in formal concept analysis. These correspondences make it understandable that – via formal concept analysis – mathematical thought may aggregate with other ways of thinking and thereby support human thought and action. (2) B. Vormbrock and R. Wille generalize in their paper from the Basic Theorem on Concept Lattices to basic theorems on algebras of semiconcepts and protoconcepts, extending the usefulness of the basic theorem on concept lattices to conceptual structures with negating operations. (3) T. Becker contributes with his paper to algebraic concept analysis by examining connections between formal concept analysis and algebraic geometry. He elaborates a theory of algebraically represented concept lattices based on notions such as algebraic varieties, coordinate algebras, and polynomial morphisms. (4) F. Dau and J. Klinger show in their contribution how formal concept analysis has been extended to "Contextual Logic," a mathematization of the traditional philosophical logic with its doctrines of concepts, judgments, and conclusions. The basic idea of this extension is to mathematize concepts by formal concepts and judgments by concept graphs whose nodes and edges are formal concepts of suitable formal contexts. (5) B. Ganter extends in his paper the known attribute logic of formal contexts to a contextual attribute logic of many-valued attributes. This allows us, in particular, to generalize the well-known attribute exploration to an attribute exploration with background knowledge. (6) P. Burmeister and R. Holzer give a survey of what has been done so far in treating incomplete knowledge using methods of formal concept analysis. In particular, they compare different algorithms for attribute explorations based on incomplete knowledge. (7) K.E. Wolff reports on a temporal concept analysis which he develops as a temporal conceptual granularity theory for movements of general objects in abstract or "real" space and time such that the notions of states, transitions, and life tracks can be defined mathematically. Basic relations to theoretical physics, mathematical system theory, automata theory, and temporal logic are discussed.

The contributions of the second part demonstrate how formal concept analysis might be applied outside mathematics. (8) U. Priss discusses in her article linguistic applications of formal concept analysis: the identification and analysis of linguistic features, the support of the automated or semi-automated construction of lexical databases for corpora, and the representation and analysis of hierarchies and classifications in lexical databases. (9) C. Carpineto and G. Romano focus in their paper on the features of formal concept analysis used to build contextual information retrieval applications as well as on its most critical aspects. The development of a formal concept analysis procedure for mining

Web results, returned by a major search engine, is envisaged as the next big challenge. (10) L. Lakhal and G. Stumme give a survey on association rule mining based on formal concept analysis. Basic ideas of applying formal concept analysis are explained by using the notion of an "iceberg concept lattice" and the specific algorithm TITANIC. (11) S.O. Kuznetsov offers a retrospective survey of the application of Galois connections in data analysis elaborated at the All-Soviet (now All-Russia) Institute for Scientific and Technical Information since 1970. He shows the connections with formal concept analysis, in particular, for the JSM method of inductive plausible reasoning. (12) R. Wille explains in his contribution how conceptual knowledge processing (based on formal concept analysis) enables effects in economic practice. This explanation is guided by the key processes of organizational knowledge management: knowledge identification, knowledge acquisition, knowledge development, knowledge distribution and sharing, knowledge usage and knowledge preservation.

The third part is concerned with applications of formal concept analysis in software engineering, including also software development for formal concept analysis. (13) T. Tilley, R. Cole, P. Becker, and P. Eklund offer a survey on formal concept analysis support for software engineering activities. This survey is based on academic papers that report the application of formal concept analysis to software engineering. The papers are classified using a framework based on the activities defined in the ISO 12207 Software Engineering standard. (14) G. Snelting gives an overview that summarizes important papers on applications of concept lattices in software analysis. He presents three methods in some detail: methods to extract classes and modules, to re-factor class hierarchies, and to infer dynamic dominators and control flow regions from program traces. (15) W. Hesse and T. Tilley focus on the use of formal concept analysis during the early phase of software development, in particular in object-oriented modelling. As a typical application, the task of finding or deriving class candidates from a given use description is considered in more detail. (16) R. Godin and P. Valtchev present an overview of work on formal concept analysis-based class hierarchy design in object-oriented software development. In particular, they discuss how to derive a concept lattice from a given class hierarchy and from the class methods and associations; and how to then turn the lattice into an improved class hierarchy. (17) P. Becker and J. Hereth Correia explain in their paper the features of the TOSCANAJ tool suite and their use in implementing conceptual information systems. TOSCANAJ as an open source project (embedded into the larger Tockit project) is offered as a starting point for creating a common base for software development for formal concept analysis.

For the basics of formal concept analysis the reader is referred to the monograph "Formal Concept Analysis: Mathematical Foundations" (Springer, 1999). The elementary definitions of a formal context and its concept lattice up to the notions used in the Basic Theorem on Concept Lattices are also presented at the beginning of the second section of the first paper in this volume.

Finally, we would like to thank all those who supported and contributed to this volume. In particular, we would like to thank all authors for their substantial contributions. Thanks also to the Deutsche Forschungsgemeinschaft for its financial support which allowed us to realize the 1st International Conference on Formal Concept Analysis.

Darmstadt, May 2005   Bernhard Ganter, Gerd Stumme, Rudolf Wille

Table of Contents

Software Engineering

Formal Concept Analysis
as Mathematical Theory
of Concepts and Concept Hierarchies

Rudolf Wille

Technische Universität Darmstadt, Fachbereich Mathematik
Schloßgartenstr. 7, D–64289 Darmstadt
wille@mathematik.tu-darmstadt.de

Abstract. *Formal Concept Analysis* has been originally developed as
a subfield of Applied Mathematics based on the mathematization of
concept and *concept hierarchy*. Only after more than a decade of development, the connections to the philosophical logic of human thought
became clearer and even later the connections to Piaget's cognitive struc-
turalism which Thomas Bernhard Seiler convincingly elaborated to a
comprehensive *theory of concepts* in his recent book [Se01]. It is the main
concern of this paper to show the surprisingly rich correspondences between Seiler's multifarious aspects of concepts in the human mind and
the structural properties and relationships of formal concepts in Formal
Concept Analysis. These correspondences make understandable, what
has been experienced in a great multitude of applications, that Formal
Concept Analysis may function in the sense of *transdisciplinary mathematics*, i.e., it allows mathematical thought to aggregate with other ways
of thinking and thereby to support human thought and action.

1 Formal Concept Analysis, Mathematics, and Logic

Formal Concept Analysis had its origin in activities of restructuring mathematics, in particular mathematical order and lattice theory. In the initial paper
[Wi82], restructuring lattice theory is explained as "an attempt to reinvigorate
connections with our general culture by interpreting the theory as concretely as
possible, and in this way to promote better communication between lattice theo-
rists and potential users of lattice theory." Since then, Formal Concept Analysis
has been developed as a subfield of *Applied Mathematics* based on the mathematization of concepts and concept hierarchies.

Only after more than a decade of development, the connections to *Philosophical Logics* of human thought became clearer, mainly through Charles Sanders
Peirce's late philosophy. Even our general understanding of mathematics did improve as pointed out in the recent paper "Kommunikative Rationalität, Logik
und Mathematik" ("Communicative Rationality, Logic, and Mathematics")
[Wi02b]. The concern of that paper is to explain and to substantiate the following thesis:

B. Ganter et al. (Eds.): Formal Concept Analysis, LNAI 3626, pp. 1–33, 2005.

> *The aim and meaning of mathematics finally lie in the fact that mathematics is able to effectively support the rational communication of humans.*

Here we only recall the key arguments founding this thesis: First, logical thinking as expression of human reason graps actual realities by the main forms of human thought: concepts, judgments, and conclusions (cf. [Ka88], p.6). Second, mathematical thinking abstracts logical human thinking for developing a cosmos of forms of potential realities (see [Pe92], p.121). Therefore, mathematics as a historically, socially and culturally detemined formation of mathematical thinking, respectively, is able to support humans in their logical thinking and hence in their rational communication. Since concepts are also prerequisites for the formation of judgments and conclusions, we can adapt the above thesis to Formal Concept Analysis as follows:

> *The aim and meaning of Formal Concept Analysis as mathematical theory of concepts and concept hierarchies is to support the rational communication of humans by mathematically developing appropriate conceptual structures which can be logically activated.*

2   Concepts and Formal Concepts

Concepts can be philosophically understood as the basic units of thought formed in dynamic processes within social and cultural environments. According to the main philosophical tradition, a concept is constituted by its *extension*, comprising all objects which belong to the concept, and its *intension*, including all attributes (properties, meanings) which apply to all objects of the extension (cf. [Wi95]). Concepts can only live in relationships with many other concepts where the *subconcept-superconcept-relation* plays a prominent role. Being a subconcept of a superconcept means that the extension of the subconcept is contained in the extension of the superconcept which is equivalent to the relationship that the intension of the subconcept contains the intension of the superconcept (cf. [Wa73], p.201).

For a mathematical theory of concepts and concept hierarchies, we obviously need a mathematical model that allows to speak mathematically about *objects*, *attributes*, and relationships which indicate that an object *has* an attribute. Such a model was introduced in [Wi82] by the notion of a "formal context" which turned out to be basic for a new area of applied mathematics: *Formal Concept Analysis*. A *formal context* is defined as a set structure $\mathbb{K} := (G, M, I)$ for which G and M are sets while I is a binary relation between G and M, i.e. $I \subseteq G \times M$; the elements of G and M are called *(formal) objects* (in German: *Gegenstände*) and *(formal) attributes* (in German: *Merkmale*), respectively, and gIm, i.e. $(g, m) \in I$, is read: the object g *has* the attribute m.

For defining the formal concepts of the formal context (G, M, I), we need the following *derivation operators* defined for arbitrary $X \subseteq G$ and $Y \subseteq M$ as follows:

$$X \mapsto X^I := \{m \in M \mid gIm \text{ for all } g \in X\},$$
$$Y \mapsto Y^I := \{g \in G \mid gIm \text{ for all } m \in Y\}.$$

The two derivation operators satisfy the following three conditions:

$$(1)\ Z_1 \subseteq Z_2 \Longrightarrow Z_1^I \supseteq Z_2^I,\ (2)\ Z \subseteq Z^{II},\ (3)\ Z^{III} = Z^I.$$

A *formal concept* of a formal context $\mathbb{K} := (G, M, I)$ is defined as a pair (A, B) with $A \subseteq G$, $B \subseteq M$, $A = B^I$, and $B = A^I$; A and B are called the *extent* and the *intent* of the formal concept (A, B), respectively. The *subconcept-superconcept-relation* is mathematized by

$$(A_1, B_1) \le (A_2, B_2) :\Longleftrightarrow A_1 \subseteq A_2\ (\Longleftrightarrow B_1 \supseteq B_2).$$

The set of all formal concepts of $\mathbb{K}$ together with the defined order relation is denoted by $\mathfrak{B}(\mathbb{K})$.

A general method of constructing formal concepts uses the derivation operators to obtain, for $X \subseteq G$ and $Y \subseteq M$, the formal concepts (X^{II}, X^I) and (Y^I, Y^{II}). For an object $g \in G$, its *object concept* $\gamma g := (\{g\}^{II}, \{g\}^I)$ is the smallest concept in $\mathfrak{B}(\mathbb{K})$ whose extent contains g and, for an attribute $m \in M$, its *attribute concept* $\mu m := (\{m\}^I, \{m\}^{II})$ is the greatest concept in $\mathfrak{B}(\mathbb{K})$ whose intent contains m. The specific structure of the ordered sets $\mathfrak{B}(\mathbb{K})$ of formal contexts $\mathbb{K}$ is clarified by the following theorem:

Basic Theorem on Concept Lattices. [Wi82] *Let* $\mathbb{K} := (G, M, I)$ *be a formal context. Then* $\mathfrak{B}(\mathbb{K})$ *is a complete lattice, called the* concept lattice of (G, M, I), *for which infimum and supremum can be described as follows:*

$$\bigwedge_{t \in T}(A_t, B_t) = (\bigcap_{t \in T} A_t, (\bigcup_{t \in T} B_t)^{II}),$$

$$\bigvee_{t \in T}(A_t, B_t) = ((\bigcup_{t \in T} A_t)^{II}, \bigcap_{t \in T} B_t).$$

In general, a complete lattice L *is isomorphic to* $\mathfrak{B}(\mathbb{K})$ *if and only if there exist mappings* $\tilde{\gamma} : G \longrightarrow L$ *and* $\tilde{\mu} : M \longrightarrow L$ *such that* $\tilde{\gamma}G$ *is* $\bigvee$*-dense in* L *(i.e.* $L = \{\bigvee X \mid X \subseteq \tilde{\gamma}G\}$), $\tilde{\mu}M$ *is* $\bigwedge$*-dense in* L *(i.e.* $L = \{\bigwedge X \mid X \subseteq \tilde{\mu}M\}$), *and* $gIm \Longleftrightarrow \tilde{\gamma}g \le \tilde{\mu}m$ *for* $g \in G$ *and* $m \in M$; *in particular,* $L \cong \mathfrak{B}(L, L, \le)$ *and furthermore:* $L \cong \mathfrak{B}(J(L), M(L), \le)$ *if the set* $J(V)$ *of all* $\bigvee$*-irreducible elements is* $\bigvee$*-dense in* L *and the set of all* $\bigwedge$*-irreducible elements is* $\bigwedge$*-dense in* L.

A formal context is best understood if it is depicted by a *cross table* as for example the formal context about bodies of waters in Fig. 1. A concept lattice is best pictured by a *labelled line diagram* as the concept lattice of our example context in Fig. 2 (see the book cover of [GW99a]). In such a diagram the name of each object g is attached to its represented object concept γg and the name of each attribute m is attached to its represented attribute concept μm. By the Basic Theorem, this labelling allows to read the extents, the intents, and the

	natural	artificial	stagnant	running	inland	maritime	constant	temporary
tarn	X		X		X		X	
trickle	X			X	X		X	
rill	X			X	X		X	
beck	X			X	X		X	
rivulet	X			X	X		X	
runnel	X			X	X		X	
brook	X			X	X		X	
burn	X			X	X		X	
stream	X			X	X		X	
torrent	X			X	X		X	
river	X			X	X		X	
channel				X	X		X	
canal		X		X	X		X	
lagoon	X		X			X	X	
lake	X		X		X		X	
mere		X	X		X		X	
plash	X		X		X			X
pond		X	X		X		X	
pool	X		X		X		X	
puddle	X		X		X			X
reservoir		X	X		X		X	
sea	X		X			X	X	

Fig. 1. Formal context partly representing the lexical field "bodies of waters"

underlying formal context from the diagram. Speaking in human logical terms, by the Basic Theorem, each concept is represented by a little circle so that its extension (intension) consists of all the objects (attributes) whose names can be reached by a descending (ascending) path from that circle. In Fig. 2, for instance, the circle vertically above the circle with the label "artificial" represents the formal concept with the extent $\{tarn, lake, pool, sea, lagoon\}$ and the intent $\{natural, stagnant, constant\}$. Furthermore, even all *attribute implications*

$$A \to B : \Longleftrightarrow A^I \subseteq B^I \text{ with } A, B \subseteq M$$

can be read from a labelled line diagram; Fig. 2, for instance, shows the attribute implication $\{artificial\} \to \{inland, constant\}$ because there are ascending paths from the circle with the label "artificial" to the circles with the labels "inland" and "constant", respectively. In the case of $M^I = \emptyset$, an implication $A \to M$ is equivalent to $A^I = \emptyset$ wherefore A is then said to be *incompatible*.

The aim of Section 2 is to give an answer to the following basic question: *How adequate is the mathematization of concepts and concept hierarchies used in Formal Concept Analysis?* For answering this question, we have to refer to a comprehensive convincing *theory of concepts*. Such a theory is presented in the book "Begreifen und Verstehen. Ein Buch über Begriffe und Bedeutungen"

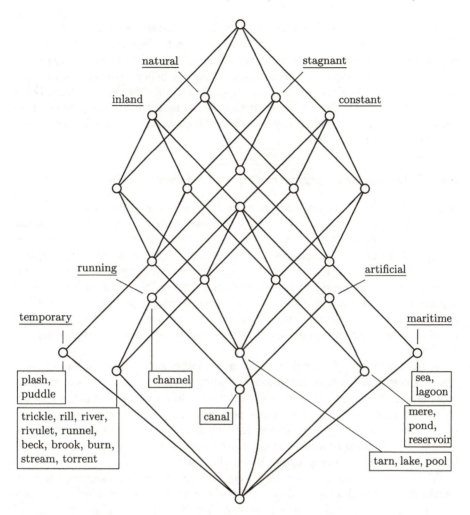

Fig. 2. Concept lattice of the formal context in Fig. 1

("Conceiving and Understanding. A Book on Concepts and Meanings") written by *Thomas Bernhard Seiler* [Se01][1]. In his book, Seiler discusses a great variety of concept theories in philosophy and psychology and concludes with his own theory which extends the concept understanding of Piaget's structure-genetic approach. In his theory, Seiler describes concepts as *cognitive structures* whose development in human mind is constructive and adaptive. Seiler elaborates his approach in twelve aspects which are briefly described in the following twelve subsections and used to review the adequacy of the mathematizations of Formal

---

[1] It might be desirable to integrate further concept theories in our discussion, but that would exceed the scope of this paper. The connections to those theories may be analyzed later

Concept Analysis. In each subsection, the first paragraph concisely summarizes Seiler's understanding of the corresponding aspect; then related notions and relationships from Formal Concept Analysis are discussed and partly concretized by at least one example. The connections between Seiler's concept theory and Formal Concept Analysis which come apparent in this way are far from being exhaustive. But they already show an astonishing multitude of correspondencies between both theories which may be taken as arguments for the adequacy of the discussed mathematizations.

2.1   Concepts Are Cognitive Acts and Knowledge Units

According to [Se01], concepts are *cognitive acts* and *knowledge units* potentially independent of language. Only if they are used to give meaning to linguistic expressions, they become so-called *word concepts* which are conventualized and incorporated. The meanings of words for an individuum presuppose conceptual knowledge of that individuum which turns linguistic expressions into signs for those concepts. *Personal concepts*[2] exceed conventional meaning with additional aspects and connotations. *Conventional concepts and meanings* are objectified and standardized contents, evolved in recurrently performed discourses. The problem arises how to explain under which conditions which knowledge aspects are actualized.

Formal concepts of formal contexts may mathematize personal and conventional concepts as units of extension and intension independent of specific concept names. They are representable in labelled line diagrams which stimulate individual *cognition acts* of creating personal and conventional concepts and knowledge. Computer programs for drawing labelled line diagrams (like ANACONDA [Vo96]) allow to indicate represented *word concepts* by attaching concept labels to the corresponding circles in the diagrams.

Mathematizations of *conventional concepts* are given, for example, through formal contexts of lexical fields in which the conventional meaning of the corresponding words are determined by so-called "noemas" (smallest elements of meaning). The formal concepts depicted in the labelled line diagram of Fig. 2 are mathematizing conventional concepts; they are derived from the formal context in Fig. 1 which originates from a mathematization of lexical fields of bodies of waters performed in [KW87].

An example based on *personal concepts* and their interrelationships is presented in Fig. 3. Its data are taken from psychology research about the development of economic concepts by young persons [Cl90]. The example reports on the outcome of interviews about *price differences* between various articles of commerce. Reasons for those differences were classified by the five characteristics "size, beauty", "use", "rarity", "production costs", and "supply/demand". The personal understanding of price differences of the 48 test persons (16 persons of age 10-11, 15, and 18-19, respectively) is represented in the line diagram of Fig. 3 by the 14 object concepts; the formal concepts with the intents

---

[2] In [Se01] personal concepts are named "idiosyncratic concepts"

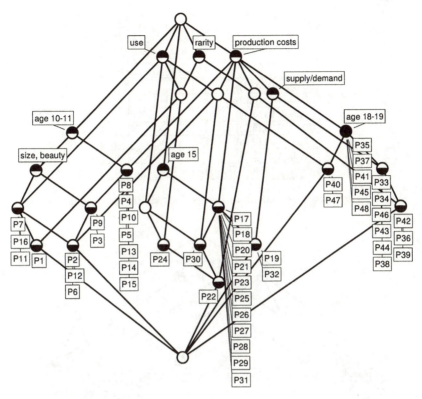

Fig. 3. Concept lattice about economic concepts of young persons

{use,rarity},{rarity, production costs}, and {supply/demand} combine two object concepts with the same characteristics from different age groups, respectively, and {use, production costs} combines even object concepts with the same characteristics from all three age groups. In particular, the development of the personal understanding over the ages becomes transparent by the labelled line diagram. Here we only mention the change from the specific characteristic "size, beauty" in the age group 10-11 to the dominance of the characteristics "production costs" and "supply/demand" in the age group 18-19. This indicates the plausible development towards the conventional meaning of the concept "money".

Concerning the mentioned problem of actualizing knowledge, labelled line diagrams as representations of concept lattices support the *actualization of knowledge aspects*. Especially, the understanding of the concepts represented by the little circles unfolds more and more when the connections of the relevant object and attribute labels with those circles are mentally established.

2.2   Concepts Are Not Categories, but Subjective Theories

According to [Se01], concepts are primarily cognitive structures and therefore elements and subsystems of our understanding and knowledge. As *naive and*

	acute	equiangular	equilateral	isosceles	oblique	obtuse	right-angled	scalene
((0,0),(1,0),(0,1))				X			X	
((0,0),(1,0),(0,2))							X	X
((0,0),(2,0),(3,1))					X	X		X
((0,0),(2,1),(4,0))				X	X	X		
((0,0),(1,2),(2,0))	X				X	X		
((0,0),(1,2),(3,0))	X						X	X
((0,0),(1,root(3)),(2,0))	X	X	X	X	X	X		

Fig. 4. Formal context with lexical attributes for triangles

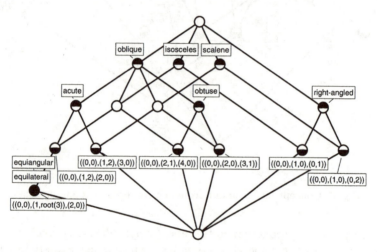

Fig. 5. Concept lattice of the formal context in Fig. 4

subjective theories, concepts contain implicit and explicit assumptions about objects and events, their conditions and causes, their characteristics, relations and functions; they are of an *abstract and idealizing nature*. They are theories which the subject creates and uses to reconstruct and to represent objects, segments, events of the surrounding world. The example in Fig. 3 indicates that young children start with creating subjective theories which slowly adapt intersubjective views and quite lately reach full conventional understanding. For conceptual subjective theories see also the example in Fig. 6 and Fig. 7.

The formal concepts of a formal context live in a *hierarchical network* of a multitude of further formal concepts. They are substructured internally by a network of subconcepts and externally in multi-relationship to further formal concepts. Thus, formal concepts are not only pairs of sets, they are part of a contextual representation of a formal theory which can be linked by inscriptions to *subjective and intersubjective theories* of human beings. As mathematical entities, formal concepts are abstract and of an idealizing nature.

In order to demonstrate a contextual mathematization of an intersubjective theory, we present a *formal theory of the lexical word concepts of triangles*. This theory is based on the formal context in Fig. 4 and conceptually unfolded in the corresponding concept lattice pictured in Fig. 5. The top element of the concept lattice represents the conventional concept of a (plane) triangle. The represented substructure of the general triangle concept is determined by the lexical attributes for triangles: *equilateral, equiangular, scalene, isosceles, oblique, acute, obtuse, right-angled*. Those attributes give rise to exactly seven object concepts for which only one generating triangle is made explicit, respectively (implicitly, there are obviously infinitely many triangles generating each of the seven object concepts). The concept lattice shows that the lexical word concepts of triangles form a *simple-implicational theory* in the sense of [Wi04] which is determined by the implications *equilateral* $\leftrightarrow$ *equiangular*, *equilateral* $\rightarrow$ *isosceles*, *equilateral* $\rightarrow$ *acute*, *acute* $\rightarrow$ *oblique*, *obtuse* $\rightarrow$ *oblique* and the incompatible subsets $\{acute, obtuse\}$, $\{acute, right - angled\}$, $\{obtuse, right - angled\}$, $\{equilateral, scalene\}$, $\{isosceles, scalene\}$, $\{oblique, right - angled\}$. Besides the seven attribute concepts, there are exactly eight consistent word concepts which can be named by combining two of the lexical attributes for triangles, for instance: *scalene obtuse triangle*. It is not surprising that the logic of the lexical word concepts of triangles is determined by implications with one-element-premise and incompatibilities; seemingly, our everyday thinking has *intersubjectively incorporated* the predominant use of logical implications with one-element premise (cf. [Wi04]).

2.3 Concepts Are Not Generally Interlinked in the Sense of Formal Logic

According to [Se01], a one-sided priority of aspects of formal logic leads to view concepts through the conventional perspective and to disregard the primarily *personal nature of concepts*. Conceptual thinking is situation and domain dependent. Personal concepts are not structured in the sense of formal logic, but they are cognitive structures which tend to amalgamate to closed and integrated systems. Although concepts are not of a formal-logic nature, they may form a *basis for logical thinking*.

Formal concepts are *mathematical entities* and not formal-logic constructs; they live in the extremely rich realm of mathematics (that allows applications as in the example of Subsection 2.6). Formal concepts are context dependent and mathematically structured in concept lattices which even tend towards more elaborated integrated systems. Formal concepts (and concept lattices) especially form the basis of *Contextual Logic* [Wi00a], a mathematization of the traditional philosophical logic based on "the three essential main functions of thinking – *concepts, judgments*, and *conclusions*" ([Ka88], p.6).

Formal concepts which mathematize *personal concepts* are derived from formal contexts which represent personal views. Fig. 6 yields an example of such a personal context which is an outcome of a *Repertory Grid* examination of an anorectic patient (see [SW93]). Since such a patient is understood to suffer from

	rational	emotional	honest	dishonest	optimistic	pessimistic	interested	uninterested	flexible	inflexible	materialistic	idealistic	not fashionable	fashionable	fond of life	depressive	purposeful	unsteady	unconstrained	constrained
Myself	X	X			X		X	X	X			X	X	X			X		X		X
Ideal	X	X		X	X		X					X	X	X		X		X		X	
Father	X	X		X	X		X					X	X	X		X		X		X	
Mother		X	X			X		X	X		X			X		X		X	X	
Sister		X	X		X		X					X	X		X			X	X	
Brother	X		X		X		X		X			X	X		X		X		X	
Otto	X		X			X	X			X	X			X	X			X		X
Anne		X	X			X		X	X		X	X			X	X		X		X
Eva	X		X			X		X	X		X	X			X		X		X	
Elke		X		X	X		X			X			X				X		X		X
Ina		X	X		X		X		X				X	X		X		X	

Fig. 6. Formal context of a repertory grid examination of an anorectic patient

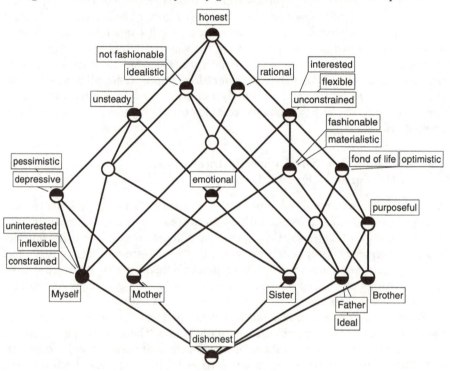

Fig. 7. Concept lattice of the first six rows of the formal context in Fig. 6

a loss of control over interpersonal relationships, the examination is performed to make disturbed relationships conscious so that she becomes able to understand those disturbances and to overcome her mental disorder. In the repertory grid examination, the patient is asked to make judgments, according to self chosen bipolar attributes, about persons (including her ideal) which she views to be important in her daily life. In Fig. 6, the bipolar attributes "rational"-"emotional", "honest"-"dishonest", etc. are represented by adjacent columns. The crosses in such pairs of adjacent columns indicate how the patient views the listed persons. To evaluate such *grids of anorectic patients*, N. Spangenberg and K. E. Wolff have successfully applied *Formal Concept Analysis* to reach meaningful line diagrams of concept lattices, as in Fig. 7 for our example. Such diagrams shall make the contents and relations of the personal views transparent and discussable for the patient and the therapist [SW88]. The diagram in Fig. 7 discloses that the patient idealizes her father and identifies herself with her mother by negative judgments; this might indicate a conflict between mother and father for which the patient takes over the responsiblility unconsciously. Making this view transparent on the basis of the line diagram may help the patient to discard the wrong responsibility (for an extensive discussion of our example see [Sp90]).

2.4   Concepts Are Domain Specific and Often Prototypical

According to [Se01], *personal concepts* originate out of internalized and transformed systems of actions and serve in particular as references to experienced situations, objects and characteristics. They are based on *implicit conceptions and experiences* which often limit strongly their conscious range and validity. The implicit core of concepts also explains the *prototypical effects* in identifying concepts.

Formal concepts result from applications of derivation operators in formal contexts which represent *domains of interest*. Therefore formal concepts are limited in their range and validity, but they might activate implicit conceptions and experiences concerning the underlying domains (as, for instance, reported in the example of Subsection 2.8). The prototypical view is reflected in the notion of *protoconcepts* [Wi00b].

As an example, we consider the formal context in Fig. 8 which is taken from the DK Eyewitness Travel Guide New Zealand [DK01]. The quite restricted context informs about leisure activities offered in the regions of Otago and Southland of New Zealand. The corresponding concept lattice in Fig. 9 shows that the eight types of leisure activities give rise to only four object concepts, i.e., the tourist locations are partitioned by the leisure activities into four classes. These classes can be characterized by four formal attributes as follows: "Observing Nature, but not Sightseeing Flights", "Sightseeing Flights, but not Jet Boating", "Jet Boating, but not Wildwater Rafting", and "Wildwater Rafting, but not Observing Nature". As prototypical for those four classes we could regard the formal objects "Oamaru", "Dunedin", "Te Anau", and "Queenstown", respectively. Consequently, the designated subcontext formed by those four formal objects and attributes has a concept lattice isomorphic to the concept lattice of the whole formal context. The prototypical nature of formal concepts of such subcontexts is mathematically grasped by the definition of a *protoconcept* of a formal context (G, M, I), that is a pair (A, B) with $A \subseteq G$ and $B \subseteq M$ satisfying the condition $A^{II} = B^{I}$ ($\iff$ $A^{I} = B^{II}$). In our example, all formal concepts of the designated subcontext are protoconcepts of the whole context.

	Hiking	Observing Nature	Sightseeing Flights	Jet Boating	Wildwater Rafting	Bungee Jumping	Parachute Gliding	Skiing
Stewart Island	X	X	X					
Fjordland NP	X	X	X					
Invercargill	X	X	X					
Milford Sound	X	X	X					
Mt. Aspiring NP	X	X	X					
Te Anau	X	X	X	X				
Dunedin	X	X	X					
Oamaru	X	X						
Queenstown	X		X	X	X	X	X	X
Wanaka	X			X	X	X	X	X
Otago Peninsula	X	X						
Haast	X	X						
Catlins	X	X						

Fig. 8. Formal context about leisure activities in Otago and Southland/NZ

Protoconcepts have been formally introduced in [Wi00b] for the development of a Boolean Concept Logic with "negation" and "opposition" as unary operations (see also [VW03]).

2.5 Concepts as Knowledge Units Refer to Reality

According to [Se01], concepts are adapted to circumstances and facts of the world arround us, but do not copy realities. The *reference of concepts to reality* is based on the cognitive content of concepts which results out of acting and perceiving confrontation with realities and is ensured by ongoing accommodations. Concepts consider things and events out of a *specific perspective* and reconstruct only those aspects and relations which follow from the specific view. To recognize something conceptually means to capture an object, an event, a characteristic in a net of previously formed *categories of experiences* and, simultaneously, to extend and to differentiate this net. These constructions are coupled with *linguistic signs* which support a constant social exchange.

Formal concepts and concept lattices are *mathematical abstractions* of concepts and concept hierarchies of human thought and may therefore be adapted contextually in mathematical terms to circumstances and facts of the world arround us. Those abstractions benefit from the rich stock of mathematical structures available in mathematics, but also extend this stock. The *inscriptions* support the discourses about the adequateness of those structures (see e.g. Subsection 2.8).

Realities are often coded in data tables which can be mathematized by *many-valued contexts* (G, M, W, I) which are set structures consisting of three sets G, M, and W, and a ternary relation $I \subseteq G \times M \times W$ such that $(g, m, w_1), (g, m, w_2) \in I$ always implies $w_1 = w_2$; the elements of G, M, and W are called *objects*,

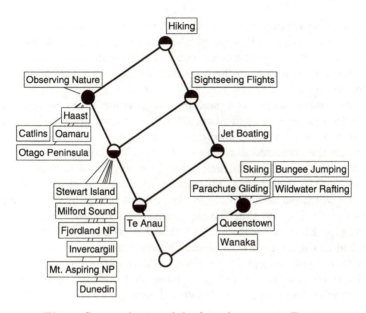

Fig. 9. Concept lattice of the formal context in Fig. 8

attributes, and *attribute values*, respectively, and $(g, m, w) \in I$ is read: the object g has the value w for the attribute m. Each $m \in M_m$ may be understood as a partial map from G into W with $m(g) = w :\Leftrightarrow (g, m, w) \in I$. To obtain formal concepts from a many-valued context (G, M, W, I), Formal Concept Analysis offers the method of *conceptual scaling* which assigns a formal context (G_m, M_m, I_m) with $m(G) \subseteq G_m$, named a *conceptual scale*, to each (many-valued) attribute $m \in M$. Such a scale views the attribute values in $m(G)$ out of a specific *purpose-oriented perspective*. In most applications, a formal context $(G, \bigcup_{m \in M} \{m\} \times M_m, J)$ is derived from the many-valued context (G, M, W, I) by the conceptual scales (G_m, M_m, I_m) $(m \in M)$ where the relation J is defined by $gJ(m, n) :\Leftrightarrow m(g) = n$ [GW89].

How challenging the confrontation with realities might be in mathematically modelling real world circumstances and facts shall be illustrated by an example from medicine. Around 1990, the Darmstadt Research Group on Formal Concept Analysis received a request to analyse *data about diabetes in children*. The original data table did contain for 111 children and 22 (many-valued) attributes as attribute values numbers from 1 to 4 which denoted segments of (not listed) measurement values. Our first analysis had as its outcome predominantly Boolean concept lattices which indicated that the represented data are highly independent. Since we could not believe that there are no essential dependencies between the attributes and attribute values in the case of diabetes, we tried to find out what was wrong with the data. We learned that the segmentation of the values of an attribute followed the principle of forming segments with equally many children to guarantee a reliable statistics. Clearly, those segments scarcely stand a chance to be medically meaningful. As a reaction to our critique, we got the original measurement values in the form of real numbers. To analyse such numerical data, one has to link the numbers to an adequate medical understanding. As non-experts for diabetes we asked a Darmstadt expert for diabetes to segment the numbers so that the segments are medically meaningful. Surprisingly, the expert was also not able to make such segmentations because each medical laboratory has its own measurement standards, but he could serve us at least with an adequate medical vocabulary for such segments. Finally, we got the segmented data corrected according to the proposed vocabulary.

Based on the corrected data, we could visualize and recognize, in particular, quite a number of dependencies. For instance, the attribute "Ph-level of the blood" got the three terms "ph. normal", "ph. pathological", and "ph. dangerous" for which we made visible in the concept lattice of the corresponding conceptual scale (Fig. 10) that "having a dangerous Ph-level" implies "having a pathodological Ph-level" of the blood (instead of listing the name of each child we attached to the circle of each object concept the number of all children who generate this object concept). The concept lattice of the aggregated conceptual scales "Ph-level" and "Coma" in Fig. 11 shows that there is a serious dependency between "having had Coma" and "having dangerous Ph-level" (cf. [SVW93]).

2.6   Concepts Are Analogous Patterns of Thought

According to [Se01], concepts do not grasp realities directly, but realities are incorporated, examined, and assimilated in an analogous manner in cognitive schemata (formed by previous experiences) which are finally adapted to the incorporated realities. In this way, the analogous character of concepts evolves so that conceptual thought can be understood as *analogous representation of realities*.

Mathematics never represents forms of realities as they are, but forms of realities give rise to abstracted mathematical forms and structures multifariously. In this way mathematics becomes constantly richer and increases its ability to assist human thought. Especially, the analogous character of concepts assimilated in cognitive schemata may be seen as the main reason that *abstractions of actual concepts* to mathematical concepts (in particular, to formal concepts in formal contexts) are so successful in supporting human discourses about the represented realities (see e.g. Subsection 2.3). "Why can concept lattices support knowledge discovery in databases?" is particularly discussed in [Wi01].

Even the analysis of geometric realities and their analogous concepts "point", "line", "circle" etc. may be supported by Formal Concept Analysis which shall be demonstrated by the *inversion in a circle*, the construction of which is sketched in Fig. 12: a point outside the circle is mapped by the inversion onto the line through the two points of contact of the two tangents through the outside point, respectively; for example, p_1 is mapped to l_1 and p_2 is mapped to l_2. By the inverse construction, each line which meets the circle in two points, but does not meet the center of the circle, is mapped to the intersection point of the tangents through the two common points of line and circle; for example, l_1 is mapped to p_1 and l_2 is mapped to p_2. Points in the circle (except the center)

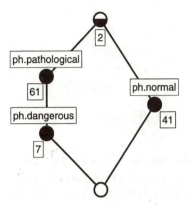

Fig. 10. Concept lattice of the Ph-level of children with diabetes

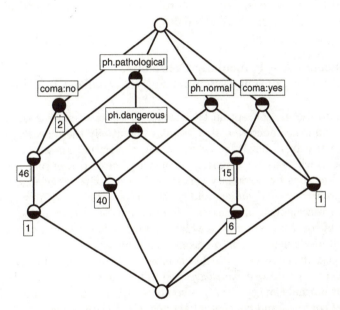

Fig. 11. Concept lattice of the Ph-level and Coma occurrence of children with diabetes

and lines outside the circle interchange by the inversion as, for example, the point q and the line m in Fig. 12, and a point on the circle interchanges with the tangent through that point. Using the set $\mathbb{R}$ of all *real numbers*, an analogous mathematical representation of the (graphic) plane by $\mathbb{R}^2$ yields a very economic conceptualization of the inversion in a circle: For the circle with radius $\sqrt{r}$, this conceptualization is based on the formal context $(\mathbb{R}^2, \mathbb{R}^2, \perp_r)$ with $(a, b) \perp_r (c, d) :\Leftrightarrow (a, b) \cdot (c, d) = r$. For each point $(u, v) \in \mathbb{R}^2$, the derivation $\{(u, v)\}^{\perp_r}$ is a line (and conversely). It follows that the derivations of the formal context $(\mathbb{R}^2, \mathbb{R}^2, \perp_r)$ represent the inversion in the circle with center $(0, 0)$ and radius

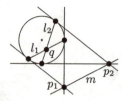

Fig. 12. Inversion in a circle

$\sqrt{r}$. Analogously, the derivations of the formal context $(\mathbb{R}^n, \mathbb{R}^n, \perp_r)$ represent the inversion in the $(n-1)$-dimensional hypersphere with center $(0, \ldots, 0)$ and radius $\sqrt{r}$, and even more general: the formal concepts of the formal context $(V, V^*, \perp_r)$ $(r \neq 0)$ where V is any finite-dimensional vector space, V^* its dual space, and $v \perp_r \varphi :\Leftrightarrow \varphi(v) = r$ represent mutually inverse antiisomorphisms between the lattices consisting of the total space V resp. V^* and of all affine subspaces not containing 0 [Wl91],[Wl99].

2.7 Concepts Are Principally Conscious, but Their Content Is Seldom Fully Actualized in Consciousness

According to [Se01], conceptual knowledge, as far as it exceeds the immediate experience and the concrete action, implies potentially the reflexive reconstruction. The conscious knowledge accompaning conceptual knowledge is *reflexive knowledge* which rests on a partial reconstruction of conceptual contents and conceptual actions by secondary concepts. The reflexive knowledge about ones own comprehension and understanding which has been explicitly realized will not be fully reactualized in all cases of later actions of corresponding concepts. A large part of conscious knowledge which we already had on our proposal remains virtual and needs new efforts to become again fully actualized.

The conscious knowledge which accompanies formal concepts – as elaborated in [GW99a] – is *reflexive knowledge* too. This knowledge is based on knowledge about Formal Concept Analysis combined with secondary knowledge about conceptual contents and conceptual actions. Even the mathematical knowledge needs efforts to make it explicit. Often large parts of potential knowledge remains virtual, in particular, when the underlying data contexts are so large that readable line diagrams of the concept lattices of those contexts cannot be established. Large data contexts as they are coded in databases may nevertheless be successfully treated by methods of Formal Concept Analysis because, in practice, one does not want to see all the information of a database at once. Therefore it is sufficient to allow specific views into the database which can be combined in such a way that a *navigation for creating knowledge* becomes possible. How this can be done shall be explained by the following example:

In 1991, members of the research group "Formale Begriffsanalyse" started a project to develop a *retrieval system* for the library of the "Center of Interdisciplinary Technology Research" (ZIT) at the TH Darmstadt which was finished in

1996 [RW00][3]. Because of the wide range of contents in interdisciplinary texts, a specific normed vocabulary was developed for satisfactory content extraction of the documents. On average, 32 catchwords from the normed vocabulary were assigned to each document which yielded a very good substitute of a content abstract for each document. These assignments, stored in a relational database, gave rise to a large cross table with 1554 documents as objects and 377 catchwords as attributes; within that table the crosses indicate which catchword is assigned to which document. From the established cross table, 137 *conceptual views* were derived with the help of experts from the respective source fields. Each conceptual view is determined by a theme and a small number of catchwords representing that theme. For instance, the conceptual view with the theme "Informatics and Knowledge Processing" got the catchwords "Formalization", "Artificial Intelligence", "Expert Systems", "Knowledge Processing", and "Hypertext". The concept lattice of this view, shown in Fig. 13 (cf. [Wi01]), is the concept lattice of the formal context represented by the five columns of the large cross table which are headed by the five listed catchwords. In Fig. 13, there are no designations of objects, but the quantities of objects in the extent of the represented concepts, respectively. For instance, the "96" attached to the circle with the label "Artificial Intelligence" indicates that there are 96 documents in the library to which the catchword "Artificial Intelligence" is assigned. Now, let us consider a researcher who is looking for *literature about expert systems dealing with traffic* and who has chosen first the conceptual view "Informatics and Knowledge Processing". The diagram in Fig. 13 gives him the information that there are 60 documents with the catchword "Expert System". To get more information about those 60 documents, particularly concerning traffic, the researcher could zoom into the circle labelled with "Expert System" with the conceptual view "Town and Traffic". Then he obtains the line diagram in Fig. 14. The diagram informs him that 9 of the 60 documents deal with "Traffic" and 4 with "Traffic" and "Means of Transportation". Since there are only few documents left, the researcher might click on those numbers to get the titles of the documents, for instance, via 4 the titles "Digital Fate", "Evolutionary Paths in the Future", "Yearbook Labour and Technology 1991", and "Cooperative Media". Via the described process, the researcher actualizes stepwise the knowledge concerning the *literature about expert systems dealing with traffic* present in the ZIT-library; in doing so, he encounters conceptual views which offer him further knowledge, for instance, that the documents dealing with expert systems and means of transportation also deal with traffic.

2.8   Concepts as Habitual and Virtual Knowledge Can Be Implicitly and Explicitly Actualized

According to [Se01], concepts have a double nature: they are actual knowledge or cognitive acts which rest on *habitual knowledge*. Since they consist of habitual

---

[3] The retrieval system of the ZIT-library was implemented with the software *TOSCANA* which allows, in general, to navigate with prepared conceptual scales (views) in relational databases (see [VW95])

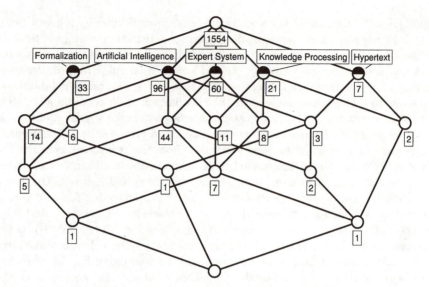

Fig. 13. Concept lattice of the conceptual view "Informatics and Knowledge Processing"

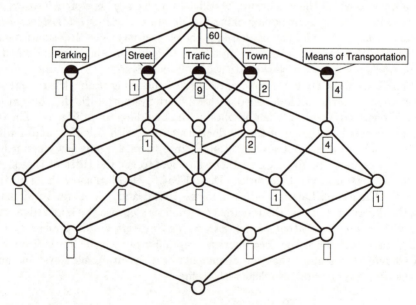

Fig. 14. Concept lattice of the conceptual view "Town and Traffic" restricted to "Expert System"

and virtual knowledge, they may be reactualized on occasion or need. But it often needs strong conceptual efforts until even some parts of the implicit knowledge become *explicit*. Content aspects and relations contained in a concept and even more so in subconcepts or superconcepts may remain implicit, although they could be determinative.

The interpretation of mathematically represented concepts and concept hierarchies builds on the *explicit and implicit knowledge* actualized by the interpreters (cf. example in Subsection 2.3). Line diagrams intelligibly presenting the conceptual relations may strongly support the interpreters. By our experiences we got the impression that the labelled diagrams may "speak" to those users who are familiar with the contents coded in the formal context; quite often, after a short glance at the diagram, users even recognize mistakes in the underlying data context. This direct support of logical thinking indicates that the *contextual and holistic nature of concepts* in the human mind are remarkably preserved by the mathematization with formal contexts and formal concepts.

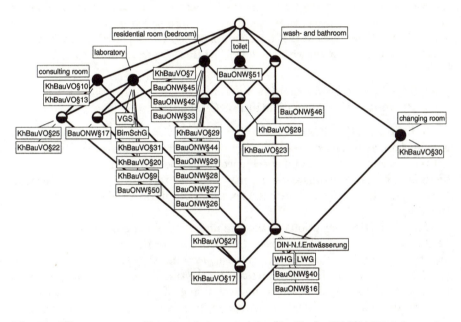

Fig. 15. Query structure "functional rooms in a hospital" of a TOSCANA information system about laws and regulations concerning building construction

The following example (cf. [Wi01]) may show that the logical connections in line diagrams of concept lattices can *stimulate background knowledge* for improving the knowledge representation; in particular, the example indicates that line diagrams of concept lattices may even stimulate critique and self-correction of the represented knowledge. In a research project for developing an information system for architects about laws, standards, and regulations concerning building constructions, we experienced again and again that line diagrams enabled the building experts to discover mistakes in the extensive data context which contributed to a considerable improvement of the data quality. An instructive case of critique and self-correction occurred in a discussion of the line diagram in Fig. 15: For testing the readability of such a diagram, a secretary in the involved ministry of building constructions, who was not working for the research project, was asked to join the discussion. After inspecting the diagram, the secretary expressed her astonishment that, in the diagram, the paragraph "BauONW§51" of the "Bauordnung Nordrhein-Westfalen" is directly attached to the circle with the label "toilet", which means that only the toilets have to be designed for handicapped people; she could not understand why the wash- and bathrooms do not have to meet requirements for handicapped people too. Even the experts became surprised when they checked again §51 and saw that only toilets are mentioned in connection with handicapped people. Only after a comprehensive discussion the experts came to the conclusion that, by superior aspects of law, §51 also applies to wash- and bathrooms. Finally, by similar reasons, the consulting rooms and the residential rooms (bedrooms) were also included so that, in the underlying cross table, three more crosses were added in the row headed by "BauONW§51" so that, in the line diagram of Fig. 15, the label "BauONW§51" moved down to the circle with the label "KhBauVO§27".

2.9   The Language as Medial System
Promotes the Actualization of Concepts

According to [Se01], although the *linkage between concept and word* is not rigid and unchangable, but on the contrary fluent within certain boundaries, a tie between specific concepts and specific words has been established in the course of time. Therefore, words as signs may activate the corresponding concepts. Language in its internalized form is a necessary condition for abstract conceptual thinking.

The linguistic inscriptions in the line diagrams of concept lattices make indeed possible the *implicit and explicit actualization* of the underlying knowledge and the interpretation of the presented conceptual structures. Thus, incorporated language expressions are necessary for the transformation of the potential-mathematical to the actual-logic understanding of the conceptual structures. How words as signs for concepts activated in discourses of language may lead the elaboration of corresponding formal concepts has been multifariously demonstrated using the Formal Concept Analysis methods of *attribute, object,* and *concept exploration* (cf. [Ga86], [Wi86], [Wi89b], [St97], [Ga00]). An idea of those explorations shall be given by the following example of an attribute exploration:

In the summer 1987, members of a *music philosophical seminar* at the TH Darmstadt were discussing how one can verbally describe musical pieces. Besides other sources, we discussed a report on an experiment which had examined how musicians assign to presented musical compositions characteristic adjectives chosen from a given list [Ba76]. Since the experiment did not convince us because of the predetermined selection of musical compositions, we carried out the attribute exploration with the eleven most discriminating adjectives of the report which were taken to be the attributes of a formal context.

In general, an *attribute exploration* is performed as an interactive procedure of questions and answers. The questions ask for the validity of attribute implications deduced from the just present formal context. The answers given should be "yes" if the implication is considered to be valid in the assumed domain, or "no" and then be justified by a counterexample taken to be a new object of the underlying formal context. A typical question in our exploration was:

> Has every musical composition with the attributes "dramatic", "lively", and "transparent" also the attributes "sprightly ", "rhythmizing", and "fast"?

The consensus of the seminar was "no", justified by the third movement of Beethoven's Moonlight Sonata as counterexample which was considered to be not sprightly, but to have the five other attributes (and additionally the attribute "strong")[4].

After four hours of intense work, the exploration was finished with the resulting context shown in Fig. 16 and the attribute implications listed in Fig. 17. The concept lattice is represented in [Wi89b]. The questions of the exploration were created by the software "ConImp", programmed by P. Burmeister [Bu00] based on the Ganter-Algorithm [Ga86] (see also [GW99a]) which guarantees that the final list of valid attribute implication forms the Duquenne-Guigues-Basis of all attribute implications of the resulting context.

2.10 Concepts Have Motivational and Emotional Qualities

According to [Se01], concepts also have a *dynamic, motivational and emotional nature*. They essentially contain an evaluating attitude. Concepts together create concentrations on points of interest which urge to deal with their objects. Such emotional states and conditions characterize not only personal concepts, but also apply to *scientific concepts*. An absolute neutrality and objectivity in the case of scientific concepts would indeed not be desirable because scientific progress lives on the dynamics of the emotions and motivations which are basic for the cognitive development of human beings.

The *dynamics of formal concepts and concept lattices* is multifarious: first, formal concepts can be generated in formal contexts out of arbitrary object and attribute sets by the derivation operators, while formal contexts themselves can be created by methods of conceptual constructions and explorations; furthermore, concept lattices can be aggregated in different ways so that, finally, *conceptual landscapes of knowledge* evolve which allow effective navigations. All this may be urged and activated by motivational and emotional qualities. It is the close connection between the potential-mathematical and the actual-logical thinking which carries over the dynamics of concepts together with motivations and emotions in both directions.

Most stimulating are *metaphorical ideas* like the idea of conceptual landscapes of knowledge. The software TOSCANA is an attempt to support the development of such landscapes by methods of Formal Concept Analysis. Quite

---

[4] The exploration was performed in German so that the English translations might not completely render the meaning of the German words ("sprightly" is here used as the translation of "munter")

	well-rounded	well-balanced	dramatic	transparent	structured thoroughly	strong	lively	sprightly	rhythmizing	fast	playful
Beethoven: Romance for violin and orchestra F-major	X	X		X	X						X
Bach: Contrapunctus I	X	X		X	X	X					
Chajkovskij: Piano concerto b flat minor, 1st movement		X			X						
Mahler: 2nd Symphony, 2nd movement					X		X	X	X	X	X
Bartok: Concert for ochestra					X	X	X	X			X
Beethoven: 9th symphony, 4th movement (presto)				X		X			X	X	
Bach: WTP 1, prelude c minor				X		X	X	X		X	
Bach: 3rd Brandenburg Concerto, 3rd movement	X	X		X	X		X	X		X	X
Ligeti: Continuum				X			X			X	X
Mahler: 9th symphony, 2nd movement (Ländler)						X	X	X		X	
Beethoven: Moonlight sonata, 3rd movement			X	X		X	X		X	X	
Hindemith: Chamber music No.1, finale				X		X	X	X	X	X	X
Bizet: Suite arlesienne	X	X		X	X	X			X		X
Mozart: Figaro, overture	X	X		X	X	X	X	X		X	X
Schubert: Wayfarer fantasy				X		X	X		X	X	
Beethoven: Spring sonata, 1st movement	X	X		X	X	X	X				X
Bach: WTP 1, fuge c minor	X	X		X	X	X	X	X			X
Shostakovich: 15th symphony, 1st movement					X	X	X	X			X
Wagner: Mastersinger, overture	X	X		X	X	X	X	X	X		X
Beethoven: String quartet op.131, final movement					X	X	X	X		X	X
Johann Strauß: Spring voice waltz	X	X		X	X	X	X	X	X	X	X
Mozart: Il Seragio, "O, how I will triumph ..."				X	X	X	X	X		X	X
Bach: Mathew's passion No.5 (choir)				X	X	X	X	X			
Brahms: Intermezzo op. 117, No.2	X	X		X	X	X			X	X	
Wagner: Ride of the valkyries				X	X	X	X	X	X		
Mozart: Magic flute, "The hell revenge rages ..."				X	X	X	X	X		X	
Mendelsohn: 4th symphony, 4th movement	X	X		X	X	X	X	X	X	X	X
Brahms: 4th symphony, 4th movement	X	X	X	X	X	X	X				
Beethoven: Great fuge op. 133	X	X	X	X	X	X	X		X		
Goretzky: Lament symphony	X	X	X	X	X	X	X				
Verdi: Requiem, dies irae	X	X	X	X	X	X	X			X	

Fig. 16. Formal context of the result of an attribute exploration in music

a number of *TOSCANA-systems* have been developed and used in practice (see, for instance, [SVW93], [Vg95], [Kf96], [EKSW00], [GH00], [KV00], [RW00], [BSWZ00], [Sc04], [Ks05]). But other TOSCANA-systems have not been established although they would be urgently needed. For instance, together with the Center of Medical Informatics at Frankfurt University, we have designed a research project for developing a medical information system for practising doctors to support their on-the-spot examinations and treatments.

Another dream is to develop a conceptually structured lexical landscape of knowledge as an extension of language thesauri. Towards this dream, basic work has been done by S. Sedelow, W. Sedelow, U. Priss, and J. Old with Roget's International Thesaurus (cf. [SS93], [Ps98], [Ol03], [Ol04], [PO04]). Here only one structural idea shall be rendered, which is inspired by the landscape metaphor: Fig. 18 represents a *restricted neighbourhood* of the word "over" by a concept lattice [PO04]. Context attributes are words of the thesaurus and context objects

1. fast, playful ⟹ lively
2. sprightly ⟹ lively, playful
3. lively, rhythmizing, playful ⟹ sprightly
4. strong, lively, fast, playful ⟹ transparent
5. structured thoroughly, strong, rhythmizing, fast ⟹ transparent
6. dramatic ⟹ strong
7. dramatic, structured thoroughly, strong, rhythmizing ⟹ transparent
8. dramatic, strong, playful ⟹ transparent, structured thoroughly, lively, fast
9. well-balanced ⟹ well-rounded, transparent, structured thoroughly
10. well-rounded ⟹ well-balanced, transparent, structured thoroughly
11. transparent, structured thoroughly, rythmizing, playful ⟹ well-rounded, well-balanced
12. well-rounded, well-balanced, transparent, structured thoroughly, fast ⟹ lively
13. transparent, structured thoroughly, lively, rhythmizing, fast ⟹ well-rounded, well-balanced
14. transparent, structured thoroughly, lively, sprightly, playful ⟹ well-rounded, well-balanced
15. well-rounded, well-balanced, transparent, structured thoroughly, lively, playful ⟹ sprightly
16. structured thoroughly, strong, rhythmizing, playful ⟹ well-rounded, well-balanced, transparent
17. well-rounded, well-balanced, dramatic, transparent, structured thoroughly, strong, rhythmizing ⟹ lively
18. well-rounded, well-balanced, dramatic, transparent, structured thoroughly, strong, lively, rhythmizing, fast ⟹ sprightly, playful
19. well-rounded, well-balanced, dramatic, transparent, structured thoroughly, strong, lively, sprightly, fast, playful ⟹ rhythmizing

Fig. 17. Basis of the implications of the formal context in Fig.16

are senses indicated by number triples denoting a category, a paragraph, and a semicolon group; a sense relates to a word if and only if the word occurs in the category, paragraph, and semicolon group denoted by the numbers of the sense. For instance, the word "through" occurs in the semicolon group 4 of the paragraph 8 of the category 105 and also in the semicolon group 1 of the paragraph 29 of the category 183. The idea of neighbourhood caused new formal operators ι^+ and ϵ^+ on formal contexts (G, M, I) defined by $\iota^+(X) := \{m \in M \mid \exists g \in X : gIm\}$ for $X \subseteq G$ and $\varepsilon^+(Y) := \{g \in G \mid \exists m \in Y : gIm\}$ for $Y \subseteq M$. These +-operators are used to create the restricted neighbourhood context of "over" with the object set $\varepsilon^+(\{over\})$ and the attribute set $\iota^+(\varepsilon^+(\{over\}))$.

2.11 Concepts Have a History and Go Through a Developmental Process

According to [Se01], each personal concept goes through a *long history of development* in which its content progressively changes. Conventional concepts on the one hand are anchored in the personal cognition, knowledge and thought, and on

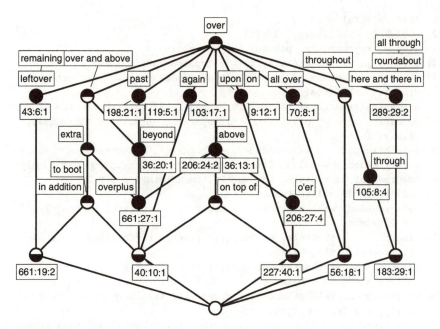

Fig. 18. Concept lattice of a restricted neighbourhood of the word "over" in Roget's International Thesaurus

the other hand are subjected to continuous social change because of their dependence on discourses. In the end, the history of development of each concept has simultaneously an *ontogenetic*, a *phylogenetic*, and a *cultural-historic* trace. Even if such developments lead to independent and structured concept systems which grant human thought self-dynamics and autonomy, highly structured concepts (even of adults) are rather the exception than the rule.

In formal contexts and their formal concepts, processes of developing concept structures and of building conceptual theories can be represented, examined, improved, and documented (cf. Fig. 3). This allows in particular productive discourses about such developments. In particular, the *developmental process of empirical theory building* can be supported by Formal Concept Analysis (cf. [SWW01]). Means of such support have been mainly elaborated on the basis of a contextual attribute logic (cf. [GW99b]). Convincing experiences have been made with the representation of local theories by small formal contexts. Larger theories can then be obtained by suitable aggregations of those contexts.

An impressive research project performed by such a process of empirical theory building is described in [MW99]. In this project, a TOSCANA-system was established as the basis for a dissertation on the theme *"Simplicity - reconstruction of a conceptual landscape in the music esthetics of the 18th century"* (see [Ma00]). 270 historical sources were exploited in their contents by a normed vocabulary of more than 400 textual attributes concerning the theme "simplicity". For the resulting formal context with the sources as formal objects and the textual attributes as formal attributes, a multitude of thematic *questioning*

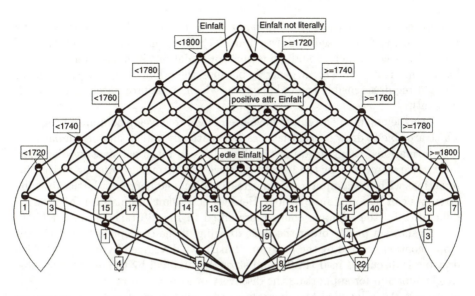

Fig. 19. Concept lattice of an aggregation of the conceptual scales "year of publication" and "literal occurrence of Einfalt"

structures were formed and tested. This thematic structuring and testing can be understood as a process of local theory building which even gives rise to larger theories by aggregating the questioning structures (also named *conceptual scales*). Finally, a rich "conceptual landscape" was established from which we only show the aggregation of the conceptual scales "year of publication" and "literal occurrence of Einfalt" in Fig. 19. Supported by the six lenticular areas representing consecutive periods of twenty years (except the last one), one can see in the diagram, for instance, that the concept of "(edle) Einfalt" is most frequent between 1780 and 1800 because "Einfalt" occurs literally in 45 and "edle Einfalt" in 22 of the 111 sources of this time period. In general, the TOSCANA-system does not only make explicit the exploitation of the contents of the 270 sources concerning simplicity, but also supports the process of categorization and theory building through the documented interplay of the thematic conceptual scales as local theories.

2.12   Concept Formation Is Not a Formalizable Automatism

According to [Se01], the process of *concept formation* always contains a creative and spontaneous impetus which, in part, is founded on accidental assimilations and, in part, uses conscious analogies and metaphoric transfers. *New concepts* often evolve from conflicting processes which cause changes in the established concept systems. Clearly, those unpredictable creations and changes of concepts and concept systems are far from being rule-based formations.

Formal concepts of given formal contexts are automatically deducible, but formal contexts which represent purpose-oriented real world relationships are usually not constructable by a *formalizable automatism* (e.g. contexts of the research project discussed in Subsection 2.11). Even intelligible line diagramsof concept lattices, which should guide concept fomations in human mind, cannot be automatically drawn, in general, because their aim of supporting human thought requires sensibly created diagrams which are appropriately readable (cf. [Wi89a]).

Nevertheless, many attempts have been made to develop computer programs which are able to draw concept lattices automatically and to support human concept formations. Being aware of the difficulties to reach satisfactory diagrams, most programmers allow the user to improve the output drawing by further interactions with the computer. An informative discussion of recent lattice drawing attempts can be found in [PHM04]. Here we only want to illustrate the drawing difficulties by the example context presented in Fig. 20 taken from [PHM04]. Fig. 21 shows a purely order-theoretic drawing of the concept lattice of the example context created by the program "LatDraw" developed by R. Freese[5]. This drawing is discussed in [PHM04] and strongly modified by the use of a suitable weight function for supporting the discovery of association rules.

In Fig. 22, a labelled line diagram of the same concept lattice is presented which was interactively established with F. Vogt's "ANACONDA" software. The purpose for drawing the diagram was to support an anwer to the question: How representative is the formal context in Fig. 20 for a *meaningful classification of animals*? First, we recognize in the diagram that eight object concepts are generated by more than one formal object. In particular, the formal concept of animals with fur is generated by six formal objects, namely by zebra, ibex, antelope, moose, mouse and pig; it is the superconcept of five object concepts with eleven generating objects altogether. All this indicates that the differentiation of the animals is quite coarse. Secondly, we can read directly from the diagram that the basic attribute implications with a one-element premise are $fur \rightarrow warmblooded$, $livestock \rightarrow warmblooded$, $pet \rightarrow warmblooded$, $warmblooded \rightarrow airbreather$, $four-legged \rightarrow airbreather$, and $nocturnal \rightarrow airbreather$. The first two implications look quite acceptable, but the third and even more so the following might be questionable for the realm of all animals. This shall be enough to demonstrate the purpose-oriented use of a labelled line diagram. Although the diagram in Fig. 22 is better readable than the one in Fig. 21 (with added object names), it is not clear whether Fig. 22 could be automatically drawn as it is or whether it could even be substantially improved. Already, the ANACONDA-drawn diagram could only be created on the basis of many years of experience in drawing concept lattices.

3   Formal Concept Analysis as Transdisciplinary Mathematics

The discussion of Seiler's twelve aspects of concepts shows that there are *close connections between concepts and formal concepts* in each of the considered as-

---

[5] See [Fr04] and `http://www.math.hawaii.edu/~ralph/LatDraw`

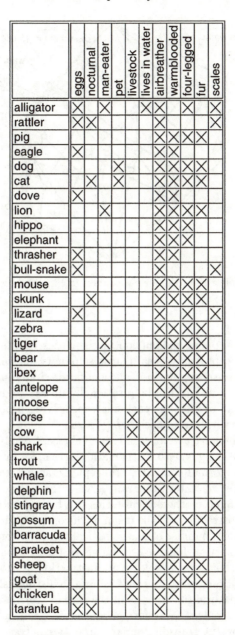

	eggs	nocturnal	man-eater	pet	livestock	lives in water	airbreather	warmblooded	four-legged	fur	scales
alligator	X		X			X	X		X		X
rattler	X	X					X				X
pig							X	X	X	X	
eagle	X						X	X			
dog				X			X	X	X	X	
cat		X		X			X	X	X	X	
dove	X						X	X			
lion			X				X	X	X	X	
hippo							X	X	X		
elephant							X	X	X		
thrasher	X						X	X			
bull-snake	X						X				X
mouse							X	X	X	X	
skunk		X					X	X	X	X	
lizard	X						X		X		X
zebra							X	X	X	X	
tiger			X				X	X	X	X	
bear			X				X	X	X	X	
ibex							X	X	X	X	
antelope							X	X	X	X	
moose							X	X	X	X	
horse					X		X	X	X	X	
cow					X		X	X	X	X	
shark			X			X					X
trout	X					X					X
whale						X	X	X			
delphin						X	X	X			
stingray	X					X					X
possum		X					X	X	X	X	
barracuda						X					X
parakeet	X			X			X	X			
sheep					X		X	X	X	X	
goat					X		X	X	X	X	
chicken	X				X		X	X			
tarantula	X	X					X				

Fig. 20. Formal context about animals

pects. Therefore the mathematization of concepts and concept hierarchies used in Formal Concept Analysis opens up the chance of supporting mathematically the logical thinking of humans. That this support can really take place has been experienced in a great multitude of applications of Formal Concept Analysis.

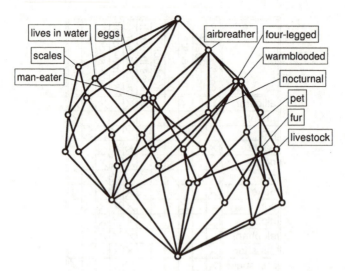

Fig. 21. Concept lattice of the formal context in Fig. 20 drawn with LatDraw

The success in applications outside mathematics results from the development of Formal Concept Analysis as "transdisciplinary mathematics". A research activity is called *transdisciplinary* if a scientific discipline uses this activity to make the disciplinary way of thinking rationally understandable, available, and applicable beyond its borders, in particular, for solving problems which cannot be managed purely within the discipline [Wi02a].

Basic transdisciplinary activities in Formal Concept Analysis are assignments of words of a natural language to mathematical terms, for instance, "data table" or "cross table" to "formal context", "concept hierarchy" to "concept lattice", "conceptual network" to "lattice diagram", "view" or "query structure" to "conceptual scale" etc. Such mathematical terms can be understood as descriptions for mathematical abstractions of the meaning of corresponding natural language words. There is quite a number of mathematical terms in Formal Concept Analysis having also a natural language meaning so that the mathematical meaning of those terms should be viewed as an abstraction of the common meaning. Sometimes the additional word "formal" is used to distinguish the mathematical meaning from the natural meaning as in the case of "formal object" - "object", "formal context" - "context", or "formal concept" - "concept".

The most important transdisciplinary activities between mathematics and the realm of common understandings lie in the conversion from the mathematical language to the common language by *interpreting mathematical meaning into logical meaning* as an expression of human reason. Such transdisciplinary activities are essential for Formal Concept Analysis. The present discussion of the many connections between Formal Concept Analysis and Seiler's comprehensive theory of concepts indicates the richness of links between mathematical and logical meanings. Basic examples of those links may be described as follows:

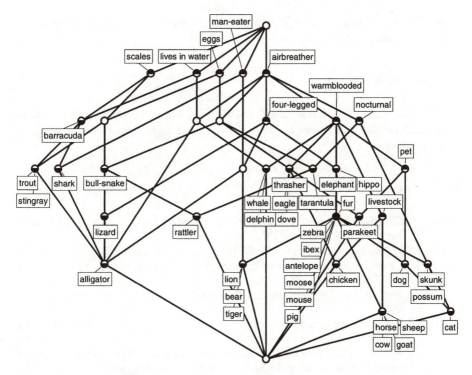

Fig. 22. Concept lattice of the formal context in Fig.20 drawn with ANACONDA

- the mathematical notion of a formal context converts to the logical meaning of a domain of interest based on object-attribute-relationships,
- the mathematical derivation of a set of formal attributes is logically viewed as the identification of all objects having all attributes of a given attribute collection,
- the mathematical order-relationship that a formal concept is less than another formal concept is logically understood as a subconcept-superconcept-relationship,
- the labelled line diagram of a concept lattice is logically considered as a hierarchical network linking nodes with object names to nodes with attribute names and thereby establishing conceptual meanings,
- formal object and attribute implications lead to the recognition of conceptual dependencies within the given domain of interest.

Although we got already many insights into relationships between mathematics and philosophical resp. everyday logic, we are far from grasping the rich network of connections between mathematical and logical meanings in human thought. A comprehensive understanding of such connections would be of great substance for further developments and applications of Formal Concept Analysis. Perhaps the most desirable aim would be to develop a *"Logical Concept Analysis"* which closely corresponds to our mathematical concept analysis and hence enables

Formal Concept Analysis to effectively support the rational comprehension and communication of human beings.

Acknowledgement

I am deeply grateful to Thomas Bernhard Seiler for all his cooperation, stimulation and critical discourses by which he has strongly supported our research over decades.

References

[Ba76]   G. Batel: *Komponenten musikalischen Erlebens.* Göttinger Musikwissenschaftliche Schriften 1976.

[BSWZ00]   K. Becker, G. Stumme, R. Wille, U. Wille, M. Zickwolff: Conceptual information systems discussed through an IT-security tool. In: R. Dieng, O. Corby (eds.): *Knowledge Engineering and Knowledge Management: Methods, Models, and Tools.* LNAI **1937**. Springer, Heidelberg 2000, 352–365.

[Bu00]   P. Burmeister: ConImp – Ein Programm zur Formalen Begriffsanalyse. In: [SW00], 25–56.

[Cl90]   A. Claar: Die Entwicklung ökonomischer Begriffe im Jugendalter. Lehr- und Forschungstexte Psychologie 37. Springer, Heidelberg 1990.

[DK01]   DK *Eyewitness Travel Guide New Zealand.* Dorling Kindersley Publishing Inc. 2001.

[EKSW00]   D. Eschenfelder, W. Kollewe, M. Skorsky, R. Wille: Ein Erkundungssystem zum Baurecht: Methoden der Entwicklung eines TOSCANA-Systems. In: [SW00], 254–272.

[Fr04]   R. Freese: Automated lattice drawing. In: P. Eklund (ed.): *Concept lattices.* LNAI **2961**. Springer, Heidelberg 2004, 112–127.

[Ga86]   B. Ganter: Algorithmen zur Formalen Begriffsanalyse. In: B. Ganter, R. Wille, K. E. Wolff (Hrsg.): *Beiträge zur Begriffsanalyse.* B.I.-Wissenschaftsverlag, Mannheim 1986, 241–254.

[Ga00]   B. Ganter: Begriffe und Implikationen. In: [SW00], 1–24.

[GW89]   B. Ganter, R. Wille: Conceptual scaling. In: F. Roberts (ed.): *Applications of combinatorics and graph theory in the biological and social sciences.* Springer-Verlag, New York 1989, 139–167.

[GW99a]   B. Ganter, R. Wille: *Formal Concept Analysis: mathematical foundations.* Springer, Heidelberg 1999.

[GW99b]   B. Ganter, R. Wille: Contextual Attribute Logic. In: W. Tepfenhart, W. Cyre (eds.): Conceptual structures: standards and practices. LNAI **1640**. Springer, Heidelberg 1999 , 377–388.

[GH00]   A. Großkopf, G. Harras: Begriffliche Erkundung semantischer Strukturen von Sprechaktverben. In: [SW00], 273–295.

[Ks05]   T. Kaiser: A TOSCANA-System for the International Regimes Database (IRD). In: H. Breitmeier, O. R. Young, M. Zürn (eds.): *Analyzing international environmental regimes: from case study to database* (to appear)

[Ka88]   I. Kant: *Logic.* Dover, New York 1988.

[Kf96]   U. Kaufmann: *Begriffliche Analyse von Daten über Flugereignisse – Imple-mentierung eines Erkundungs- und Analysesystems mit TOSCANA.* Diplomarbeit, FB4, TU Darmstadt, 1996.

[KW87]   U. Kipke, R. Wille: Formale Begriffsanalyse erläutert an einem Wortfeld. LDV-Forum **5** (1987), 31–36.

[KV00]   B. Kohler-Koch, F. Vogt: Normen- und regelgeleitete internationale Kooperationen. In: [SW00], 325–340.

[KW94]   W. Kollewe, M. Skorsky, F. Vogt, R. Wille: TOSCANA – ein Werkzeug zur begrifflichen Analyse und Erkundung von Daten. In: R. Wille, M. Zickwolff (Hrsg.): *Begriffliche Wissensverarbeitung – Grundfragen und Aufgaben.* B.I.-Wissenschaftsverlag, Mannheim 1994, 267–288.

[Ma00]   K. Mackensen: *Simplizität – Genese und Wandel einer musikästhetischen Kategorie des 18. Jahrhunderts.* Bärenreiter, Kassel 2000.

[MW99]   K. Mackensen, U. Wille: Qualitative text analysis supported by conceptual data systems. *Quality & Quantity* **33** (1999), 135–156.

[Ol03]   J. Old: The semantic structure of Roget's, a whole-language thesaurus. PhD Dissertation. Indiana University, Bloomington 2003.

[Ol04]   J. Old: Unlocking the semantics of Roget's thesaurus using Formal Concept Analysis. In: P. Eklund (ed.): *Concept lattices.* LNAI **2961**. Springer, Heidelberg 2004, 244–251.

[Pe92]   Ch. S. Peirce: *Reasoning and the logic of things.* Edited by K. L. Ketner; with an introduction by K. L. Ketner and H. Putnam. Havard University Press, Cambridge 1992.

[PHM04]   A. Pogel, T. Hanan, L. Miller: Visualization of concept lattices using weight functions. In: H. D. Pfeiffer, K. E. Wolff, H. S. Delugach (eds.): *Conceptual structures at work.* Shaker Verlag, Aachen 2004, 1–14.

[Pr00]   S. Prediger: Mathematische Logik in der Wissensverarbeitung: Historisch-philosophische Gründe für eine Kontextuelle Logik. *Mathematische Semesterberichte* 47 (2000), 165–191.

[Ps98]   U. Priss: *Relational Concept Analysis: Semantic structures in dictionaries and lexical databases.* Dissertation, TU Darmstadt. Shaker Verlag, Aachen 1996.

[PO04]   U. Priss, J. Old: Modelling lexical databases with Formal Concept Analysis. *Journal of Universal Computer Science* **10** (2004), 967–984.

[RW00]   T. Rock, R. Wille: Ein TOSCANA-Erkundungssystem zur Literatursuche. In: [SW00], 239–253.

[SVW93]   P. Scheich, M. Skorsky, F. Vogt, C. Wachter, R. Wille: Conceptual data systems. In: O. Opitz, B. Lausen, R. Klar (eds.): *Information and classification.* Springer, Heidelberg 1993, 72–84.

[Sc04]   S. Schmidt: *Ein TOSCANA-System über die Besteuerung von Einkünften aus Kapitalvermögen (Anlage KAP zur EST2003).* Diplomarbeit, FB4, TU Darmstadt 2004.

[SS93]   S. Sedelow, W. Sedelow: The Concept concept. *Proc. Fifth Int. Conf. on Computing and Information.* Sudbury, Ontario 1993, 339–343.

[Se01]   Th. B. Seiler: *Begreifen und Verstehen. Ein Buch über Begriffe und Bedeutungen.* Verlag Allgemeine Wissenschaft, Mühltal 2001.

[Sp90]   N. Spangenberg: *Familienkonflikte eßgestörter Patientinnen. Eine empirische Untersuchung mit der Repertory Grid Technik.* Habilitationsschrift, Universität Gießen 1990.

[SW88]   N. Spangenberg, K. E. Wolff: Conceptual grid evaluation. In: H. H. Bock
          (ed.): *Classification and related methods of data analysis*. Elsevier, Amster-
          dam 1988, 577–580.
[SW93]   N. Spangenberg, K. E. Wolff: Datenreduktion durch die Formale Begriffs-
          analyse von Repertory Grids. In: J. W. Scheer, A. Catina (eds.): *Einführung
          in die Repertory Grid-Technik*. Bd. 2: *Klinische Forschung und Praxis*. Hu-
          ber, Bern 1993, 38–54.
[SWW01]   S. Strahringer, R. Wille, U. Wille: Mathematical support for empirical the-
          ory building. In: H. S. Delugach, G. Stumme (eds.): *Conceptual structures:
          broadening the base*. LNAI 2120. Springer, Heidelberg 2001, 169–186.
[St97]   G. Stumme: *Concept exploration: knowledge acquisition in conceptual
          knowledge systems*. Dissertation, TU Darmstadt. Shaker Verlag, Aachen
          1997.
[SW00]   G. Stumme, R. Wille (Hrsg.): *Begriffliche Wissensverarbeitung: Methoden
          und Anwendungen*. Springer, Heidelberg 2000.
[Vg95]   N. Vogel: *Ein begriffliches Erkundungssystem für Rohrleitungen*. Diplomar-
          beit, FB4, TU Darmstadt, 1995.
[Vo96]   F. Vogt: *Formale Begriffsanalys mit C^{++}: Datenstrukturen und Algorith-
          men*. Springer, Heidelberg 1996.
[VW95]   F. Vogt, R. Wille: TOSCANA – A graphical tool for analyzing and exploring
          data. In: R. Tamassia, I. G. Tollis (eds.): *Graph drawing '94*. LNCS **894**.
          Springer, Heidelberg 1995, 226–233.
[VW03]   B. Vormbrock, R. Wille: Semiconcept and protoconcept algebras: the basic
          theorems. This volume.
[Wa73]   H. Wagner: Begriff. In: H. Krings, H. M. Baumgartner, C. Wild (eds.):
          Handbuch philosophischer Grundbegriffe. Kösel, München 1973, 191–209.
[Wi82]   R. Wille: Restructuring lattice theory: an approach based on hierarchies of
          concepts. In: I. Rival (ed.): Ordered sets. Reidel, Dordrecht-Boston 1982,
          445–470.
[Wi86]   R. Wille: Bedeutungen von Begriffsverbänden. In: B. Ganter, R. Wille,
          K. E. Wolff (Hrsg.): *Beiträge zur Begriffsanalyse*. B.I.-Wissenschaftsverlag,
          Mannheim 1986, 161–211.
[Wi89a]   R. Wille: Lattices in data analysis: how to draw them with a computer. In:
          I. Rival (Ed.): *Algorithms and order*. Kluwer, Dordrecht 1989, 33–58.
[Wi89b]   R. Wille: Knowledge acquisition by methods of Formal Concept Analysis. In:
          E. Diday (ed.): *Data analysis and learning symbolic and numeric knowledge*.
          Nova Science Publisher, New York–Budapest 1989, 365–380.
[Wi95]   R. Wille: Begriffsdenken: Von der griechischen Philosophie bis zur künst-
          lichen Intelligenz heute. Dilthey-Kastanie, Ludwig-Georgs-Gymnasium,
          Darmstadt 1995, 77–109.
[Wi00a]   R. Wille: Contextual Logic summary. In: G. Stumme (ed.): *Working with
          conceptual structures: Contributions to ICCS 2000*. Shaker-Verlag, Aachen
          2000, 265–276.
[Wi00b]   R. Wille: Boolean Concept Logic. In: B. Ganter, G. W. Mineau (eds.):
          Conceptual structures: logical, linguistic, and computational issues. LNAI
          1867. Springer, Heidelberg 2000, 317–331.
[Wi01]   R. Wille: Why can concept lattices support knowledge discovery in
          databases? In: E. Mephu Nguifo et al. (eds.): ICCS 2001 Workshop on Con-
          cept Lattice-Based Theory, Methods and Tools for Knowledge Discovery in
          Databases. Stanford University 2001, 7–20; also in: *Journal of Experimental
          and Theoretical Artificial Intelligence* **14** (2002), 81–92.

[Wi02a]   R. Wille: Transdisziplinarität und Allgemeine Wissenschaft. In: H. Krebs, U. Gehrlein, J. Pfeifer, J. C. Schmidt (Hrsg.): *Perspektiven Interdisziplinärer Technikforschung: Konzepte, Analysen, Erfahrungen.* Agenda-Verlag, Münster 2002, 73–84.

[Wi02b]   R. Wille: Kommunikative Rationalität und Mathematik. In: S. Prediger, F. Siebel, K. Lengnink (Hrsg.): *Mathematik und Kommunikation.* Verlag Allgemeine Wissenschaft, Mühltal 2002, 181–195.

[Wi04]   R. Wille: Truncated distributive lattices: conceptual structures of simple-implicational theories. Order **20** (2004), 229–238.

[Wl91]   U. Wille: Eine Axiomatisierung bilinearer Kontexte. Mitt. Math. Sem. Gießen **200** (1991), 71–112.

[Wl99]   U. Wille: Characterization of ordered bilinear contexts. Journal of Geometry **64** (1999), 167–207.

Semiconcept and Protoconcept Algebras: The Basic Theorems

Björn Vormbrock and Rudolf Wille

Technische Universität Darmstadt, Fachbereich Mathematik,
Schloßgartenstr. 7, D–64289 Darmstadt
{vormbrock,wille}@mathematik.tu-darmstadt.de

Abstract. The concern of this paper is to elaborate a basic understanding of *semiconcepts* and *protoconcepts* as notions of Formal Concept Analysis. First, semiconcepts and protoconcepts are motivated by their use for effectively describing formal concepts. It is shown that one can naturally operate with those units of description, namely with operations which constitute algebras of semiconcepts and algebras of protoconcepts as so-called *double Boolean algebras*. The main results of this paper are the two basic theorems which characterize *semiconcept* resp. *protoconcept algebras* as pure resp. fully contextual double Boolean algebras whose related Boolean algebras are complete and atomic. Those theorems may, for instance, be applied to check whether line diagram representations of semiconcept and protoconcept algebras are correct.

1 Semiconcepts and Protoconcepts

Formal Concept Analysis has been formally enriched by introducing the notions of *semiconcept* and *protoconcept*. The concern of this paper is to elaborate a basic understanding of those notions. First, semiconcepts and protoconcepts are motivated by their use for effectively describing formal concepts. It is shown that one can naturally operate with those units of description, namely with operations which constitute algebras of semiconcept and algebras of protoconcept as so-called *double Boolean algebras*. The main results of this paper are the two basic theorems which characterize *semiconcept* resp. *protoconcept algebras* as pure resp. fully contextual double Boolean algebras whose related Boolean algebras are complete and atomic.

How to describe and define *concepts* properly is a basic question of the philosophical doctrine of concepts. Since complete declarations of the extension and the intension of a concept are seldom possible, concepts are usually described by sets of *prototypic objects* and *characteristic attributes*, respectively (cf. [La87],[Sch90], [Fo98]). Of course, concepts allow quite different descriptions for which an often cited example is given by the description terms "equilateral triangle" and "equiangular triangle". In general, *concept descriptions* should be rich enough to support the described concepts for fulfilling their role as basic units of thought and knowledge (cf. [Se01],[Wi04a]). Inspite of their richness, concept descriptions can only be understood on the basis of suitable *background knowledge*.

B. Ganter et al. (Eds.): Formal Concept Analysis, LNAI 3626, pp. 34–48, 2005.

In *Formal Concept Analysis* [GW99a], background knowledge is mathematized by *formal contexts* representing object-attribute-relationships. Concepts are then mathematized by *formal concepts* within those formal contexts; such a formal concept consists of an *extent*, mathematizing the original concept extension, and of an *intent*, mathematizing the original concept intension. This mathematization yields a general method of concept descriptions which constitute the intent and the extent of a formal concept by applying the corresponding *derivation operators* of the formal context to suitable sets of objects and of attributes, respectively. In the following, we discuss how formal concepts can be sufficiently described by the derivation method applied to "generating" pairs consisting of an object set and an attribute set.

Let us recall that a *formal context* has been introduced in [Wi82] as a set structure $\mathbb{K} := (G, M, I)$ for which G and M are sets while I is a binary relation between G and M, i.e. $I \subseteq G \times M$; the elements of G and M are called *objects* and *attributes*, respectively, and gIm, i.e. $(g, m) \in I$, is read: the object g *has the attribute* m. The *derivation operators* of $\mathbb{K}$ are defined as follows ($X \subseteq G$, $Y \subseteq M$):

$$X \mapsto X' := \{m \in M \mid gIm \text{ for all } g \in X\},$$
$$Y \mapsto Y' := \{g \in G \mid gIm \text{ for all } m \in Y\}.$$

Obviously, the two derivation operators satisfy the following three conditions:

$$(1)\ Z_1 \subseteq Z_2 \Longrightarrow Z_1' \supseteq Z_2', \ (2)\ Z \subseteq Z'', \ (3)\ Z''' = Z'.$$

Now, a *formal concept* of $\mathbb{K}$ is defined as a pair (A, B) with $A \subseteq G$, $B \subseteq M$, $A = B'$, and $B = A'$; A and B are called the *extent* and the *intent* of the formal concept (A, B), respectively. Because of condition (3), (X'', X') for $X \subseteq G$ and (Y', Y'') for $Y \subseteq M$ are always formal concepts, and each formal concept can be obtained by each of those derivation constructions. The *subconcept-superconcept-relation* is mathematized by

$$(A, B) \leq (C, D) :\Longleftrightarrow A \subseteq C \ (\Longleftrightarrow B \supseteq D).$$

The set of all formal concepts of $\mathbb{K}$ together with the defined order relation is a complete lattice, called the *concept lattice* of $\mathbb{K}$ and denoted by $\mathfrak{B}(\mathbb{K})$.

Only for small contexts, complete listings of the formal concepts with their extents and intents are possible, but even in those cases it might not be clear how to make effective use of such listings. If the formal context is infinite then complete listings of its extents and intents are obviously impossible, but sufficient concept descriptions might be still available. This shall be demonstrated by the formal context $(\mathbb{N}, M_{\mathbb{N}}, I_{\mathbb{N}})$ with $\mathbb{N} := \{1, 2, 3, \ldots\}$, $M_{\mathbb{N}} := \{$ *even, odd, prime, square, sum of two squares*$\}$ where $I_{\mathbb{N}}$ indicates which natural number has which property out of $M_{\mathbb{N}}$. Since the attribute set is finite, the context $(\mathbb{N}, M_{\mathbb{N}}, I_{\mathbb{N}})$ can only have finitely many formal concepts. As shown in [GW99b], the natural numbers $1, 2, 3, 4, 5, 25, 100$ suffice to describe the 18 formal concepts of $(\mathbb{N}, M_{\mathbb{N}}, I_{\mathbb{N}})$; for instance, the pair $(\{5, 25\}, \{$ *odd, sum of two squares*$\})$ describes the concept of all odd numbers which are sums of two squares, i.e., $\{5, 25\}$ is a prototypic object set for the odd-and-(sum of two square)-number-concept in the scope of the attribute set $M_{\mathbb{N}}$.

In an arbitrary context $\mathbb{K} := (G, M, I)$, a pair (A, B) with $A \subseteq G$ and $B \subseteq M$ is called a $\sqcap$-*semiconcept* if $A' = B$, which means that A is a *prototypic object set* for the formal concept (A'', B) of $\mathbb{K}$. Dually, a pair (C, D) with $C \subseteq G$ and $D \subseteq M$ is called a $\sqcup$-*semiconcept* if $D' = C$, which means that D is a *characteristic attribute set* for the formal concept (C, D'') of $\mathbb{K}$. Such semiconcepts have been first considered in the development of conceptual knowledge systems [LW91]. A typical occurrence of semiconcepts is already demonstrated by the subcontext $(\{1, 2, 3, 4, 5, 25, 100\}, M_{\mathbb{N}}, I \cap (\{1, 2, 3, 4, 5, 25, 100\} \times M_{\mathbb{N}}))$ of the number context $(\mathbb{N}, M_{\mathbb{N}}, I_{\mathbb{N}})$: The 18 formal concepts of the subcontext are $\sqcap$-semiconcepts of $(\mathbb{N}, M_{\mathbb{N}}, I_{\mathbb{N}})$, which function as descriptions of the 18 formal concepts of $(\mathbb{N}, M_{\mathbb{N}}, I_{\mathbb{N}})$, respectively.

Clearly, one is even interested in concept descriptions by pairs with a prototypic object set and a characteristic attribute set, in particular, if the given formal context has an infinite object set and an infinite attribute set. Such a pair (A, B) should at least be a *preconcept* of the concerning context $\mathbb{K} := (G, M, I)$ which, in general, is defined by $A \subseteq G$, $B \subseteq M$, and $A \subseteq B'$ or equivalently by $B \subseteq A'$ [SW86]. With repect to the *order* $\subseteq^2$ between preconcepts defined by

$$(A, B) \subseteq^2 (C, D) :\Longleftrightarrow A \subseteq C \text{ and } B \subseteq D,$$

the formal concepts of $\mathbb{K}$ are exactly the maximal preconcepts of $\mathbb{K}$. Thus, a preconcept (A, B) indicates uniquely a formal concept if and only if there is exactly one formal concept greater than or equal to (A, B). How that formal concept can be constructed by (A, B) is answered by the following lemma:

Lemma 1 *A preconcept* (A, B) *of a formal context* $\mathbb{K} := (G, M, I)$ *is less than or equal to exactly one formal concept* (C, D) *of* $\mathbb{K}$ *if and only if* $(B', A') = (C, D)$.

Proof: Obviously, $(A, B) \subseteq^2 (A'', A')$ and $(A, B) \subseteq^2 (B', B'')$. Therefore, if there is only one formal concept (C, D) above (A, B), it follows $(A'', A') = (B', B'')$ and hence $(B', A') = (C, D)$. Conversely, if $(B', A') = (C, D)$ and if $(A, B) \subseteq^2 (E, F)$ for some formal concept (E, F) of $\mathbb{K}$, then we obtain $F \subseteq A'$ and $E \subseteq B'$ by condition (1), i.e. $(E, F) \subseteq^2 (B', A')$ which forces $(E, F) = (C, D)$. Thus, there is exactly one formal concept above (A, B). $\square$

Lemma 1 motivates to introduce the notion of a *protoconcept* of a formal context $\mathbb{K} := (G, M, I)$ defined as a preconcept (A, B) of $\mathbb{K}$ for which (B', A') is a formal concept of $\mathbb{K}$ (cf. [Wi00a]). Semiconcepts are obviously the protoconcepts which have an extent or an intent as one of its components. Protoconcepts may be understood as mathematizations of units of thought which are constituted as concepts in restricted contexts in such a way that they extend uniquely to concepts in appropriate extensions of those contexts. The example in Fig.1 shall demonstrate this understanding of protoconcepts: A music beginner is first

	d	F	a	C	e	G	d–f	f–a	a–c	c–e	e–g	g–b	b–d
c		×	×	×					×	×			
d	×					×	×						×
e			×	×	×					×	×		
f	×	×					×	×					
g				×	×	×					×	×	
a	×	×	×					×	×				
b					×	×						×	×
c′		×	×	×					×	×			
d′	×					×	×						×
e′			×	×	×					×	×		
f′	×	×					×	×					
g′				×	×	×					×	×	
a′	×	×	×					×	×				
b′					×	×						×	×
c″		×	×	×					×	×			

Fig. 1. A context of 2- and 3-harmonies of the diatonic scale; the attributes are the 3-harmonies d-minor, F-major, a-minor, C-major, e-minor, G-major and the 2-harmonies d-f-minor, f-a-major, a-c-minor, c-e-major, e-g-minor, g-b-major, b-d-minor

learning harmonies within the basic diatonic scale c,d,e,f,g,a,b,c': for instance, the G-major 3-harmony is understood to be represented by the notes g,b,d and the e-minor 3-harmony by the notes e,g,b; the g-b-major 2-harmony is then recognized as the common part of those two harmonies. In this way the beginner acquires all the (formal) concepts of the 8×6 - subcontext in the upper left of Fig.1. Later on, similar conceptualizations lead to an extended understanding of harmonies on larger diatonic scales as, for instance, represented by the whole context in Fig.1. Since the formal concepts of the 8×6 - subcontext are protoconcepts of the whole context, the extended understanding is compatible with the first understanding of the presented diatonic harmonies. The concept lattices of the two contexts are even isomorphic which becomes visible in Fig.2.

Interestingly, it can be shown that every formal context has extensions in which each of its formal concepts is a proper protoconcept. For proving this, we use the *direct product* of formal contexts $\mathbb{K} := (G, M, I)$ and $\mathbb{K}^* := (G^*, M^*, I^*)$ which is defined as follows:

$$\mathbb{K} \times \mathbb{K}^* := (G \times G^*, M \times M^*, \nabla) \text{ with } (g, g^*)\nabla(m, m^*) : \iff gIm \text{ or } g^*I^*m^*$$

Proposition 1 *Let* $\mathbb{K} := (G, M, I)$ *and* $\mathbb{K}^* := (G^*, M^*, I^*)$ *be formal contexts with* $\emptyset \neq I^* \neq G^* \times M^*$ *and let* $(g^*, m^*) \in G^* \times M^* \setminus I^*$ *for which* $g^*I^*n^*$ *always implies* $(n^*)^{I^*} = G^*$ *and* $h^*I^*m^*$ *always implies* $(h^*)^{I^*} = M^*$. *Then* $g \mapsto (g, g^*)$ *for* $g \in G$ *and* $m \mapsto (m, m^*)$ *for* $m \in M$ *describe a context isomorphism from* $\mathbb{K}$ *onto the subcontext* $\mathbb{K}_{(g^*,m^*)} := (G \times \{g^*\}, M \times \{m^*\}, \nabla \cap (G \times \{g^*\}) \times (M \times \{m^*\}))$ *of* $\mathbb{K} \times \mathbb{K}^*$ *whose formal concepts are proper protoconcepts of* $\mathbb{K} \times \mathbb{K}^*$. *If* $|G^*| = 1$ *resp.* $|M^*| = 1$ *then the formal concepts of the subcontext* $\mathbb{K}_{(g^*,m^*)}$ *are* $\sqcup$-*semiconcepts resp.* $\sqcap$-*semiconcepts of* $\mathbb{K} \times \mathbb{K}^*$.

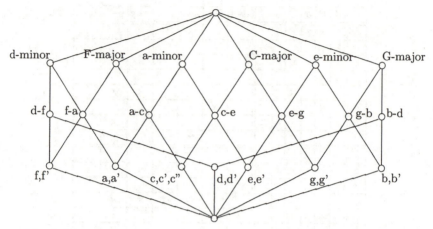

Fig. 2. Line diagram of the concept lattice of the formal context in Fig.1

Proof: Since $(g^*, m^*) \notin I^*$, we have $gIm \iff (g, g^*)\nabla(m, m^*)$; hence the maps $g \mapsto (g, g^*)$ and $m \mapsto (m, m^*)$ form a context isomorphism (α, β) from $\mathbb{K}$ onto the subcontext $\mathbb{K}_{(g^*, m^*)}$ of $\mathbb{K} \times \mathbb{K}^*$. Now, let (A, B) be a formal concept of $\mathbb{K}$. Then

$$(A \times (G^* \setminus (m^*)^{I^*}) \cup G \times (m^*)^{I^*}, \ B \times (M^* \setminus (g^*)^{I^*}) \cup M \times (g^*)^{I^*})$$

is a formal concept of $\mathbb{K} \times \mathbb{K}^*$. Thus, $(\alpha(A), \beta(B))$ is a protoconcept of $\mathbb{K} \times \mathbb{K}^*$ which is not a formal concept of $\mathbb{K} \times \mathbb{K}^*$ because $|G^*| > 1$ or $|M^*| > 1$. If $|G^*| = 1$ then $\alpha(A) = \beta(B)^\nabla$ so that $(\alpha(A), \beta(B))$ is a $\sqcup$-semiconcept of $\mathbb{K} \times \mathbb{K}^*$. If $|M^*| = 1$ then $\beta(B) = \alpha(A)^\nabla$ so that $(\alpha(A), \beta(B))$ is a $\sqcap$-semiconcept of $\mathbb{K} \times \mathbb{K}^*$.   □

2   Double Boolean Algebras

Originally, protoconcepts have been introduced in [Wi00a] for the mathematical development of a *Boolean Concept Logic*. The crucial question was how to define suitable operations of negation in conceptual structures. The problem is that, in a formal context $\mathbb{K} := (G, M, I)$, the complement of an extent need not to be an extent again. Therefore the Boolean negation on the powerset of G cannot be directly transformed to a negation operation on the concept lattice of $\mathbb{K}$. This becomes possible if the set $\mathfrak{B}(\mathbb{K})$ of all formal concepts of $\mathbb{K}$ is extended to the set $\mathfrak{P}(\mathbb{K})$ of all protoconcepts of $\mathbb{K}$. In the end, six fundamental operations have been defined on $\mathfrak{P}(\mathbb{K})$:

$$(A_1, B_1) \sqcap (A_2, B_2) := (A_1 \cap A_2, (A_1 \cap A_2)')$$
$$(A_1, B_1) \sqcup (A_2, B_2) := ((B_1 \cap B_2)', B_1 \cap B_2)$$
$$\neg(A, B) := (G \setminus A, (G \setminus A)')$$
$$\lnot(A, B) := ((M \setminus B)', M \setminus B)$$
$$\bot := (\emptyset, M)$$
$$\top := (G, \emptyset)$$

The set $\mathfrak{P}(\mathbb{K})$ together with the operations $\sqcap, \sqcup, \neg, \lrcorner, \bot$, and $\top$ is called the *protoconcept algebra* of $\mathbb{K}$ and is denoted by $\mathfrak{P}(\mathbb{K})$; the operations are named *"meet"*, *"join"*, *"negation"*, *"opposition"*, *"nothing"*, and *"all"*. For the structural analysis of the protoconcept algebra $\underline{\mathfrak{P}}(\mathbb{K})$, it is useful to define additional operations on $\mathfrak{P}(\mathbb{K})$:

$$\mathfrak{x} \sqcupplus \mathfrak{y} := \neg(\neg\mathfrak{x} \sqcap \neg\mathfrak{y}) \text{ and } \mathfrak{x} \sqcapplus \mathfrak{y} := \lrcorner(\lrcorner\mathfrak{x} \sqcup \lrcorner\mathfrak{y}),$$
$$\top := \neg\bot \text{ and } \bot := \lrcorner\top.$$

The semiconcepts of $\mathbb{K}$ form a subalgebra $\underline{\mathfrak{H}}(\mathbb{K})$ of $\underline{\mathfrak{P}}(\mathbb{K})$ which is called the *semiconcept algebra* of $\mathbb{K}$. The set $\mathfrak{H}_\sqcap(\mathbb{K})$ of all $\sqcap$-semiconcepts is closed under the operations $\sqcap, \sqcupplus, \neg, \bot$, and $\top$; therefore, $\underline{\mathfrak{H}}_\sqcap(\mathbb{K}) := (\mathfrak{H}_\sqcap(\mathbb{K}), \sqcap, \sqcupplus, \neg, \bot, \top)$ is a Boolean algebra isomorphic to the Boolean algebra of all subsets of G. Dually, the set $\mathfrak{H}_\sqcup(\mathbb{K})$ of all $\sqcup$-semiconcepts is closed under the operations $\sqcapplus, \sqcup, \lrcorner, \bot$, and $\top$; therefore, $\underline{\mathfrak{H}}_\sqcup(\mathbb{K}) := (\mathfrak{H}_\sqcup(\mathbb{K}), \sqcapplus, \sqcup, \lrcorner, \bot, \top)$ is a Boolean algebra antiisomorphic to the Boolean algebra of all subsets of M. Furthermore, $\mathfrak{B}(\mathbb{K}) = \mathfrak{H}_\sqcap(\mathbb{K}) \cap \mathfrak{H}_\sqcup(\mathbb{K})$, and $(\mathfrak{B}(\mathbb{K}), \sqcap, \sqcup)$ is the concept lattice of $\mathbb{K}$ of which $\sqcap$ and $\sqcup$ are the meet and join operation, respectively. The general order relation $\sqsubseteq$ of $\mathfrak{P}(\mathbb{K})$, which coincides on $\mathfrak{B}(\mathbb{K})$ with the subconcept-superconcept-order $\leq$, is defined by

$$(A_1, B_1) \sqsubseteq (A_2, B_2) :\Longleftrightarrow A_1 \subseteq A_2 \text{ and } B_1 \supseteq B_2.$$

The following theorem about equations in protoconcept algebras states an analogue of Stone's result that the equational axioms of the Boolean algebras form a basis for all equations which are valid in all powerset algebras:

Theorem 1 [Wi00a] *A basis of all equations which are valid in all protoconcept algebras is given by the following equations which are the equational axioms of the so-called* double Boolean algebras*:*

1a) $(x \sqcap x) \sqcap y = x \sqcap y$

1b) $(x \sqcup x) \sqcup y = x \sqcup y$

2a) $\quad x \sqcap y = y \sqcap x$

2b) $\quad x \sqcup y = y \sqcup x$

3a) $x \sqcap (y \sqcap z) = (x \sqcap y) \sqcap z$

3b) $x \sqcup (y \sqcup z) = (x \sqcup y) \sqcup z$

4a) $x \sqcap (x \sqcup y) = x \sqcap x$

4b) $x \sqcup (x \sqcap y) = x \sqcup x$

5a) $x \sqcap (x \sqcupplus y) = x \sqcap x$

5b) $x \sqcup (x \sqcapplus y) = x \sqcup x$

6a) $x \sqcap (y \sqcupplus z) = (x \sqcap y) \sqcupplus (x \sqcap z)$

6b) $x \sqcup (y \sqcapplus z) = (x \sqcup y) \sqcapplus (x \sqcup z)$

7a) $\neg\neg(x \sqcap y) = x \sqcap y$

7b) $\lrcorner\lrcorner(x \sqcup y) = x \sqcup y$

8a) $\neg(x \sqcap x) = \neg x$

8b) $\lrcorner(x \sqcup x) = \lrcorner x$

9a) $x \sqcap \neg x = \bot$

9b) $x \sqcup \lrcorner x = \top$

10a) $\neg\bot = \top \sqcap \top$

10b) $\lrcorner\top = \bot \sqcup \bot$

11a) $\neg\top = \bot$

11b) $\lrcorner\bot = \top$

12) $(x \sqcap x) \sqcup (x \sqcap x) = (x \sqcup x) \sqcap (x \sqcup x)$.

For a *double Boolean algebra* which is an algebra $\underline{D} := (D, \sqcap, \sqcup, \neg, \lrcorner, \bot, \top)$ of type $(2, 2, 1, 1, 0, 0)$ satisfying the equations 1a) to 11a), 1b) to 11b), and 12) of Theorem 1, further operations are defined as in the case of protoconcept algebras:

$$x \sqcup y := \neg(\neg x \sqcap \neg y) \text{ and } x \sqcap y := {}^{\lrcorner}({}^{\lrcorner}x \sqcup {}^{\lrcorner}y),$$
$$\top := \neg \bot \text{ and } \bot := {}^{\lrcorner}\top.$$

Clearly, each protoconcept algebra is a double Boolean algebra. Semiconcept algebras satisfy the following additional condition:

$$13) \; x \sqcap x = x \text{ or } x \sqcup x = x.$$

A double Boolean algebra $\underline{D}$ satisfying the condition 13) is called *pure*, because it is only the union of the two subsets $D_\sqcap := \{x \in D \mid x \sqcap x = x\}$ and $D_\sqcup := \{x \in D \mid x \sqcup x = x\}$ which both carry a Boolean structure, i.e., $\underline{D}_\sqcap := (D_\sqcap, \sqcap, \sqcup, \neg, \bot, \top)$ and $\underline{D}_\sqcup := (D_\sqcup, \sqcap, \sqcup, {}^{\lrcorner}, \bot, \top)$ are Boolean algebras. A detailed investigation of the structure of semiconcept algebras and double Boolean algebras is presented in [HLSW00]. For introducing an order on a double Boolean algebra, we imitate the order definition for protoconcept algebras:

$$x \sqsubseteq y : \Longleftrightarrow \; x \sqcap y = x \sqcap x \text{ and } x \sqcup y = y \sqcup y$$

On double Boolean algebras, the relation $\sqsubseteq$ is a quasiorder, i.e., it is reflexive and transitive, but not necessarily antisymmetric. The following lemma is basic for understanding this quasiorder $\sqsubseteq$:

Lemma 2 *For an element x in a double Boolean algebra $\underline{D}$, $x_\sqcap := x \sqcap x$ is the largest element in $D_\sqcap$ below x, i.e. $y \sqsubseteq x_\sqcap$ for all $y \in D_\sqcap$ with $y \sqsubseteq x$, and $x_\sqcup := x \sqcup x$ is the smallest element in $D_\sqcup$ above x, i.e. $y \sqsupseteq x_\sqcup$ for all $y \in D_\sqcup$ with $y \sqsupseteq x$.*

Proof: First of all, $x_\sqcap \sqsubseteq x$ because $x_\sqcap \sqcap x = x_\sqcap \sqcap x_\sqcap = x_\sqcap$ and $x_\sqcap \sqcup x = x \sqcup x$. Now, let $y \in D_\sqcap$ with $y \sqsubseteq x$, i.e. $y \sqcap x = y \sqcap y = y$ and $y \sqcup x = x \sqcup x$. Then $y \sqcap x_\sqcap = y \sqcap y$ and $y \sqcup x_\sqcap = (y \sqcap y) \sqcup x_\sqcap = (y \sqcap x_\sqcap) \sqcup x_\sqcap = x_\sqcap \sqcup x_\sqcap$ and hence $y \sqsubseteq x_\sqcap$. The dual claim follows dually. $\qquad\Box$

A double Boolean algebra $\underline{D}$ is said to be *complete* if the Boolean algebras $\underline{D}_\sqcap$ and $\underline{D}_\sqcup$ are complete. The existing infimum resp. supremum of a subset A of $D_\sqcap$ are denoted by $\sqcap A$ resp. $\sqcup A$ and, dually, of a subset B of $D_\sqcup$ by $\sqcap B$ resp. $\sqcup B$. In general, we define $\sqcap C := \sqcap \{c_\sqcap \mid c \in C\}$ and $\sqcup C := \sqcup \{c_\sqcup \mid c \in C\}$ for arbitrary subsets C of D. Clearly, semiconcept algebras and protoconcept algebras are examples of complete double Boolean algebras.

3   The Basic Theorem on Semiconcept Algebras

In Formal Concept Analysis, the basic theorems characterize abstractly basic structures which are derived from formal contexts. For those theorems, the Basic Theorem on Concept Lattices (see [Wi82], [GW99a]) is paradigmatic. Besides the multifarious use of this theorem in theoretic developments, it is frequently applied in practice, in particular by non-mathematicians. A typical aim is to check whether a lattice representation really presents the correct concept lattice. In the case of a labelled line diagram of a finite concept lattice, the Basic Theorem yields the following check list which guarantees a correct diagram:

1. each circle from which exactly one line segment descends must have an object label,
2. each circle from which exactly one line segment ascends must have an attribute label,
3. there is an ascending path of line segments from a circle with an object label to a circle with an attribute label if and only if that object has that attribute in the given formal context (this includes the case where the object label and the attribute label are attached to the same circle).

Using this check list, it can be easily seen that the line diagram of Fig.2 represents the concept lattice of the upper-left-8×6-subcontext in the of the formal context in Fig.1.

In this and the next section, the basic theorems on semiconcept and protoconcept algebras are presented (further basic theorems can be found in [Ha92] on lattices of topologically closed concepts, in [Wi95] on concept trilattices, in [Wl99] on ordered bilinear contexts, in [Dö99] on coherence networks of the concept lattices of a multicontext, and in [Wi03] on lattices of conceptual contents).

The Basic Theorem on Semiconcept Algebras. *For a context* $\mathbb{K} := (G, M, I)$, *the semiconcept algebra* $\underline{\mathfrak{H}}(\mathbb{K})$ *is a complete pure double Boolean algebra whose Boolean algebras* $\underline{\mathfrak{H}}_{\sqcap}(\mathbb{K})$ *and* $\underline{\mathfrak{H}}_{\sqcup}(\mathbb{K})$ *are atomic. The (arbitrary) meet and join of* $\underline{\mathfrak{H}}(\mathbb{K})$ *are given by*

$$\bigsqcap_{t \in T}(A_t, B_t) = (\bigcap_{t \in T} A_t, (\bigcap_{t \in T} A_t)') \quad and \quad \bigsqcup_{t \in T}(A_t, B_t) = ((\bigcap_{t \in T} B_t)', \bigcap_{t \in T} B_t).$$

In general, a complete pure double Boolean algebra $\underline{D}$ *whose Boolean algebras* $\underline{D}_{\sqcap}$ *and* $\underline{D}_{\sqcup}$ *are atomic, is isomorphic to* $\underline{\mathfrak{H}}(\mathbb{K})$ *if and only if there exist a bijection* $\tilde{\gamma}$ *from* G *onto the set* $A(\underline{D}_{\sqcap})$ *of all atoms of* $\underline{D}_{\sqcap}$ *and a bijection* $\tilde{\mu}$ *from* M *onto the set* $C(\underline{D}_{\sqcup})$ *of all coatoms of* $\underline{D}_{\sqcup}$ *such that* $gIm \iff \tilde{\gamma}(g) \sqsubseteq \tilde{\mu}(m)$ *for all* $g \in G$ *and* $m \in M$. *In particular, for any complete pure double Boolean algebra* $\underline{D}$ *whose Boolean algebras are atomic, we get* $\underline{D} \cong \underline{\mathfrak{H}}(A(\underline{D}_{\sqcap}), C(\underline{D}_{\sqcup}), \sqsubseteq)$, *i.e., the semiconcept algebras are up to isomorphism the complete pure double Boolean algebras* $\underline{D}$ *whose Boolean algebras* $\underline{D}_{\sqcap}$ *and* $\underline{D}_{\sqcup}$ *are atomic.*

Proof: Using Theorem 1, it is straightforward to check that every semiconcept algebra is a pure double Boolean algebra. Since $\underline{\mathfrak{H}}_{\sqcap}(\mathbb{K})$ and $\underline{\mathfrak{H}}_{\sqcup}(\mathbb{K})$ are isomorphic to the powerset algebra $\underline{\mathfrak{P}}(G)$ and the dual of the powerset algebra $\underline{\mathfrak{P}}(M)$, respectively, they are complete atomic Boolean algebras. Because of $(A, \overline{B})_{\sqcap} = (A, A')$ and $(A, B)_{\sqcup} = (B', B)$ for each semiconcept (A, B), we obtain

$$\bigsqcap_{t \in T}(A_t, B_t) = \inf_{\underline{\mathfrak{H}}_{\sqcap}(\mathbb{K})}\{(A_t, A'_t) \mid t \in T\} = (\bigcap_{t \in T} A_t, (\bigcap_{t \in T} A_t)'),$$

$$\bigsqcup_{t \in T}(A_t, B_t) = \sup\nolimits_{\mathfrak{H}_{\sqcup}(\mathbb{K})}\{(B'_t, B_t) \mid t \in T\} = ((\bigcap_{t \in T} B_t)', \bigcap_{t \in T} B_t).$$

Now, let $\varphi : \mathfrak{H}(\mathbb{K}) \to \underline{D}$ be an isomorphism. Then we define $\tilde{\gamma}(g) := \varphi(\{g\}, \{g\}')$ and $\tilde{\mu}(m) := \varphi(\{m\}', \{m\})$. Since $\mathcal{A}(\underline{\mathfrak{H}_{\sqcap}}(\mathbb{K})) = \{(\{g\}, \{g\}') \mid g \in G\}$ and $\mathcal{C}(\underline{\mathfrak{H}_{\sqcup}}(\mathbb{K}))$
$= \{(\{m\}', \{m\}) \mid m \in M\}$, it follows $\mathcal{A}(\underline{D_{\sqcap}}(\mathbb{K})) = \{\tilde{\gamma}(g) \mid g \in G\}$ and $\mathcal{C}(\underline{D_{\sqcup}}(\mathbb{K})) = \{\tilde{\mu}(m) \mid m \in M\}$. Thus, $\tilde{\gamma}$ is a bijection from G onto $\mathcal{A}(\underline{D_{\sqcap}})$ and $\tilde{\mu}$ is a bijection from M onto $\mathcal{C}(\underline{D_{\sqcup}})$. Furthermore, $gIm \iff g \in \{m\}'$ and $m \in \{g\}' \iff (\{g\}, \{g\}') \sqsubseteq (\{m\}', \{m\}) \iff \tilde{\gamma}(g) \sqsubseteq \tilde{\mu}(m)$ for all $g \in G$ and $m \in M$.

Conversely, we assume the existence of the bijections $\tilde{\gamma}$ and $\tilde{\mu}$ with the required properties. Then we define two maps $\varphi_{\sqcap} : \underline{\mathfrak{H}_{\sqcap}}(\mathbb{K}) \to \underline{D_{\sqcap}}$ and $\varphi_{\sqcup} : \underline{\mathfrak{H}_{\sqcup}}(\mathbb{K}) \to \underline{D_{\sqcup}}$ by $\varphi_{\sqcap}(A, A') := \bigsqcup\{\tilde{\gamma}(g) \mid g \in A\}$ and $\varphi_{\sqcup}(B', B) := \bigsqcap\{\tilde{\mu}(m) \mid m \in B\}$. Since $\underline{D_{\sqcap}}$ and $\underline{D_{\sqcup}}$ are complete atomic Boolean algebras, $\varphi_{\sqcap}$ and $\varphi_{\sqcup}$ are isomorphisms onto those Boolean algebras. For an arbitrary $\sqcap$-semiconcept (A, A') of $\mathbb{K}$, let $x := \varphi_{\sqcap}(A, A')$ and $y := \varphi_{\sqcup}(A'', A')$. Lemma 2 yields that $x_{\sqcup} = \bigsqcap\{b \in \underline{D_{\sqcup}} \mid x \sqsubseteq b\} = \bigsqcap\{c \in \mathcal{C}(\underline{\mathfrak{H}_{\sqcup}}(\mathbb{K})) \mid x \sqsubseteq c\} = y$ because of the equivalence $gIm \iff \tilde{\gamma}(g) \sqsubseteq \tilde{\mu}(m)$. Thus,

$$(\varphi_{\sqcap}(A, A'))_{\sqcup} = \varphi_{\sqcup}((A, A')_{\sqcup}) \text{ and } (\varphi_{\sqcup}(B', B))_{\sqcap} = \varphi_{\sqcap}((B', B)_{\sqcap})$$

because, for a $\sqcup$-semiconcept (B', B) of $\mathbb{K}$, we obtain dually $y_{\sqcap} = x$ if $y := \varphi_{\sqcup}(B', B)$ and $x := \varphi_{\sqcap}(B', B'')$. If (A, B) is even a formal concept of $\mathbb{K}$, we have $x = y_{\sqcap} = x_{\sqcup\sqcap} = x_{\sqcap\sqcup} \in D_{\sqcap} \cap D_{\sqcup}$ by the equation (12) in Theorem 1. It follows that $\varphi_{\sqcap}(A, B) = x = x_{\sqcup} = y = \varphi_{\sqcup}(A, B)$. Thus, $\varphi_{\sqcap}$ and $\varphi_{\sqcup}$ coincide on $\mathfrak{B}(\mathbb{K})(= \mathfrak{H}(\mathbb{K})_{\sqcap} \cap \mathfrak{H}(\mathbb{K})_{\sqcup})$ and therefore $\varphi(A, A') := \varphi_{\sqcap}(A, A')$ for $A \subseteq G$ and $\varphi(B', B) := \varphi_{\sqcup}(B', B)$ for $B \subseteq M$ defines a bijection φ from $\mathfrak{H}(\mathbb{K})$ onto $\underline{D}$. φ preserves the operations of double Boolean algebras which can be seen as follows: Since $\varphi_{\sqcap}$ and $\varphi_{\sqcup}$ are isomorphisms between the corresponding Boolean algebras and since $\varphi_{\sqcap}((A, A')_{\sqcap}) = (\varphi_{\sqcap}(A, A'))_{\sqcap}$ and $\varphi_{\sqcap}((B', B)_{\sqcap}) = (\varphi_{\sqcup}(B', B))_{\sqcap}$, we get

$$\varphi\bigsqcap_{t \in T}(A_t, B_t) = \varphi_{\sqcap}\bigsqcap_{t \in T}(A_t, B_t)_{\sqcap} = \bigsqcap_{t \in T}\varphi_{\sqcap}((A_t, B_t)_{\sqcap})$$
$$= \bigsqcap_{t \in T}(\varphi(A_t, B_t))_{\sqcap} = \bigsqcap_{t \in T}(\varphi(A_t, B_t));$$
$$\varphi(\neg(A, B)) = \varphi_{\sqcap}(\neg((A, B)_{\sqcap})) = \neg(\varphi_{\sqcap}((A, B)_{\sqcap}))$$
$$= \neg((\varphi(A, B))_{\sqcap}) = \neg(\varphi(A, B));$$
$$\varphi(\emptyset, M) = \varphi_{\sqcap}(\emptyset, M) = \bot.$$

Dually, we obtain that φ preserves joins $\bigsqcup$, opposition $\lrcorner$, and the top element $\top$.

Finally, let $\underline{D}$ be a complete pure double Boolean algebra whose Boolean algebras are atomic, let $\tilde{\gamma}$ be the identity on $\mathcal{A}(\underline{D_{\sqcap}})$, and let $\tilde{\mu}$ be the identity on $\mathcal{C}(\underline{D_{\sqcup}})$. Then the already proved second part of the basic theorem yields directly the claimed isomorphy $\underline{D} \cong \mathfrak{H}(\mathcal{A}(\underline{D_{\sqcap}}), \mathcal{C}(\underline{D_{\sqcup}}), \sqsubseteq)$. $\qquad\square$

By an example, we show how the Basic Theorem on Semiconcept Algebras may be used to check a *line diagram representation* of a semiconcept algebra.

	Guest House	Hotel	First Class	Luxus
Kempinski		×	×	×
Hilton		×	×	
ibis		×		
ISSY	×			

Fig. 3. A context $\mathbb{K}^h$ of some hotels in Dresden

For this, we consider the formal context $\mathbb{K}^h := (G^h, M^h, I^h)$ in Fig.3 whose semiconcept algebra is presented by the labelled line diagram in Fig.4. In this figure, the semiconcept algebra $\mathfrak{H}(\mathbb{K}^h)$ is first of all drawn as the ordered set $(\mathfrak{H}(\mathbb{K}^h), \sqsubseteq)$. The $\sqcap$-semiconcepts of the Boolean algebra $\mathfrak{H}_\sqcap(\mathbb{K}^h)$ are represented by the 16 circles in which the lower half is blackened and the $\sqcup$-semiconcepts of the Boolean algebra $\mathfrak{H}_\sqcup(\mathbb{K}^h)$ are represented by the 16 circles in which the upper half is blackened. Therefore the formal concepts of the concept lattice $\mathfrak{B}(\mathbb{K}^h)$ are presented by the 5 fully blackened circles. The formal concepts give rise to the indicated partition into five equivalence classes where two semiconcepts are equivalent if and only if they generate the same formal concept. In such an equivalence class, the $\sqcap$-semiconcepts are below and the $\sqcup$-semiconcepts are above the unique formal concept in that class. The bijection $\tilde{\gamma}$ becomes visible by the object names attached to the atoms of the Boolean algebra $\mathfrak{H}_\sqcap(\mathbb{K}^h)$ and the bijection $\tilde{\mu}$ is visible by the attribute names attached to the coatoms of the Boolean algebra $\mathfrak{H}_\sqcup(\mathbb{K}^h)$. Finally, there is an ascending path of line segments from a circle with an object label to a circle with an attribute label if and only if the object has the attribute according to the formal context $\mathbb{K}^h$. After checking all of this, we know by the Basic Theorem on Semiconcept Algebras that the labelled line diagram in Fig.4 correctly represents the semiconcept algebra of the formal context $\mathbb{K}^h$ given in Fig.3.

4   The Basic Theorem on Protoconcept Algebras

For formulating the Basic Theorem on Protoconcept Algebras, we introduce the following notions: A double Boolean algebra $\underline{D}$ is called *contextual* if its quasiorder $\sqsubseteq$ is antisymmetric, i.e. the relation $\sqsubseteq$ is a (partial) order on $\underline{D}$. A contextual double Boolean algebra $\underline{D}$ is said to be *fully contextual* if, in addition, for each $x \in \underline{D}_\sqcap$ and $y \in \underline{D}_\sqcup$ with $x_\sqcup = y_\sqcap$ there is a unique $z \in \underline{D}$ with $z_\sqcap = x$ and $z_\sqcup = y$.

The Basic Theorem on Protoconcept Algebras. *For a context* $\mathbb{K} := (G, M, I)$, *the protoconcept algebra* $\mathfrak{P}(\mathbb{K})$ *of* $\mathbb{K}$ *is a complete fully contextual double Boolean algebra whose Boolean algebras* $\mathfrak{H}_\sqcap(\mathbb{K})$ *and* $\mathfrak{H}_\sqcup(\mathbb{K})$ *are atomic. The (arbitrary) meet and join of* $\mathfrak{P}(\mathbb{K})$ *are given by*

$$\bigsqcap_{t \in T}(A_t, B_t) = (\bigcap_{t \in T} A_t, (\bigcap_{t \in T} A_t)') \quad and \quad \bigsqcup_{t \in T}(A_t, B_t) = ((\bigcap_{t \in T} B_t)', \bigcap_{t \in T} B_t).$$

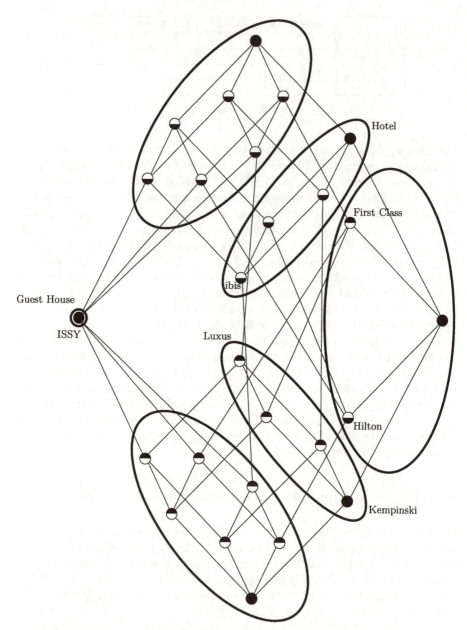

Fig. 4. Line diagram of the semiconcept algebra of the formal context $\mathbb{K}^h$ in Fig.3

In general, a complete fully contextual double Boolean algebra $\underline{D}$ whose Boolean algebras $\underline{D}_\sqcap$ and $\underline{D}_\sqcup$ are atomic, is isomorphic to $\mathfrak{P}(\mathbb{K})$ if and only if there exist a bijection $\tilde{\gamma}$ from G onto the set $\mathcal{A}(\underline{D}_\sqcap)$ of all atoms of $\underline{D}_\sqcap$ and a bijection $\tilde{\mu}$ from M onto the set $\mathcal{C}(\underline{D}_\sqcup)$ of all coatoms of $\underline{D}_\sqcup$ such that $gIm \iff \tilde{\gamma}(g) \sqsubseteq \tilde{\mu}(m)$ for all $g \in G$ and $m \in M$. In particular, for any complete fully *contextual double Boolean algebra $\underline{D}$ whose Boolean algebras are atomic, we get $\underline{D} \cong \mathfrak{P}(\mathcal{A}(\underline{D}_\sqcap), \mathcal{C}(\underline{D}_\sqcup), \sqsubseteq)$, i.e., the protoconcept algebras are up to isomorphism the complete fully contextual double Boolean algebras $\underline{D}$ whose Boolean algebras $\underline{D}_\sqcap$ and $\underline{D}_\sqcup$ are atomic.*

Proof: Using Theorem 1, it is straightforward to check that every protoconcept algebra $\mathfrak{P}(\mathbb{K})$ is a complete contextual double Boolean algebra whose Boolean algebras $\mathfrak{P}_\sqcap(\mathbb{K})$ and $\mathfrak{P}_\sqcup(\mathbb{K})$ are atomic. For semiconcepts (A, A') and (B', B) with $(A, \overline{A'})_\sqcup = (B', \overline{A'}) = (B', B)_\sqcap$, (A, B) is the unique protoconcept with $(A, B)_\sqcap = (A, A')$ and $(A, B)_\sqcup = (B', B)$; hence $\mathfrak{P}(\mathbb{K})$ is even fully contextual. The descripton of arbitrary meets and joins can be justified as in the proof of the Basic Theorem on Semiconcept Algebras because each proper join and meet of $\mathfrak{P}(\mathbb{K})$ always results in $\mathfrak{H}(\mathbb{K})$.

Since $\mathfrak{H}(\mathbb{K})$ is a subalgebra of $\mathfrak{P}(\mathbb{K})$ and $\underline{D}_p := \underline{D}_\sqcap \cup \underline{D}_\sqcap$ is a complete pure subalgebra of the complete double Boolean algebra $\underline{D}$, the Basic Theorem on Semiconcept Algebras can be applied to obtain that $\mathfrak{H}(\mathbb{K}) \cong \underline{D}_p$ if and only if there exist a bijection $\tilde{\gamma}$ from G onto $\mathcal{A}(\underline{D}_\sqcap)$ and a bijection $\tilde{\mu}$ from M onto $\mathcal{C}(\underline{D}_\sqcup)$ such that $gIm \iff \tilde{\gamma}(g) \sqsubseteq \tilde{\mu}(m)$. An isomorphism $\varphi : \mathfrak{P}(\mathbb{K}) \to \underline{D}$ restricts to an isomorphism from $\mathfrak{H}(\mathbb{K})$ onto $\underline{D}_p$. Then the Basic Theorem on Semiconcept Algebras guarantees the existence of the bijections $\tilde{\gamma}$ and $\tilde{\mu}$ with the desired properties. Now, we assume conversely the existence of such bijections $\tilde{\gamma}$ and $\tilde{\mu}$. Then, by the Basic Theorem on Semiconcept Algebras, there exists an isomorphism φ from $\mathfrak{H}(\mathbb{K})$ onto $\underline{D}$. Let (A, B) be a protoconcept of $\mathbb{K}$ which is not a semiconcept. Then (A, A') and (B', B) are proper $\sqcap$- and $\sqcup$-semiconcepts, respectively, with $(A, A') = (A, B)_\sqcap$, $(B', B) = (A, B)_\sqcup$, and $(A, A')_\sqcup = (B', A') = (B', B)_\sqcap$. For $x := \varphi(A, A')$ and $y := \varphi(B', B)$, it follows $x_\sqcup = \varphi(B', A') = y_\sqcap$. Therefore, there is a unique $z(A, B) \in \underline{D}$ with $z(A, B)_\sqcap = x$ and $z(A, B)_\sqcup = y$. Thus, we can extend φ to a bijection $\hat{\varphi} : \mathfrak{P}(\mathbb{K}) \to \underline{D}$ with $\hat{\varphi}(A, B) = z(A, B)$ for all protoconcepts (A, B) which are not semiconcepts. Using $(\hat{\varphi}(A, B))_\sqcap = \hat{\varphi}(A, A') = \hat{\varphi}((A, B)_\sqcap)$ and its dual, the equations of Theorem 1 yield a proof that $\hat{\varphi}$ is even an isomorphism.

Finally, let $\underline{D}$ be a complete fully contextual double Boolean algebra whose Boolean algebras are atomic, let $\tilde{\gamma}$ be the identity on $\mathcal{A}(\underline{D}_\sqcap)$, and let $\tilde{\mu}$ be the identity on $\mathcal{C}(\underline{D}_\sqcup)$. Then the already proved second part of the basic theorem yields directly the claimed isomorphy $\underline{D} \cong \mathfrak{P}(\mathcal{A}(\underline{D}_\sqcap), \mathcal{C}(\underline{D}_\sqcup), \sqsubseteq)$. □

For graphically representing a protoconcept algebra, a labelled line diagram of its semiconcept algebra (as shown in Section 3) is sufficient. In such a diagram, a protoconcept (A, B) which is not a semiconcept is depicted in the equivalence class of the corresponding formal concept (B', A') by the two circles which represent the semiconcepts (A, A') and (B', B), respectively. Conversely, the two representing circles of semiconcepts (A, A') and (B', B) in the equivalence class of a formal concept (B', A') depict a proper protoconcept, namely (A, B). Of course, one could add a single circle for representing (A, B) with links downwards to the circle of (A, A') and upwards to the circle to (B', B), but this might negatively effect the readability of the diagram.

5   Further Developments

Further research on semiconcept and protoconcept algebras is mainly motivated by the aim to mathematically support the development of *Conceptual Knowledge Processing* [Wi94]. This aim has as consequence that semiconcept and protoconcept algebras are first of all investigated in the scope of *Contextual Logic* [Wi00b], in particular of *"Boolean Concept Logic"* whose role as part of Contextual Logic is outlined in [Wi00a]. The results of Boolean Concept Logic are basic for the *Boolean Judgment Logic* which aims at a comprehensive theory of formal judgments mathematically represented by (semi- and proto-)concept graphs of power context families (cf. [KV03],[DK04]). A semantic approach to the investigation of semiconcept and protoconcept graphs is presented in [Wi01] and [Wi02a]. The interplay of syntax and semantics of semiconcept graphs is studied in [Kl01] and [Kl02]. Algebras of distinctive judgments are analysed in [Wi02b]. For applying that and further research in Contextual Boolean Logic, there are presently the two main fields of data analysis and information systems; for both fields the most promising software would be a suitable adaptation of TOSCANA, the successful program system of Formal Concept Analysis (cf. [EGSW00]).

The mathematical theory of semiconcept algebras has substantially started in [HLSW00] where, in particular, is proved that there are equations valid in all semiconcept algebras, but not in all protoconcept algebras. The main result of [Wi00a] is that protoconcept algebras generate the equationally defined class of all double Boolean algebras (cf. Theorem 1). This result motivates to attack the word problem of double Boolean algebras which, in particular, sets the task of determining the free double Boolean algebras. In this scope, the question arises whether there are useful normal forms for algebraic terms concerning double Boolean algebras. Solutions to those problems and tasks will lead to further questions about the algorithmic treatment of those solutions. Structural knowledge which might help to master those tasks consists in decomposition theorems and effective descriptions of congruence relations of double Boolean algebras; basic results have already been proved in [Vo03]. Further research has been started on the more general preconcept algebras and the corresponding generalized double Boolean algebras in [Wi04b].

References

[DK04]   F. Dau, J. Klinger: From Formal Concept Analysis to Contextual Logic. This volume.

[Dö99]   S. Dörflein: Coherence networks of concept lattices. Dissertation, TU Darmstadt. Shaker Verlag, Aachen 1999.

[EGSW00]   P. Eklund, B. Groh, G. Stumme, R. Wille: A contextual-logic extension of TOSCANA. In: B. Ganter, G. W. Mineau (eds.): Conceptual structures: logical, linguistic and computational issues. LNAI **1867**. Springer, Heidelberg 2000, 453-467.

[Fo98]   J. A. Fodor: *Concepts: where cognitive science went wrong.* Oxford University Press, New York 1998.

[GW99a]   B. Ganter, R. Wille: *Formal Concept Analysis: mathematical foundations*. Springer, Heidelberg 1999; German version: Springer, Heidelberg 1996.

[GW99b]   B. Ganter, R. Wille: Contextual Attribute Logic. In: W. Tepfenhart, W. Cyre (eds.): *Conceptual structures: standards and practices*. LNAI **1640**. Springer, Heidelberg 2000, 377–388.

[Ha92]   G. Hartung: Topological representation of lattices via formal concept analysis. *Algebra Universalis* **29** (1992), 273–299.

[HLSW00]   C. Herrmann, P. Luksch, M. Skorsky, R. Wille: Algebras of semiconcepts and double Boolean algebras. In: *Contributions to General Algebra* **13**. Verlag Johannes Heyn, Klagenfurt 2001, 175–188.

[Kl01]   J. Klinger: Simple semiconcept graphs: a Boolean approach. In: H. Delugach, G. Stumme (eds.): *Conceptual structures: broadening the base*. LNAI **2120**. Springer, Heidelberg 2001, 101–114.

[Kl02]   J. Klinger: Semiconcept graphs with variables. In: U. Priss, D. Corbett, G. Angelova (eds.): *Conceptual structures: integration and interfaces*. LNAI **2393**. Springer, Heidelberg 2002, 369–381.

[KV03]   J. Klinger, B. Vormbrock: Contextual Boolean Logic: how did it develop? In: B. Ganter, A. de Moor (eds.): *Using conceptual structures. Contributions to ICCS 2003*. Shaker Verlag, Aachen 2003, 143–156.

[La87]   G. Lakoff: Cognitive models and prototype theory. In: U. Neisser (ed.): *Concepts and conceptual developments: ecological and intellectual factors in categorization*. Cambridge University Press, Cambridge 1987, 63–100.

[LW91]   P. Luksch, R. Wille: A mathematical model for conceptual knowledge systems. In: H. H. Bock, P. Ihm (eds.): *Classification, data analysis, and knowledge organisation*. Springer, Heidelberg 1991, 156–162.

[Sch90]   E. Schröder: *Algebra der Logik*. Bd.1. Leipzig 1890; published again by Chelsea Publ. Comp., New York 1966.

[Se01]   Th. B. Seiler: *Begreifen und Verstehen. Ein Buch über Begriffe und Bedeutungen*. Verlag Allgemeine Wissenschaft, Mühltal 2001.

[SW86]   J. Stahl, R. Wille: Preconcepts and set representations of contexts. In: W. Gaul, M. Schader (eds.): *Classification as a tool of research*. North-Holland, Amsterdam 1986, 431–438.

[Vo03]   B. Vormbrock: Congruence relations on double Boolean algebras. *Algebra Universalis* (submitted)

[Wi82]   R. Wille: Restructuring lattice theory: an approach based on hierarchies of concepts. In: I. Rival (ed.): *Ordered sets*. Reidel, Dordrecht 1982, 445–470.

[Wi94]   R. Wille: Plädoyer für eine philosophische Grundlegung der Begrifflichen Wissensverarbeitung. In: R. Wille, M. Zickwolff (eds.): *Begriffliche Wissensverarbeitung: Grundfragen und Aufgaben*. B.I.-Wissenschaftsverlag, Mannheim 1994, 11–25.

[Wi95]   R. Wille: The basic theorem of Triadic Concept Analysis. *Order* **12** (1995), 149–158.

[Wi00a]   R. Wille: Boolean Concept Logic. In: B. Ganter, G. W. Mineau (eds.): *Conceptual structures: logical, linguistic, and computational issues*. LNAI **1867**. Springer, Heidelberg 2000, 317–331.

[Wi00b]   R. Wille: Contextual Logic summary. In: G. Stumme (ed.): *Working with Conceptual Structures. Contributions to ICCS 2000*. Shaker, Aachen 2000, 265–276.

[Wi01]   R. Wille: Boolean Judgment Logic. In: H. Delugach, G. Stumme (eds.): *Conceptual structures: broadening the base*. LNAI **2120**. Springer, Heidelberg 2001, 115–128.

[Wi02a]   R. Wille: Existential concept graphs of power context families. In: U. Priss, D. Corbett, G. Angelova (eds.): *Conceptual structures: integration and interfaces*. LNAI **2393**. Springer, Heidelberg 2002, 382–395.

[Wi02b]   R. Wille: The contextual-logic structure of distinctive judgments. In: G. Angelova, U. Priss, D. Corbett (eds.): *Foundations and applications of conceptual structures. Contributions to ICCS 2002*. Bulgarian Academy of Sciences, Sofia 2002, 92–101.

[Wi03]   R. Wille: Conceptual content as information - basics for Conceptual Judgment Logic. In: A. de Moor, W. Lex, B. Ganter (eds.): *Conceptual structures for knowledge creation and communication*. LNAI **2746**. Springer, Heidelberg 2003,1-5.

[Wi04a]   R. Wille: Formal Concept Analysis as mathematical theory of concepts and concept hierarchies. This volume.

[Wi04b]   R. Wille: Preconcept algebras and generalized double Boolean algebras. In: P. Eklund (ed.): *Concept lattices*. LNAI **2961**. Springer, Heidelberg 2004, 1–13.

[Wl99]   U. Wille: Characterization of ordered bilinear contexts. *Journal of Geometry* **64** (1999), 167–207.

Features of Interaction
Between Formal Concept Analysis
and Algebraic Geometry*

Tim Becker

Institute for Medical Biometry, Informatics and Epidemiology,
Sigmund-Freud-Str. 25, D-53105 Bonn
becker@imbie.meb.uni-bonn.de

Abstract. This paper contributes to Algebraic Concept Analysis by examining connections between Formal Concept Analysis and Algebraic Geometry. The investigations are based on **polynomial contexts** (over a field K in n variables) which are defined by $\mathbb{K}^{(n)} := (K^n, K[x_1, \ldots, x_n], \perp)$ where $a \perp f :\Leftrightarrow f(a) = 0$ for $a \in K^n$ and any polynomial $f \in K[x_1, \ldots, x_n]$. Important notions of Algebraic Geometry such as algebraic varieties, coordinate algebras, and polynomial morphisms are connected to notions of Formal Concept Analysis. That allows to prove many interrelating results between Algebraic Geometry and Formal Concept Analysis, even for more abstract notions such as affine and projective schemes.

1 Introduction

The following paper formulates results from Algebraic Geometry in the language of Formal Concept Analysis. We are able to determine which classical results from Algebraic Geometry follow already from the concept analysis consideration. Furthermore, we get a new way of interpretation and gain additional insight in the classical results. However, Formal Concept Analysis will also benefit from this investigation since it can take over several algebraic notions. We will proceed from a generalization of the Galois correspondence arising from the fundamental relation $\perp$ in Algebraic Geometry, where we have $a \perp f$ if and only if $f(a) = 0$. Here f is a polynomial in n variables over a given field K and a is an element from the affine space K^n over K. In it notions such as affine subspaces and parallelity are defined. The final version of the generalized relation yields concept lattices with additional properties that correspond to the notion of a scheme in Algebraic Geometry. In this way we get a new understanding of schemes and make them accessible by the methods of Formal Concept Analysis. Of course, we are able to treat less sophisticated notions as well.

The goal of this paper cannot be to reinvent (parts of) Algebraic Geometry. Many questions treated here arise only from a concept analysis point of view.

---

* This paper is an adapted version of the first part of [Be99]

B. Ganter et al. (Eds.): Formal Concept Analysis, LNAI 3626, pp. 49–80, 2005.
© Springer-Verlag Berlin Heidelberg 2005

Nevertheless, we will get a deeper insight into several features of the theory. A good example of this is the description of the classical sheaf construction in terms of Formal Concept Analysis. On the other hand, construction principles from Algebraic Geometry will find their way into the language of formal contexts. For instance, the glueing of contexts along "open" sets, introduced in section 7, is a new construction principle which can be understood as a generalization of the semi-product of contexts. Furthermore, the classical sheaf construction shows possibilities for introducing "general objects" not only in Algebraic Geometry, but also for a large class of usual formal contexts. Basic notions and results of Formal Concept Analysis used in this paper can be found in [GW99]. Basics of Algebraic Geometry are presented in the next section.

2 Basics of Algebraic Geometry

Algebraic Geometry is the mathematical discipline that arose from the investigation of the solutions of algebraic equations in the affine space K^n over a given field K. Such solution sets are called (affine) **algebraic varieties**. An algebraic variety V can be written in the form $V := \{a \in K^n \mid \forall i \in I : f_i(a) = 0\}$ where the $f_i \in K[x_1, \ldots, x_n]$ are polynomials in n variables. A useful introduction to Algebraic Geometry (and Commutative Algebra) is [Ku80]. A recent update of this book, with focus on Algebraic Geometry, is [Ku97]. Further information can be found in [Br89]. Let us start by giving someexamples of algebraic varieties. In order to be able to draw them, we work in the real plane. For instance, the variety defined by the polynomial $f := x^2 + y^2 - 1$ is just the circle of radius 1 with centre at the origin shown in Figure 3.

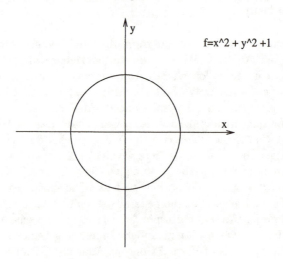

Fig. 1. A circle variety

The equation $(x^2 + y^2)^3 - 4x^2y^2 = 0$ yields the **four-leaved rose** which is also defined by the polar equation $r = sin(2\theta)$.

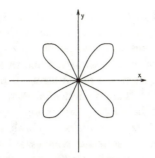

Fig. 2. The four-leaved rose

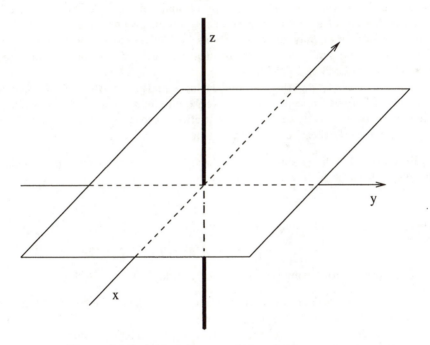

Fig. 3. An algebraic variety consisting of two components

Just as in the case of vector spaces which can be interpreted as algebraic varieties defined by linear polynomials, each algebraic variety has a dimension. The algebraic variety V depicted below which is defined by $xz = 0$ and $yz = 0$ has dimension 2. Note that this algebraic variety is the union of two components (xy-plane and z-axis), each of which has no further finite decomposition into algebraic varieties. The existence of such a decomposition into **irreducible** components is a characteristic feature of algebraic varieties as we will see later.

It can be easily seen that a variety $V := V(f_1, \ldots f_n)$ depends only on the ideal generated by the polynomials $f_1, \ldots, f_n$. On the other hand, the set $I(V)$ of all polynomials vanishing on a given algebraic variety V is an ideal. Thus the question arises how the ideal $I(V(I))$ vanishing on the variety determined by I can be described. In the case of an algebraically closed field K, the answer is given by Hilbert's Nullstellensatz (cited below).

Algebraic Geometry not only deals with affine varieties but also with projective varieties which are solution sets of homogenous polynomials in a projective space. The consideration of those varieties is adequate because they allow a smoother theory. They can be applied, for instance, in the theory of elimination of algebraic equations [CLO92]. Nowadays, algebraic varieties are often replaced by the more general notion of a **scheme**, which was introduced by Grothendieck [GD60]. Schemes are also described in [Ha77]. We will investigate them in Section 7. The most recent development in Algebraic Geometry are so-called **Gröbner bases** which are very suitable for computational purposes. A Gröbner basis is a generating set of an ideal with especially useful properties. It helps to find explicit solutions of polynomial equations if they do exist, and to determine if a given polynomial f belongs to an ideal or not. Gröbner bases are treated carefully in [CLO92] and [BW91].

Now, in order to describe the underlying situation in Algebraic Geometry in terms of Formal Concept Analysis, we would like to define a formal context whose extents are exactly the algebraic varieties in the affine space K^n and whose intents are exactly those ideals which are determined by algebraic varieties.

Definition 1 Let K be a field and let $K[x_1, \ldots, x_n]$ be the ring of polynomials in n variables over K. Then the formal context

$$\mathbb{K}^{(n)} := (K^n, K[x_1, \ldots, x_n], \perp),$$

where for $a \in K^n$ and a polynomial f we have

$$a \perp f : \Longleftrightarrow f(a) = 0,$$

is called **polynomial context** (over K in n variables).   ◇

We wish to determine the corresponding concept lattice. For $A \subseteq K^n$ it can be easily seen that $A^\perp = \{f \in K[x_1, \ldots, x_n] \mid \forall a \in A : f(a) = 0\}$ is closed under addition and under multiplication with arbitrary polynomials, i.e., $A^\perp$ is an ideal[1]. According to Hilbert's Basissatz (cf. [Ku80], I.2), every ideal in the polynomial ring is finitely generated. Therefore we have $A^\perp = < f_1, \ldots, f_m >$ with a finite number of polynomials. From basic results of Formal Concept Analysis it follows that every extent V of $\mathbb{K} := \mathbb{K}^{(n)}$ is of the form $A^{\perp\perp}$ for some $A \subseteq K^n$. Hence, every extent of $\mathbb{K}$ has the form $V := < f_1, \ldots, f_m >^\perp = \{a \in K^n \mid f_i(a) = 0, i = 1, \ldots, m\} = I^\perp$.[2] Here I is the ideal generated by the f_i, and we see that these extents are exactly the **algebraic varieties** contained in K^n. It remains to determine the intents of the polynomial context. We know that these are exactly the sets of the form $I^{\perp\perp}$ for some ideal I (respectively $I(V(I))$ in the classical notation).

---

[1] In Algebraic Geometry one usually writes $I(A)$ instead of $A^\perp$

[2] This equation reads $V = V(f_1, \ldots, f_m) = V(I)$ in the classical notation, which means that "V" is written for the derivation of polynomials

Example

a) Let us consider the polynomial ring $K[x]$ in one variable. Let $I = <x^2>$. Then we have $I^{\perp\perp} = <x>$. This phenomenon is, of course, due to the fact that for a polynomial f the equation $f^m(a) = 0$ is equivalent to $f(a) = 0$. The following theorem will show that, if K is algebraically closed, this is the only reason why $I^{\perp\perp}$ can become larger than I.

b) Let $I \subseteq \mathbb{R}[x]$ be the ideal generated by $x^2 + 1$. In this case $I^{\perp}$ is empty and so $I^{\perp\perp}$ is the whole ring.

c) Consider the polynomials $f = x^2y + 2$ and $g = x^2(y+1) + 3$. Both polynomials have real zeroes, but the only zeroes they have in common are $(i, 2)$ and $(-i, 2)$. Thus in $(\mathbb{R}^2, \mathbb{R}[x, y], \perp)$ we have $<f, g>^{\perp\perp} = \mathbb{R}[x, y]$.

Theorem 1 (Hilbert's Nullstellensatz)

Let K be an algebraically closed field. Then

$$I^{\perp\perp} = \sqrt{I} \qquad (I(V(I)) = \sqrt{I})$$

holds where the **radical** *of I is defined as follows:*
$\sqrt{I} := \{f \in K[x_1, \ldots, x_n] \mid f^m \in I$ *for some positive integer m*$\}$. *In particular, $I^{\perp}$ is not empty if I is a proper ideal, i.e., $I^{\perp}$ is empty only if $1 \in I$.*

The **proof** can be found in ([Ku80], I.3).

Now, in the case of an algebraically closed field, we know from Hilbert's Nullstellensatz that the formal concepts of the basic context are exactly the pairs (V, I) where V is an algebraic variety and $I := V^{\perp}$ is the corresponding reduced ideal. (An ideal is called **reduced** if it is equal to its radical).

Corollary 1 *Let K be an algebraically closed field. The set of all algebraic varieties in K^n and the set of all reduced ideals in $K[x_1, \ldots, x_n]$ form complete lattices which are dually isomorphic. In particular, the mapping $V \mapsto V^{\perp}$ which assigns to every algebraic variety its corresponding ideal is an anti-isomorphism between the lattice of all algebraic varieties in K^n and the lattice of all reduced ideals in $K[x_1, \ldots, x_n]$; it is the mapping which sends every extent of $\mathbb{K}^{(n)}$ to its corresponding intent.*

For our analysis of interactions between Formal Concept Analysis and Algebraic Geometry we introduce further basic notions from Algebraic Geometry:

Definition 2 A variety $V \neq \emptyset$ is called **irreducible** if it is not the union of two proper subvarieties. ◊

Theorem 2 *a) V is irreducible if and only if $V^{\perp}$ is a prime ideal.*
b) The system of all algebraic varieties forms the system of closed sets of a topology on K^n, the so-called **"Zariski"** *topology.*
c) Every algebraic variety V has a unique finite decomposition into irreducible varieties of the form $V = V_1 \cup \ldots \cup V_m$.

For the **proof** see again ([Ku80], I.1+I.2). Note that the results above hold even when K is not algebraically closed.

Example   Consider $V := < xy >^{\perp} = \{(a,b) \,|\, a = 0 \text{ or } b = 0\}$. V is the union of the x-axis and the y-axis, which are its irreducible components. Indeed, the ideals of the x-axis and the y-axis are $< y >$ and $< x >$, respectively, which are prime.

In a polynomial context the union of two extents is again an extent. In general, only the intersection of extents is again an extent. Formal contexts with the special property that the union of any finite number of extents (resp. intents) is again an extent (resp. intent) are called **extent-topological** (resp. **intent-topological**). Formal contexts which are both extent-topological and intent-topological are treated, for instance, in [HKS99].

Theorem 3 *Let $a := (a_1, \dots, a_n) \in K^n$. Then the set $\{a\}$ is an algebraic variety with the ideal $\{a\}^{\perp} = < x_1 - a_1, \dots, x_n - a_n >$, which is a maximal ideal. Moreover, if K is algebraically closed, the mapping $a \mapsto a^{\perp}$ is a bijection between the points of the affine space K^n and the maximal ideals of $K[x_1, \dots, x_n]$.*

Proof   Obviously $< x_1 - a_1, \dots, x_n - a_n >^{\perp} = \{a\}$ holds, and so $\{a\}$ is an algebraic variety. It is clear that $a^{\perp} \supseteq < x_1 - a_1, \dots, x_n - a_n >$ and that $< x_1 - a_1, \dots, x_n - a_n >$ is a maximal ideal: consider the homomorphism α : $K[x_1, \dots, x_n] \longrightarrow K$ given by $x_i \mapsto a_i$, $i = 1, \dots, n$. We have $< x_1 - a_1, \dots, x_n - a_n > \in \; kernel \, \alpha$. Each polynomial $f \in a^{\perp}$ is a K-linear combination of the polynomials $(x_i - a_i)$, $i = 1, \dots, n$, since we can apply the Taylor formula. We conclude $K[x_1, \dots, x_n] / < x_1 - a_1, \dots, x_n - a_n > \cong K$ and $< x_1 - a_1, \dots, x_n - a_n >$ is a maximal ideal. However, $\{a\}^{\perp}$ has to be maximal under those ideals which have a nonempty set of zeroes since $\{a\}$ is a minimal non-empty variety. This proves the first statement. If K is algebraically closed and if $\mathfrak{m}$ is a maximal ideal then by Hilbert's Nullstellensatz, $\mathfrak{m}^{\perp}$ must be a single point and so $\mathfrak{m}$ must be of the form $< x_1 - a_1, \dots, x_n - a_n >$. $\qquad\qquad\square$

Note that the latter statement is not true if K is not algebraically closed. For instance, if $K = \mathbb{R}$ the maximal ideal $I = < x^2 + 1 >$ does not occur in the one-to-one correspondence.

3   Subcontexts of Polynomial Contexts

Substructures occur naturally within mathematical theories and applications. A **subcontext** of a formal context (G, M, I) is a formal context (H, N, J) where $H \subseteq G$ and $N \subseteq M$ are subsets and where the relation J is derived from I via $J := I \cap (H \times N)$. With respect to polynomial contexts, subcontexts with a given algebraic variety V and with an unchanged set of attributes are of special interest. These contexts are isomorphic to contexts in which the set of attributes is replaced by the coordinate algebra $K[V]$ of V. There is a functorial correspondence between algebraic varieties and coordinate algebras, which is the basis of the interplay between Algebraic Geometry and Commutative Algebra. This situation has an impact on polynomial concept analysis as well.

Let f be a polynomial with unique factorization $f = cf_1^{\alpha_1} \ldots f_m^{\alpha_m}$. We define $f_{red} := f_1 \ldots f_m$. Then $f_{red}^{\perp} = f^{\perp}$ holds. Thus, we can partially clarify our basic context by taking $\{f \in K[x_1, \ldots, x_n] \mid f = f_{red}\} \cup \{0, 1\}$ as our set of attributes instead of $K[x_1, \ldots, x_n]$. If K is algebraically closed, we know from the Nullstellensatz that this context is fully clarified. From now on, we will consider the polynomial context or, when it is more appropriate, this partially clarified version without explicitly mentioning in each case which one we are speaking of. It always should be clear from the situation (or be of no importance).

Now, let V be an algebraic variety. We consider the subcontext $(V, K[x_1, \ldots, x_n], \perp)$. The following observation is needed. Its proof can be found in ([GW99], p.98).

Lemma 1 *Let (G, M, I) be a formal context and let (H, M, J) be a subcontext, i.e., $H \subseteq G$ and $J = I \cap (H \times M)$. Then every intent of the subcontext is an intent of the supercontext.*

In our situation this means that all intents of $(V, K[x_1, \ldots, x_n], \perp)$ are reduced ideals. Moreover, these ideals must contain $V^{\perp}$ because they are of the form $A^{\perp}$ for some subset $A \subseteq V$. On the other hand, for any reduced ideal I containing $V^{\perp}$, the derivation in the polynomial context and in its subcontext $(V, K[x_1, \ldots, x_n], \perp)$ are the same. Consequently, the mapping which assigns to every extent of $(V, K[x_1, \ldots, x_n], \perp)$ the corresponding intent is an inclusion reversing bijection between the set of all subvarieties of V and all reduced ideals in $K[x_1, \ldots, x_n]$ containing $V^{\perp}$. Let us clarify the context $(V, K[x_1, \ldots, x_n], \perp)$. $f^{\perp} = g^{\perp}$ holds if and only if for all $a \in V$ we have the equivalence $f(a) = 0 \iff g(a) = 0$. Thus we can partially clarify our context if we take one representative from each equivalence class of the equivalence relation $\equiv$ given by $f \equiv g :\iff f - g \in V^{\perp}$, because then we have $f(a) = -g(a)$ for all $a \in V$, in particular $f(a) = 0$ if and only if $g(a) = 0$. Now, let $K[V] := K[x_1, \ldots, x_n]/V^{\perp}$ be the quotient ring consisiting of the elements $f + V^{\perp}$. $K[V]$ is called the **coordinate algebra** of V. We have $f + V^{\perp} = g + V^{\perp}$ if and only if $f - g \in V^{\perp}$. Let $\bar{f}$ denote the coset represented by f. Let $\mathbb{K}_V$ be the context $(V, K[V], \perp_V)$ where

$$a \perp_V \bar{f} :\iff f(a) = 0.$$

Now, it is clear from our previous considerations that $\perp_V$ is well-defined and that the pair of mappings (α, β), where $\alpha := id_V$ and where β is given by $f \longrightarrow \bar{f}$, is a context isomorphism between the partially clarified version of the context $(V, K[x_1, \ldots, x_n], \perp)$ and $(V, K[V], \perp_V)$. Thus, the concepts of $\mathbb{K}_V$ are of the form $(W, \beta(I))$ where (W, I) is a formal concept of $(V, K[x_1, \ldots, x_n], \perp)$. It is known from the homomorphism theorem of rings that every ideal of $K[V]$ is of the form $\bar{I} = I/V$ for some ideal $I \in K[x_1, \ldots, x_n]$.

Corollary 2 *The set of all subvarieties of a given algebraic variety V and the set of all reduced ideals in $K[V]$ form complete lattices which are dually isomorphic. In particular, the mapping which assigns to each subvariety $W \subseteq V$ its ideal $W^{\perp_V}$ is an antiisomorphism between the lattice of all subvarieties of V and the* *lattice of all reduced ideals of $K[V]$; it is the mapping which assigns to every extent of $\mathbb{K}_V$ the corresponding intent.*

Proof The extents of $\mathbb{K}_V$ are exactly the extents of $(V, K[x_1, \ldots, x_n], \perp)$ which are exactly the subvarieties W of V. By definition, $W^{\perp_V} = W^\perp / V^\perp$, which is an ideal of $K[V]$. If $J = I/V^\perp$ is any ideal, then the mapping β tells us that $J^{\perp\perp} = \sqrt{I}/V^\perp$. By the definition of the multiplication in the ring $K[V]$, we obtain immediately that the latter expression is equal to the radical of J in $K[V]$. □

4   Isomorphisms of Algebraic Varieties

This section shows how isomorphisms between algebraic varieties and between the corresponding concept lattices and contexts imply each other. Similar questions with respect to graphs and matrices are settled in [Xi93].

Let $V \subseteq K^n$ and $W \subseteq K^m$ be algebraic varieties. A **morphism** between V and W is a mapping φ from V to W for which polynomials $f_1, \ldots, f_m \in K[x_1, \ldots, x_n]$ exist such that, for all $a := (a_1, \ldots, a_n) \in V$, the equation $\varphi(a) = (f_1(a), \ldots, f_m(a))$ holds. (Keep in mind that $(f_1(a), \ldots, f_m(a))$ must be in W!) Two algebraic varieties V and W are **isomorphic** if there are morphisms φ from V to W and ψ from W to V which are inverses of each other. In this case φ is called an **isomorphism**.

Example   We claim that the variety $Q := \langle z - xy \rangle^\perp \subseteq \mathbb{R}^3$ is isomorphic to $\mathbb{R}^2$. Indeed, the morphisms $\alpha : \mathbb{R}^2 \longrightarrow Q$, given by $(x, y) \longrightarrow (x, y, xy)$, and π from Q to $\mathbb{R}^2$, given by projecting to the first two components, are inverses of each other. For all $(a, b, c) \in Q$ we have $\alpha \circ \pi(a, b, c) = \alpha(a, b) = (a, b, ab)$, which shows that $\alpha \circ \pi$ is the identity on Q because (a, b, c) satisfies the defining equation $z = xy$ of Q. Obviously, $\pi \circ \alpha$ is the identity on $\mathbb{R}^2$. Note that the corresponding coordinate algebras $\mathbb{R}[x, y]$ and $\mathbb{R}[Q]$ are isomorphic as $\mathbb{R}$-algebras. One computes or concludes from the following theoretical results that the mapping $\tilde{\alpha}(f) := f \circ \alpha$, which means that for $f \in \mathbb{R}[Q]$ we have $\alpha(f) = f(x, y, xy)$, is an isomorphism of rings from $\mathbb{R}[Q]$ to $\mathbb{R}[x, y]$ which is the identity on $\mathbb{R}$.

Theorem 4 *Let $V \subseteq K^n$ and $W \subseteq K^m$ be algebraic varieties. If V and W are isomorphic, so are the concept lattices $\underline{\mathfrak{B}}(\mathbb{K}_V)$ and $\underline{\mathfrak{B}}(\mathbb{K}_W)$.*

Proof Let α be an isomorphism from V to W. Then the mapping β from $K[V]$ to $K[W]$ given by $\beta(f) := f \circ \alpha^{-1}$ is an isomorphism of K-algebras (cf. [CLO92], 5.4), in particular a bijective mapping from $K[V]$ to $K[W]$. Thus, it is clear that the pair (α, β) is an isomorphism between the contexts $\mathbb{K}_V$ and $\mathbb{K}_W$. In particular, $\underline{\mathfrak{B}}(\mathbb{K}_V)$ and $\underline{\mathfrak{B}}(\mathbb{K}_W)$ are isomorphic. □

Remark The converse is not true, as the following example shows: consider the curve $V := \langle y^5 - x^2 \rangle$ in $\mathbb{R}^2$. There is a one-to-one morphism from V to $\mathbb{R}$ given by projecting V onto the x-axis, which shows that the concept lattices of $\mathbb{K}_V$

and $(\mathbb{R}, \mathbb{R}[x], \perp)$ are isomorphic. (Subvarieties are exactly the finite subsets of the respective varieties). It can be shown that the two varieties are not isomorphic. Perhaps the easiest way to see this is to apply the next theorem and to show that the corresponding coordinate algebras are not isomorphic. However, note that the inverse of the projection π is not a morphism and so π^{-1} cannot be used to define an isomorphism of contexts because the assignment $f \mapsto f \circ \pi^{-1}$ will not give a mapping from $\mathbb{R}[V]$ to $\mathbb{R}[x]$. A reason why we do not want the two varieties to be isomorphic is the fact that the origin is a singularity of the curve V, whereas $\{0\}$ is not a singularity of the variety $\mathbb{R}$.

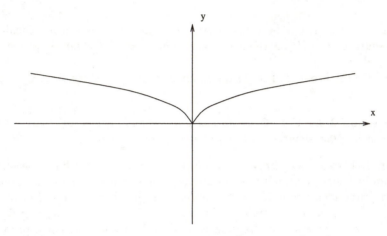

Fig. 4. The algebraic variety defined by $f = y^5 - x^2$

Theorem 5 *Let K be a field and let $V \subseteq K^n$ and $W \subseteq K^m$ be algebraic varieties. Then the following three statements are equivalent:*
a) The varieties V and W are isomorphic.
b) $K[V]$ and $K[W]$ are isomorphic as K-algebras.
c) $\mathbb{K}_V$ and $\mathbb{K}_W$ are isomorphic via an isomorphism (α, β) which satisfies the additional condition that $f(a) = c$ is always equivalent to $\beta(f)(\alpha(a)) = c$ for all $c \in K$.

Proof The equivalence of a) and b) is a standard result from Algebraic Geometry (cf. [CLO92], 5.4).
a) $\Longrightarrow$ c) is clear from the proof of Theorem 4.
c) $\Longrightarrow$ b) We show that β is an isomorphism of K-algebras. Since (α, β) is an isomorphism of contexts, β is bijective. According to our assumption we have $f(a) = (\beta(f))(\alpha(a))$ for all $a \in V$. In particular, β is the identity on K. Multiplicativity follows from the equations $(\beta(fg))(\alpha(a)) = fg(a) = f(a)g(a) = (\beta(f))(\alpha(a))$ and $(\beta(g))(\alpha(a)) = (\beta(f)\beta(g))(\alpha(a))$ and the additivity from $(\beta(f+g))(\alpha(a)) = (f + g)(a) = f(a) + g(a) = (\beta(f))(\alpha(a)) + (\beta(g))(\alpha(a)) = (\beta(f) + \beta(g))(\alpha(a))$, which hold for all $a \in V$. $\qquad \square$

Now, we wish to apply our results to the dimension of an algebraic variety.

Definition 3 The **dimension** of an algebraic variety V is the supremum of the lengths[3] of chains $V_0 \subseteq V_1 \ldots \subseteq V_m = V$ of distinct irreducible subvarieties of V.
$\diamond$

Proposition 1 *Isomorphic varieties have the same dimension.*

Proof The concept lattices of $\mathbb{K}_V$ and $\mathbb{K}_W$ are isomorphic when V and W are isomorphic.
$\square$

Definition 4 a) Let R be a noetherian ring and let P be a prime ideal of R. The **height** of P, $h(P)$, is the supremum of all lengths of chains of prime ideals contained in P.
b) The **dimension** of R is the supremum over all $h(P)$ where $P \subseteq R$ is a prime ideal.
$\diamond$

Proposition 2 *Let $V \subseteq K^n$ be an algebraic variety. Then $dim V = n - h(V^\perp)$. In particular, the dimension of any algebraic variety is finite.*

Proof It is clear that $dim V = dim K[x_1, \ldots, x_n] - h(V^\perp)$ holds because we have an inclusion reversing bijection between irreducible varieties contained in V and prime ideals containing $V^\perp$. It remains to show that the dimension of $K[x_1, \ldots, x_n]$ is n. The chain of inclusions $\{0\} \subseteq < x_1 > \subseteq < x_1, x_2 > \subseteq \ldots \subseteq < x_1, \ldots, x_n >$ shows that the dimension must be at least n. For a proof that $dim K[x_1, \ldots, x_n]$ is indeed equal to n see ([Ku80], II.3.4).
$\square$

Example Consider the algebraic variety $W := < xz, yz >^\perp \subseteq K^3$. (Figure 1.5 in Section 2). Its irreducible components are the x, z-plane and the y-axis. It can be easily seen that the dimension of an arbitrary algebraic variety is equal to the maximum of the dimensions of its irreducible components. Now, in our example the dimension of the x, z-plane is two, because its ideal $< y >$ has height one. Hence, the dimension of W is two as well (as one would expect).

It can be shown that in the case where V is a linear subspace of K^n, i.e., V is defined by linear polynomials, the dimension of V as an algebraic variety is equal to its dimension as a vector space. In this way, the dimension of algebraic varieties is a natural generalization of the usual notion of dimension for vector spaces.

Remark The question arises, whether the existence of an isomorphism between contexts $\mathbb{K}_V$ and $\mathbb{K}_W$ already implies that V and W are isomorphic. We know that the corresponding concept lattices are isomorphic in this case. In particular, the dimensions of V and W are the same and, if $V = V_1 \cup \ldots \cup V_k$ is the decomposition of V into irreducible varieties, then the decomposition of W has the

---

[3] A chain of the form $V_0 \subseteq \ldots V_m = V$ has length m

form $W = W_1 \cup \ldots \cup W_k$. How could we show that $(V, K[V], \perp) \cong (W, K[W], \perp)$ implies $V \cong W$? What makes the problem difficult is the fact that, even if $V \cong W$ and if (α, β) is an isomorphism from $(V, K[V], \perp)$ to $(W, K[W], \perp)$, neither α nor β have to be morphisms. We have, for instance, $(K, K[x], \perp) \cong (K, K[x], \perp)$ via (α, β), where $\alpha(1) := 2, \alpha(2) := 1, \alpha := id_K$ else, and β is defined on irreducible polynomials as follows: $\beta|_K := id_K$, $\beta(x+1) := x+2, \beta(x+2) := x+1, \beta(x+c) := x+c$ else,$\beta(f^m) := (\beta(f))^m$. (We assume K to be algebraically closed). α is not a polynomial mapping and β is not additive. This shows that we must construct an isomorphism $\tilde{\beta}$ from β. One idea is to use the standard method used in Algebraic Geometry to construct a morphism from W to V starting from a morphism from $K[V]$ to $K[W]$. (In fact, it can be shown that there is a contra-variant functor from the category of algebraic varieties over K to the category of finitely generated K-algebras ([Ha77], I.3)). Let $V \subseteq K^n$ and $W \subseteq K^m$ be algebraic varieties and let (α, β) be an isomorphism of the corresponding contexts. Consider the congruence classes $\bar{x}_1, \ldots, \bar{x}_n$ belonging to the variables $x_1, \ldots, x_n$. The image of $\bar{x}_i$, $i = 1, \ldots, n$, under β is represented by a polynomial f_i in the variables $x_1, \ldots, x_m$. We can use these images to define a morphism $\tilde{\alpha} : W \longrightarrow K^n$ by sending $(a_1, \ldots, a_m)$ to $(f_1(a_1, \ldots, a_m), \ldots, f_n(a_1, \ldots, a_m))$. On the other hand, we obtain a mapping $\tilde{\beta} : K[V] \longrightarrow K[W]$ by sending $f \in K[W]$ to $f \circ \tilde{\alpha}$, but only if $\tilde{\alpha}(W) \subseteq V$. In this case we can ask if $\tilde{\alpha}$ and $\tilde{\beta}$ are isomorphisms. The problem is that, once it is known that $\tilde{\alpha}$ is an isomorphism, it is easy to prove that $\tilde{\beta}$ is an isomorphism as well, and vice versa. (In fact, this is how Theorem 5 is proved). However, we have to argue without this knowledge, as it is not clear whether $\tilde{\alpha}$ maps to V. If we consider again our example $(K, K[x], \perp) \cong (K, K[x], \perp)$ with (α, β) as before, $\tilde{\alpha}$ and $\tilde{\beta}$ turn out to be the identity functions because we have $\beta(x) = x$. In the remaining part of this section we will elaborate conditions that guarantee that the method just described can be applied.

The definition of algebraic varieties does not depend on affine coordinate transformations. This is justified by the following lemmas.

Definition 5 Let $y_1, \ldots, y_n \in K[x_1, \ldots, x_n]$. An **affine coordinate transformation** on $\{x_1, \ldots, x_n\}$ is a mapping $\sigma : \{x_1, \ldots, x_n\} \longrightarrow K[y_1, \ldots, y_n]$ with $\sigma(x_i) := \sum_{k=1}^{n} a_{ik} y_k + b_i$, $i = 1, \ldots n$ where $det(a_{ik}) \neq 0$. $\Diamond$

Lemma 2 Let σ be an affine coordinate transformation on $\{x_1, \ldots, x_n\}$. Then there is a context isomorphism $(\rho, \hat{\sigma})$ from $(K^n, K[x_1, \ldots, x_n], \perp)$ to $(K^n, K[y_1, \ldots, y_n], \perp)$ where $\hat{\sigma}$ is defined by $\hat{\sigma}(f(x_1, \ldots, x_n)) := f(\sigma(x_1), \ldots, \sigma(x_n))$.

Proof Let A be the matrix (a_{ik}) and consider the system of linear equations $A(y_1, \ldots, y_n) = (c_1 - b_1, \ldots, c_n - b_n)$ for a given point $c := (c_1, \ldots, c_n) \in K^n$. (Here $b_1, \ldots, b_n$ are as in Definition 5). It has a unique solution $\tilde{c} := (\tilde{c}_1, \ldots, \tilde{c}_n)$ and we obtain a bijective mapping ρ of the affine space by sending c to $\tilde{c}$. We compute $\tilde{\sigma}(f(x_1, \ldots, x_n))(\rho(c_1, \ldots, c_n)) = f(\sigma(x_1), \ldots, \sigma(x_n))(\rho(c_1, \ldots, c_n)) = f(\sum a_{1i} y_i + b_1, \ldots, \sum a_{ni} y_i + b_n)(\tilde{c}_1, \ldots, \tilde{c}_n) = f(c_1, \ldots, c_n)$. In particular, $c \perp f$ holds if and only if $\rho(c) \perp \hat{\sigma}(f)$. $\square$

Lemma 3 *Let $V \subseteq K^n$ and $W \subseteq K^m$ be algebraic varieties. Let (α, β) : $(V, K[x_1, \ldots, x_n], \perp) \longrightarrow (W, K[x_1, \ldots, x_m], \perp)$ be a context isomorphism and let σ be an affine coordinate transformation on $\{x_1, \ldots, x_n\}$. Then the contexts $(\rho(V), \hat{\sigma}(K[x_1, \ldots, x_n]), \perp)$ and $(W, K[x_1, \ldots, x_m], \perp)$ are isomorphic (Here ρ and $\hat{\sigma}$ are defined as before).*

Proof We obtain an isomorphism if we define $\tilde{\beta}(\hat{\sigma}(f)) := \beta(f)$ and $\tilde{\alpha}(\rho(a_1, \ldots, a_n)) := \alpha(a_1, \ldots, a_n)$. □

Now let us return to the situation described before. Let $V \subseteq K^n$ and $W \subseteq K^m$ be algebraic varieties and let $(\alpha, \beta) : \mathbb{K}_V \longrightarrow \mathbb{K}_W$ be a context isomorphism. Let $\beta(\overline{x}_i) =: \overline{f_i}$ with $f_i \in K[x_1, \ldots, x_m]$, $i = 1, \ldots n$, and let $\beta^{-1}(x_j) =: \overline{g_j}$ with $g_j \in K[x_1, \ldots, x_n]$, $j = 1, \ldots m$. Moreover, let $\widetilde{\alpha^{-1}} : W \longrightarrow K^n$ be given by $(a_1, \ldots, a_m) \mapsto f_1(a_1, \ldots, a_m), \ldots, f_n(a_1, \ldots, a_m))$ and let $\tilde{\alpha} : V \longrightarrow K^m$ be given by $(b_1, \ldots, b_n) \mapsto (g_1(b_1, \ldots, b_n), \ldots, g_m(b_1, \ldots, b_n))$.

Theorem 6 *Let $(\alpha, \beta) : \mathbb{K}_V \longrightarrow \mathbb{K}_W$ be a context isomorphism and let $\tilde{\alpha}$ and $\widetilde{\alpha^{-1}}$ as before. Then, if $\tilde{\alpha}(W) \subseteq V$ and $\widetilde{\alpha^{-1}}(V) \subseteq W$, the algebraic varieties V and W are isomorphic.*

Proof Because of the preceding lemmas we may assume that both V and W contain the origin and that $\beta(0) = 0$. Since $\tilde{\alpha}(W)$ is a subset of V, the definition $\tilde{\beta} : K[V] \longrightarrow K[W]$, $\tilde{\beta}(f) := f \circ \tilde{\alpha}$ yields a well-defined K-homomorphism from $K[V]$ to $K[W]$. In the same way $\widetilde{\beta^{-1}} : K[W] \longrightarrow K[V]$ given by $\widetilde{\beta^{-1}}(p) := p \circ \widetilde{\alpha^{-1}}$ is a K-homomorphism. If we can show that $\tilde{\beta}$ is bijective, then $\tilde{\beta}$ is an isomorphism, which means that $K[V]$ and $K[W]$ and consequently V and W are isomorphic. For this purpose we show that $\widetilde{\beta^{-1}} \circ \tilde{\beta} : K[V] \longrightarrow K[V]$ and $\tilde{\beta} \circ \widetilde{\beta^{-1}} : K[W] \longrightarrow K[W]$ are bijective, because then, by elementary properties of mappings, we conclude that $\tilde{\beta}$ is bijective (and $\widetilde{\beta^{-1}}$ as well). For reasons of symetry it is enough to show that $\widetilde{\beta^{-1}} \circ \tilde{\beta} : K[V] \longrightarrow K[V]$ is bijective. Since $\widetilde{\beta^{-1}} \circ \tilde{\beta}$ is a K-homomorphism it is enough to show that $\widetilde{\beta^{-1}} \circ \tilde{\beta}(\{x_1, \ldots, x_n\}) = \{f_1, \ldots, f_n\}$, where the polynomials f_j are such that $< f_1, \ldots, f_n >$ is an maximal ideal of $K[V]$. To prove this we show that $C := \{\widetilde{\beta^{-1}} \circ \tilde{\beta}(x_1), \ldots, \widetilde{\beta^{-1}} \circ \tilde{\beta}(x_n)\}^\perp = \{0\}$. By definition, we have $\widetilde{\beta^{-1}} \circ \tilde{\beta}(x_i) = f_i(g_1(b_1, \ldots, b_n), \ldots, g_m(b_1, \ldots, b_n))$. Hence, we have $(b_1, \ldots, b_n) \in C$ if and only if $f_i(g_1(b_1, \ldots, b_n), \ldots, g_m(b_1, \ldots, b_n)) = 0$ for all $i \in I$. Since (α, β) is an isomorphism of contexts we conclude that $g_j((b_1, \ldots, b_n)) = 0$ for all $j \in \{1, \ldots, m\}$. Appplying again the fact that (α, β) is an isomorphism of contexts, we conclude $(b_1, \ldots, b_n) = 0$. □

5   Projective Varieties

Projective varieties are located in projective space and are defined by homogenous polynomials. This situation can be treated analogously to the affine case.

Definition 6 Let $(x_0, \ldots, x_n)$ and $(y_0, \ldots, y_n)$ be elements of $K^{n+1} \setminus \{0\}$. We define $(x_0, \ldots, x_n)$ and $(y_0, \ldots, y_n)$ to be equivalent if there is a $\lambda \in K \setminus \{0\}$ such that $(x_0, \ldots, x_n) = \lambda(y_0, \ldots, y_n)$. The set $\mathbb{P}^n(K)$ of all such equivalence classes of $K^{n+1} \setminus \{0\}$ is referred to as the n-dimensional **projective space** over K. Each $(n+1)$-tuple $(x_0, \ldots, x_n)$ of elements of K defines a point in $\mathbb{P}^n(K)$ and we say that $(x_0, \ldots, x_n)$ are **homogenous coordinates** of p. ◇

Remark   Let $U_0 := \{(x_0, \ldots, x_n) \in \mathbb{P}^n(K) \mid x_0 \neq 0\}$. It can be easily seen that the mapping $\phi : (a_1, \ldots, a_n) \mapsto (1, a_1, \ldots, a_n)$ is a bijection from the affine space K^n onto the subset U_0 of the n-dimensional projective space over K. Thus, we can think of $\mathbb{P}^n(K)$ consisting of the affine space K^n plus the **hyperplane at infinity**, namely the set of points with $x_0 = 0$. For example, the projective line $\mathbb{P}^1(\mathbb{C})$ consists of $\mathbb{C}$ and the point ∞, represented by $(0, a)$ for arbitrary $a \in \mathbb{C} \setminus \{0\}$.

Consider a point $p \in \mathbb{P}^n(K)$ and a polynomial $f \in K[x_0, \ldots, x_n]$. Now, $f(p)$ depends on what homogenous coordinates of p we choose: if $f := x_0 + x_1{}^2$ and if $p := (1, 1)$, we have $f(1, 1) = 2$. However, $(2, 2)$ are homogenous coordinates of p as well and $f(2, 2) = 6$. Therefore we make the following definition:

Definition 7 The context

$$(\mathbb{P}^n(K), K[x_0, \ldots, x_n,] \perp)$$

where $p \perp f := \Longleftrightarrow f(a_0, \ldots, a_n) = 0$ for **all** homogenous coordinates $(a_0, \ldots, a_n)$ of p is called the **projective polynomial context** (over K in n variables) and is sometimes denoted as $\mathbb{K}_{proj}^{(n)}$. The exents of $\mathbb{K}_{proj}$ are called **projective varieties**. ◇

To determine the intents of this context we first observe that, for any subset A of the projective space, $A^\perp$ is an ideal. To proceed we have to recall further notions and results from Commutative Algebra.

Definition 8 A polynomial $f \in K[x_0, \ldots, x_n]$ is **homogenous** if all its monomials have the same total degree. An ideal is **homogenous** if it is generated by homogenous polynomials. If f is an arbitrary polynomial we can write f uniquely as the sum of homogenous polynomials, which are called the **homogenous components** of f. ◇

Lemma 4 *An ideal $I \subseteq K[x_0, \ldots, x_n]$ is homogenous if and only if, for each $f \in I$, all the homogenous components of f are in I.*

For a **proof** see ([Ku80, I.5]).

Lemma 5 *Let K be an infinite field. $A^\perp$ is an homogenous ideal for all $A \subseteq \mathbb{P}^n(K)$ and every projective variety V is of the form $V = \{g_1, \ldots, g_l\}^\perp$, where the g_i are homogenous polynomials.*

Proof We already know that $A^\perp$ is an ideal. Choose $f \in A^\perp$ and $a := (a_0, \ldots, a_n)$ $\in A$. Let $f = f_0 + \cdots + f_k$ be the decomposition of f into homogenous polynomials, where the total degree of f_i is i or f_i is zero. For all $\lambda \in K \setminus \{0\}$ we have the following equality: $f(\lambda(a_0, \ldots, a_n)) = f_0(a_0, \ldots, a_n) + \lambda f_1(a_0, \ldots, a_n) + \cdots + \lambda^k f_k(a_0, \ldots, a_n)$. By the choice of a and f, the expression on the left vanishes for all values of λ. But the expression on the right can only vanish for all λ if every $f_i(a_0, \ldots, a_n)$ vanishes, which is seen if one considers the expression on the right as a polynomial in λ (Here we need that the underlying field is infinite). Now every f_i vanishes on $(a_0, \ldots, a_n)$. However, since the f_i are homogenous they will vanish on all other homogenous coordinates of a as well. Thus, $f_i \in A^\perp$ for all i, which shows that $A^\perp$ is an homogenous ideal. To prove the second assertion recall that by Hilbert's Basissatz every ideal $I \subseteq K[x_0, x_1, \ldots, x_n]$ is finitely generated. If I is homogenous and if f is one of the generators, then all its homogenous components are in I and we can take them as new generators replacing f. Hence, I is generated by a finite number of homogenous polynomials. $\qquad\square$

Theorem 7 *Let K be algebraically closed and let $I \subseteq K[x_0, x_1, \ldots, x_n]$ be a homogenous ideal which does not contain a power of $< x_0, \ldots, x_n >$. Then $I^{\perp\perp} = \sqrt{I}$ holds.*

Proofs can be found in all books on Algebraic Geometry listed in this paper, for instance ([Ku97], p.52).

Similar to the case of affine varieties we see that the lattice of all projective varieties in $\mathbb{P}^n(K)$ is dually isomorphic to the lattice of all reduced homogenous ideals in $K[x_0, \ldots, x_n]$, the ideal $< x_0, \ldots, x_n >$ is excluded. $< x_0, \ldots, x_n >^{\perp\perp}$ is the whole ring because $< x_0, \ldots, x_n >^\perp$ is empty. This is the reason why $< x_0, \ldots, x_n >$ is sometimes called the **irrelevant ideal** of $K[x_0, \ldots, x_n]$.

6   Modifying the Affine Space

We have seen that the analysis of a given variety V is closely connected to the analysis of the coordinate algebra $K[V]$ of V. A basic method of Algebraic Geometry is to investigate more general classes of rings within a setting that generalizes the initial situation. This procedure will also lead to the possibility of counting multiplicities on algebraic varieties, as it will be described in 6.6.

Definition 9 Let A be a commutative ring with 1. The **spectrum** of A, denoted by $Spec\, A$, is the set of all prime ideals of A. $\qquad\diamond$

In Algebraic Geometry the polynomial context is often replaced by the formal context $(Spec\, K[x_1, \ldots, x_n], K[x_1, \ldots, x_n], \ni)$ and, analogously, the context $\mathbb{K}_V$ is replaced by $(Spec\, K[V], K[V], \ni)$. This is justified by the following theorem:

Theorem 8 *Let V be an algebraic variety over an algebraically closed field K. Then the contexts $(V, K[V], \perp)$ and $(\operatorname{Spec} K[V], K[V], \ni)$ have isomorphic concept lattices. Moreover, they have the same intents. Two contexts $(V, K[V], \perp)$ and $(W, K[W], \perp)$ are isomorphic if and only if $(\operatorname{Spec} K[V], K[V], \ni)$ and $(\operatorname{Spec} K[W], K[W], \ni)$ are isomorphic.*

Usually, the theorem is proven by showing $I^{\ni\ni} = \sqrt{I}$ directly for any ideal I (cf. [Ku80], I.4). We will proceed in a different way. We model the situation of spectra for formal contexts in general. The set of objects is replaced by a set of **general objects** in the following way: for each $\cup$-irreducible extent we take its corresponding intent as a new object. For "many" formal contexts the concept lattice remains unchanged under this procedure.

Definition 10 Let $\mathbb{K} := (G, M, I)$ be a formal context. An extent of $\mathbb{K}$ is called $\cup$-**irreducible**, if it is not a union of finitely many proper subextents. Let $\mathcal{U}_{irr}$ be the set of all such extents. Let $\hat{G} := \{B \subseteq M \mid B = A^I \text{ for some } A \in \mathcal{U}_{irr}\}$, let $\hat{\mathbb{K}} := (\hat{G}, M, \ni)$, and let $\tilde{\mathbb{K}} := (\mathcal{U}_{irr}, M, \triangle)$ where $A \triangle m :\iff m \in A^I$. $\Diamond$

Lemma 6 *Let $\mathbb{K}$ be a formal context. If each extent of $\mathbb{K}$ is the union of $\cup$-irreducible extents, then the concept lattices $\underline{\mathfrak{B}}(\mathbb{K})$, $\underline{\mathfrak{B}}(\hat{\mathbb{K}})$, and $\underline{\mathfrak{B}}(\tilde{\mathbb{K}})$ are isomorphic with identical intents. Conversely, if $\underline{\mathfrak{B}}(\mathbb{K})$ and $\underline{\mathfrak{B}}(\hat{\mathbb{K}})$ are isomorphic, then every extent of $\mathbb{K}$ is the union of $\cup$-irreducible extents.*

Proof The definition $\alpha : \mathcal{U}_{irr} \longrightarrow \hat{G}$, $A \mapsto A^I$, $\beta := id_M$ yields an isomorphism (α, β) between the contexts $\tilde{\mathbb{K}}$ and $\hat{\mathbb{K}}$ which shows that the corresponding concept lattices are isomorphic with identical intents, because β is the identity. Let $\mathfrak{A} \subseteq \mathcal{U}_{irr}$. Then $\mathfrak{A}^{\triangle} = \{m \in M \mid \forall A \in \mathfrak{A} : m \in A^I\} = \{m \in M \mid m \in (\bigcap_{A \in \mathfrak{A}} A^I)\} = \{m \in M \mid m \in (\bigcup_{A \in \mathfrak{A}} A)^I\} = (\bigcup_{A \in \mathfrak{A}} A)^I$. Thus, the intents of $\tilde{K}$ are exactly the sets of the form $(\bigcup_{A \in \mathfrak{A}} A)^I$ for some $\mathfrak{A} \subseteq \mathcal{U}_{irr}$. However, the extents of $\mathbb{K}$ are of the form $\bigcup_{A \in \mathfrak{A}} A$ for some $\mathfrak{A} \subseteq \mathcal{U}_{irr}$. Passing to the corresponding intent shows that $\mathbb{K}$ and $\tilde{\mathbb{K}}$ have the same intents, and the proof of the first assertion is completed.

For the converse note that if $\mathbb{K}$ has an extent which is not a union of $\cup$−irreducible extents, the corresponding intent will not occur as an intent of the context $\tilde{\mathbb{K}}$ as the above reasoning shows. This proves the second statement since $\tilde{\mathbb{K}}$ and $\hat{\mathbb{K}}$ are always isomorphic. $\square$

Proof of Theorem 8 Consider the context $(V, K[V], \perp)$ where V is an algebraic variety. A subvariety W of V is irreducible if and only if the ideal $W^{\perp} \subseteq K[V]$ is prime. According to Hilbert's Nullstellensatz, every prime ideal is an intent. Additionally, for every algebraic variety we have a (finite) decomposition into irreducible subvarieties, according to the results of Section 4. Now, since finite unions of algebraic varieties are again algebraic varieties, V is irreducible as an algebraic variety if and only if it is $\cup$−irreducible as an extent of $(V, K[V], \perp)$, except for the empty set: by definition, $\emptyset$ is not an irreducible variety, the reason being that its derivation is the whole ring, which is not a prime ideal. So the previous lemma tells us that the contexts $(V, K[V], \perp)$ and $(Spec\,K[V] \cup \{K[V]\}, K[V], \ni)$ have isomorphic concept lattices and that they have the sameintents. However, $K[V]$ is in relation with every attribute and therefore part of every extent. This shows that we can delete the object $K[V]$, without changing the concept lattice and the intents.

Now, let $(\alpha, \beta) : (Spec\,K[V], K[V], \ni) \longrightarrow (Spec\,K[W], K[W], \ni)$ be an isomorphism. We define $\hat{\alpha}$ from V to W by $\hat{\alpha}(a) := b$ if $\alpha(\overline{\mathfrak{m}}_a) = \overline{\mathfrak{m}}_b$, where $\overline{\mathfrak{m}}_a$ and $\overline{\mathfrak{m}}_b$ are the maximal ideals belonging to a and b, respectively. We obtain an isomorphism $(\hat{\alpha}, \beta) : (V, K[V], \perp) \longrightarrow (W, K[W], \perp)$ since we have $a \perp f \Longleftrightarrow f \in \overline{\mathfrak{m}}_a \Longleftrightarrow \beta(f) \in \overline{\mathfrak{m}}_b \Longleftrightarrow b \perp \beta(f) \Longleftrightarrow \alpha(a) \perp \beta(f)$.

If on the other hand $(\alpha, \beta) : (V, K[V], \perp) \longrightarrow (W, K[W], \perp)$ is an isomorphism we can define an isomorphism $(\tilde{\alpha}, \beta) : (Spec\,K[V], K[V], \ni) \longrightarrow (Spec\,K[W], K[W], \ni)$ as follows: if $P \in Spec\,K[V]$ is equal to $\bigcap \{\overline{\mathfrak{m}}_a \,|\, a \in P^{\perp}\}$, we define $\tilde{\alpha}(P) := \bigcap \{\overline{\mathfrak{m}}_{\alpha(a)} \,|\, a \in P^{\perp}\}$. We conclude $\tilde{\alpha}(P) \ni \beta(f) \Longleftrightarrow \beta(f) \in \bigcap \{\overline{\mathfrak{m}}_a \,|\, a \in P^{\perp}\} \Longleftrightarrow \beta(f) \in \overline{\mathfrak{m}}_{\alpha(a)}$ for all $a \in P^{\perp} \Longleftrightarrow \beta(f) \perp \alpha(a)$ for all $a \in P^{\perp} \Longleftrightarrow f \perp a$ for all $a \in P^{\perp} \Longleftrightarrow f \in P^{\perp\perp} = P$. $\qquad\square$

Corollary 3 *Let K be algebraically closed.*
a) For any ideal $I \subseteq K[V]$ we have $I^{\ni\ni} = \sqrt{I}$, i.e., $\sqrt{I}$ is the intersection of all prime ideals containing I.
b) Let $A \subseteq Spec\,K[V]$, $A \neq \emptyset$. A is $\cup-$irreducible if and only if $A^{\ni}$ is prime.
c) The system of all extents of $(Spec\,K[V], K[V], \ni)$ forms a topology on $Spec\,K[V]$.
d) If K is not algebraically closed, then $I^{\perp\perp}$ is equal to the intersection of all prime ideals P containing I which satisfy $P^{\perp\perp} = P$.

Proof Theorems 2 and 8 and Lemma 6. $\qquad\square$

Usually, one understands the spectrum as follows: each point $a \in K^n$ corresponds to exactly one maximal ideal $\mathfrak{m}_a$ in $Spec\,K[x_1, \dots, x_n]$. In this way, we can consider the maximal ideals as "points" of $Spec\,K[x_1, \dots, x_n]$. For each irreducible variety V, its ideal $\mathfrak{p}_V := V^{\perp}$ is a prime ideal. For such "points" we have $(\mathfrak{p}_V)^{\ni\ni} = \{\mathfrak{p} \,|\, \mathfrak{p} \supseteq \mathfrak{p}_V\}$. In particular, we have $a \in V$ if and only if $(\mathfrak{p}_V)^{\ni\ni} \ni \mathfrak{m}_a \supseteq \mathfrak{p}_V$. If we identify V and $(\mathfrak{p}_V)^{\ni\ni}$ we can consider the extents of the spectrum as varieties. Now, every irreducible "variety" V is the extent g^{II} of some **object concept** (g^{II}, g^I), indeed it is the object concept of $\mathfrak{p}_V$. $\mathfrak{p}_V$ is called a **generic point** of the variety V. Hence, he spectrum consists of closed points (maximal ideals) which are in one-to-one correspondence with the points of the affine space K^n, and additionally of generic points, whose topological closure can be regarded as an irreducible algebraic variety. It should be mentioned that one reason that renders the introduction of general objects especially interesting in Algebraic Geometry is the fact that every point is an extent such that it can still be identified within the new context. In more general cases, information about single objects may be lost. For instance, for formal contexts (G_1, M, I) and (G_2, N, J), the contexts $(\hat{G}_1, M, \ni)$ and $(\hat{G}_2, N, \ni)$ can be isomorphic even if (G_1, M, I) and (G_2, N, J) are not isomorphic.

Remark   We used Lemma 6 and Hilbert's Nullstellensatz to prove that for any ideal $I \in K[V]$ the radical of I is equal to the intersection of all prime ideals containing I. The latter statement can be proven directly for every commutative ring (cf. [Ku80], I.4). Therefore we could try to start from this result to prove the Nullstellensatz. The problem is, that we would need to know that every prime ideal belongs to the set $\hat{G}$ constructed in Definition 10, which means that we had to show that every prime ideal is an intent without using the Nullstellensatz. It seems that this is not possible.

Example   In the next section we will introduce the notion of an affine scheme. As a motivation we wish to give an example which can be found in ([Ku80], I.4). It explains a simplified version of schemes. It corresponds to the state we have reached so far. As indicated in the beginning of this section, we need not concentrate on coordinate algebras. For a ring R, the tuple $(Spec\,R, R)$ is called an affine scheme (In the next chapter we will replace R with a more complicated object). For an ideal $I \subseteq R$ the pair $(Spec\,R/I, R/I)$ is called a closed subscheme. Note that the mapping from $Spec\,R$ to $Spec\,R/I$ which sends $\mathfrak{p}$ to $\bar{\mathfrak{p}}$ induces a homomorphism from $I^{\ni} = \{\mathfrak{p} \,|\, \mathfrak{p} \supseteq I\}$ onto $Spec\,R/I$. Consider $R := K[x,y]$ where K is algebraically closed. $Spec\,(R/ < x^2 >)$ and $Spec\,(R/ < x >)$ can topologically be identified with the y-axis. Yet $(Spec\,(R/ < x >), R/ < x >)$ is a proper closed subscheme of $(Spec\,(R/ < x^2 >), R/ < x^2 >)$. $(Spec\,(R/ < x^2 >), R/ < x^2 >)$ can be regarded as a y-axis that must be counted twice.

The necessity of counting algebraic varieties several times occurs in the following example: the algebraic variety $V := < xz, yz >$ is the union of the x, y-plane and the z-axis. The origin lies on both irreducible components of V. Consider the plane $W := < x - z >$. The intersection of V and W is equal to the y-axis. Now let $(Spec\,(K[x, y, z]/ < xy, yz >), K/ < xy, yz >)$ the affine scheme corresponding to V and $(Spec\,(K[x, y, z]/ < x - z >), K[x, y, z]/ < x - z >)$ be the affine scheme corresponding to W. Since $I_1^{\ni} \cap I_2^{\ni} = (I_1 + I_2)^{\ni}$ holds for ideals in $K[x, y, z]$, the intersection of these affine schemes is equal to $(Spec\,(K[x, y, z]/(< xy, yz > + < x - z >), K[x, y, z]/(< xy, yz > + < x - z >))$. We compute that $K[x, y, z]/(< xy, yz > + < x - z >) \cong K[x, y]/(x^2, xy)$, which shows that the intersection of the two affine schemes can be regarded as a y-axis with an origin that must be counted twice, which matches the fact that the origin lies on two irreducible components of V.

Here we have encountered an example of a formal context where it is useful not to clarify the set of attributes. On the contrary, the set of attributes is enlarged in such a way that it carries an algebraic structure which describes certain intrinsic properties of the underlying data. This idea is carried on in the next section.

7   Affine Schemes

In Algebraic Geometry, the idea of the spectrum of a ring is further generalized. To $Spec\,A$ we associate a sheaf of rings on A. A construction of this kind will be called an **affine scheme**. We want to interpret affine schemes as formal contexts

with certain additional properties, which will lead to a better understanding of
the passage from algebraic varieties over spectra to affine schemes. To do this,
we must go through several notions from the theory of sheaves.

Definition 11 Let X be a topological space. A **sheaf** of rings $\mathcal{O}$ on X is constituted as follows:
(a) for each open set $U \subseteq X$ there is a commutative ring $\mathcal{O}(U)$ with '1',[4]
(b) for each inclusion of open sets $V \subseteq U$ there is a ring homomorphism[5] ρ_{UV}
from $\mathcal{O}(U)$ to $\mathcal{O}(V)$ called **restriction map** subject to the conditions
(1) $\mathcal{O}(\emptyset) = 0$,
(2) $\rho_{UU} = id_{\mathcal{O}(U)}$,
(3) if $W \subseteq V \subseteq U$, then $\rho_{VW} \rho_{UV} = \rho_{UW}$,
(4) if $U = \bigcup_i V_i$ is an open covering and if $s \in \mathcal{O}(U)$ is an element with
$s|_{V_i} := \rho_{UV_i}(s)$ equal to zero for all i, then s is zero,
(5) if $U = \bigcup_i V_i$ is an open covering and if $s_i \in \mathcal{O}(V_i)$ are elements such that
$s_i|_{V_i \cap V_j} = s_j|_{V_i \cap V_j}$ for all i, j, then there is an element $s \in \mathcal{O}(U)$ such that
$s|_{V_i} = s_i$ for all i. (According to (4), s must be unique!).
Elements $s \in \mathcal{O}(U)$ are referred to as **sections** over U.   ◊

Definition 12 Let $x \in X$. The **stalk** $\mathcal{O}_x$ of $\mathcal{O}$ at x is defined as the direct
limit over all rings $\mathcal{O}(U)$ where U is an open set containing x, via the restriction
maps. Thus the elements of $\mathcal{O}_x$ are of the form $< s, U >$, where $s \in \mathcal{O}(U)$ and
U is an open neighbourhood of x, and we have $< s, U > = < t, V >$ if and only
if there is an open set $W \subseteq U \cap V$ such that the sections of s and t over W are
the same.   ◊

Definition 13 a) If $\mathcal{F}$ and $\mathcal{G}$ are sheaves on a topological space X, a **morphism**
ϕ from $\mathcal{F}$ to $\mathcal{G}$ consists of a ring homomorphism $\phi_U : \mathcal{F}(U) \longrightarrow \mathcal{G}(U)$ for each
open set U, such that whenever $V \subseteq U$ is an inclusion of open sets, the diagram

$$
\begin{array}{ccc}
& \phi_U & \\
\mathcal{F}(U) & \longrightarrow & \mathcal{G}(U) \\
\rho \downarrow & & \downarrow \rho' \\
\mathcal{F}(V) & \longrightarrow & \mathcal{G}(V) \\
& \phi_V &
\end{array}
$$

is commutative, where ρ and ρ' are the restriction maps in $\mathcal{F}$ and $\mathcal{G}$, respectively.
An **isomorphism** is a morphism which has a two-sided inverse which is again
a morphism.
b) Let $f : X \longrightarrow Y$ be a continuous map between topological spaces. For any
sheaf $\mathcal{F}$ on X, we define the **direct image** sheaf $f_* \mathcal{F}$ on Y by $(f_* \mathcal{F})(V) :=$
$\mathcal{F}(f^{-1}(V))$ for any open set $V \subseteq Y$. For any sheaf $\mathcal{G}$ on Y, we define the
inverse image sheaf $f_{-1} \mathcal{G}$ on X to be the sheaf associated (cf. remark after
Corollary 4) to the presheaf $U \mapsto lim_{V \supseteq f(U)} \mathcal{G}(V)$, $U \subseteq X$ open. (The limit is
taken over all open sets V of Y containing $f(U)$).

---

[4] A ring in the sense of this paper is always commutative and has a '1'
[5] A ring homomorphism always sends '1' to '1'

c) If Z is a subset of X carrying the induced topology, if $i : Z \longrightarrow X$ is the inclusion map, and if $\mathcal{F}$ is a sheaf on X, then we call $i_{-1}\mathcal{F}$ the **restriction** of $\mathcal{F}$ to Z, often denoted by $\mathcal{F}|_Z$. For any point $p \in Z$ the stalk of $\mathcal{F}|_Z$ at p is just $\mathcal{F}_p$.   ◊

There is a standard way of associating a sheaf to a given commutative ring A, which we present next. We want to consider schemes, which consist of a topological space and an associated sheaf, as a generalization of a polynomial context. Therefore the reader may think of a coordinate algebra playing the role of A.

Definition 14 Let A be a commutative ring with 1. Let $Spec\,A$ be the set of all prime ideals of A. The extents of the context $(Spec\,A, A, \ni)$ are the closed sets of a topology on $Spec\,A$. We have proven this in the case where A is the coordinate algebra $K[V]$ of an algebraic variety V. We also have $I^{\ni\ni} = \sqrt{I}$ for any ideal $I \subseteq A$. For a proof see ([Ku80], I.4). For this reason we sometimes refer to the extents of $(Spec\,A, A, \ni)$ as varieties.

We wish to define a sheaf of rings $\mathcal{O} := \mathcal{O}_{Spec\,A}$ on $Spec\,A$. For a prime ideal $\mathfrak{p} \in Spec\,A$, let $A_\mathfrak{p}$ be the localization of A at $\mathfrak{p}$, i.e., the ring of fractions $T^{-1}A$ with $T := A - \mathfrak{p}$. The elements of $A_\mathfrak{p}$ are of the form $\frac{f}{g}$, $f \in A$, $g \in T$, and we have $\frac{f}{g} = \frac{m}{n}$ if and only if there is an element $t \in T$ with $(fn - gm)t = 0$. $A_\mathfrak{p}$ is a local ring with maximal ideal $\mathfrak{p}A_\mathfrak{p}$. Now, for an open set U, let $\mathcal{O}(U)$ be the ring of elements $(r_\mathfrak{p})_\mathfrak{p} \in \prod_{\mathfrak{p}\in U} A_\mathfrak{p}$ subject to the condition

$(\star)$ for all $\mathfrak{p} \in U$, there are elements $a \in A$ and $g \in A$ such that $\mathfrak{p} \in D(g) \subseteq U$, and such that for all $\mathfrak{q} \in D(g)$ we have $r_\mathfrak{q} = \frac{a}{g}$ in $A_\mathfrak{q}$. (Here $D(g)$ is the open complement of $g^\ni$).

If we take the restriction maps in the usual sense, we obtain a sheaf of rings on $Spec\,A$. $(Spec\,A, \mathcal{O}_{Spec\,A})$ is called the **spectrum** of A; $Spec\,A$ is its topological space, $\mathcal{O}_{Spec\,A}$ is its **structure sheaf**.   ◊

Theorem 9 *Let A be a commutative ring and $\mathcal{O}$ be its structure sheaf.*
a) For each prime ideal $\mathfrak{p} \in A$, its stalk $\mathcal{O}_\mathfrak{p}$ is isomorphic to the local ring $A_\mathfrak{p}$.
b) For each element $f \in A$ the ring $\mathcal{O}(D(f))$ is isomorphic to A_f, the ring of fractions with powers of f as denominators.
c) In particular, the case $f = 1$ yields $\mathcal{O}(Spec\,A) \cong A$.

For a **proof** see ([Ha77], II.2.2).

We wish to interpret the spectrum of a ring as a formal context. A slight modification is in order: For $U \subseteq Spec\,A$, let $\overline{\mathcal{O}(U)} := \bigcup\{\mathcal{O}(V) \mid V \subseteq U,\ V \text{ open}\}$ and let $\overline{\mathcal{O}}$ be the union over all rings of sections $\mathcal{O}(U)$. Let $(r_\mathfrak{p})_{\mathfrak{p}\in U} \in \overline{\mathcal{O}}$ where $r_\mathfrak{p} = \frac{a_{(\mathfrak{p})}}{f_{(\mathfrak{p})}}$ is a representation of $r_\mathfrak{p}$ such that $\mathfrak{p} \in D(f_{(\mathfrak{p})}) \subseteq U$ and such that for all $\mathfrak{q} \in D(f_{(\mathfrak{p})})$ we have $r_\mathfrak{q} = \frac{a_{(\mathfrak{p})}}{f_{(\mathfrak{p})}}$ in $A_\mathfrak{q}$.
Define $\mathfrak{q} \triangle (r_\mathfrak{p})_{\mathfrak{p}\in U} :\iff f_{(\mathfrak{p})} \in \mathfrak{q}$ for all $\mathfrak{p} \in U$.

Theorem 10 *Let A be a commutative ring and let $(Spec\,A, \mathcal{O}_{SpecA})$ be its spectrum. Then $\mathfrak{B}(SpecA, A, \ni)$ and $\mathfrak{B}(SpecA, \overline{\mathcal{O}}, \triangle)$ are isomorphic. Moreover, the systems of extents are the same. For closed sets $Y \subseteq Spec\,A$, we have $Y^{\triangle} = \overline{\mathcal{O}}(U)$ where U is the complement of Y.*

Proof Let $Y \subseteq Spec\,A$ be closed.

(i) First we show $\overline{\mathcal{O}(Y^c)} \subseteq Y^{\triangle}$. Let $(r_{\mathfrak{p}})_{\mathfrak{p}} \in \mathcal{O}(U)$ where $U \subseteq Y^c$. We have to show $(r_{\mathfrak{p}})_{\mathfrak{p}} \triangle \mathfrak{q}$ for all $\mathfrak{q} \in Y$, which means $f_{(\mathfrak{p})} \in \mathfrak{q}$ for all $\mathfrak{q} \in Y$. Let $\tilde{\mathfrak{p}} \in U$, where $U \subseteq Y^c$. Let $r_{\tilde{\mathfrak{p}}} = \frac{a}{f}$ with $D(f) \subseteq U$ and $r_{\hat{\mathfrak{p}}} = \frac{a}{f}$ for all $\hat{\mathfrak{p}} \in D(f)$. Since $D(f) \subseteq U \subseteq Y^c$, i.e. $\{\mathfrak{p} \mid f \not\subseteq \mathfrak{p}\} \subseteq Y^c$, we obtain $f \in \mathfrak{q}$ for all $\mathfrak{q} \in Y$.

(ii) We show the inclusion $\overline{\mathcal{O}(V^c)} \supseteq Y^{\triangle}$. Let $(r_{\mathfrak{p}})_{\mathfrak{p}} \in Y^{\triangle}$, hence $f_{(\mathfrak{p})} \in \mathfrak{q}$ for all $\mathfrak{q} \in Y$ $(\star)$. We must show that $(r_{\mathfrak{p}})_{\mathfrak{p}}$ is in some $\mathcal{O}(U)$ where U is contained in the complement of Y. Suppose $(r_{\mathfrak{p}})_{\mathfrak{p}}$ is in some $\mathcal{O}(U)$ where U is not contained in Y^c. In this case let $\tilde{\mathfrak{p}} \in U \cap Y$. Then we have for $r_{\tilde{\mathfrak{p}}} = \frac{a}{f}$ that $f \notin \tilde{\mathfrak{p}}$ because of the definition of $A_{\tilde{\mathfrak{p}}}$. This is a contradiction to $(\star)$.

(iii) Now, we prove $A^{\ni\ni} \subseteq A^{\triangle\triangle}$. We have $A^{\triangle} = \{(r_{\mathfrak{p}})_{\mathfrak{p}} \mid \forall \mathfrak{q} \in A : f_{(\mathfrak{p})} \in \mathfrak{q}\}$ and $A^{\triangle\triangle} = \{\tilde{\mathfrak{q}} \mid f_{(\mathfrak{p})} \in \tilde{\mathfrak{q}}$ for all $(r_{\mathfrak{p}})_{\mathfrak{p}}$ with $f_{(\mathfrak{p})} \in \mathfrak{q}$ for all $\mathfrak{q} \in A\}$. Now $\hat{\mathfrak{q}} \in A^{\ni\ni}$ is equivalent to $\hat{\mathfrak{q}} \supseteq \bigcap\{\mathfrak{q} \mid \mathfrak{q} \in A\}$, which implies $\hat{\mathfrak{q}} \in A^{\triangle\triangle}$ for $\hat{\mathfrak{q}} \in A^{\ni\ni}$.

(iv) Finally we show $A^{\ni\ni} \supseteq A^{\triangle\triangle}$. This assertion follows from the equality $(A^{\ni\ni})^{\triangle\triangle} = A^{\ni\ni}$, because then we have $A^{\ni\ni} = (A^{\ni\ni})^{\triangle\triangle} \supseteq A^{\triangle\triangle}$. Therefore let $Y := A^{\ni\ni}$ be a closed set. We must show that $Y^{\triangle\triangle} = Y$. One inclusion is trivial. The other inclusion $Y^{\triangle\triangle} \subseteq Y$ is equivalent to $\overline{\mathcal{O}(Y^c)}^{\triangle} \subseteq Y$. Let $\mathfrak{q} \in \overline{\mathcal{O}(Y^c)}^{\triangle}$, which means that for all $U \subseteq Y^c$ and all $(r_{\mathfrak{p}})_{\mathfrak{p}} \in \mathcal{O}(U)$ we have $f_{(\mathfrak{p})} \in \mathfrak{q}$. Now, if $\mathfrak{q}$ was not an element of Y, we would have $\mathfrak{q} \in U = Y^c$ and $f_{(\mathfrak{q})} \in \mathfrak{q}$, which is impossible. $\qquad\square$

Example Let $K := \mathbb{F}_3 = \{0, 1, 2\}$ be the field with three elements. We wish to demonstrate the passage from the polynomial context over $\mathbb{F}_3$ in one variable to the corresponding sheaf context.

a) The clarified version of $(K, K[x], \perp)$ is

$\perp$	0	1	x	$x+1$	$x+2$	x^2+2	x^2+x	x^2+2x
0	×	×					×	×
1	×			×	×			×
2	×		×		×	×		

and we obtain a boolean concept lattice since every finite set is an algebraic variety.

b) If we consider $Spec\,K[x] = \{0, <x>, <x-1>, <x-2>\}$ we obtain the context

$\ni$	0	1	x	$x+1$	$x+2$	x^2+2	x^2+x	x^2+2x
$\{0\}$	×							
$<x>$	×	×					×	×
$<x-1>$	×			×	×			×
$<x-2>$	×		×			×	×	

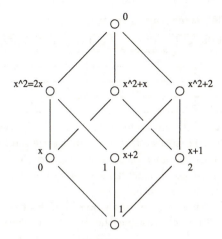

Fig. 5. $\mathfrak{B}(K, K[x], \perp)$

which is almost isomorphic to $(K, K[x], \perp)$, except for the existence of the zero ideal as an additional object. It is the generic point of the whole space which was missing in $(K, K[x], \perp)$.

c) Finally, we intend to elaborate the context $(Spec\,K[x], \overline{\mathcal{O}}, \triangle)$ in order to illustrate Theorem 10. The open sets are $U_1 := \emptyset$, $U_2 := Spec\,K[x]$, $U_3 := \{\{0\}, <x-1>, <x-2>\}$, $U_4 := \{\{0\}, <x>, <x-1>\}$, $U_5 := \{\{0\}, <x>, <x-2>\}$, $U_6 := \{\{0\}, <x>\}$, $U_7 := \{\{0\}, <x-1>\}$, $U_8 := \{\{0\}, <x-2>\}$. By definition, $\mathcal{O}(U_1)$ is the 0-ring. $\mathcal{O}(U_6)$ is the set of all tuples $(\frac{a_1}{g_1}, \frac{a_2}{g_2})$ with $\frac{a_1}{g_1} \in K[x]_{\{0\}}, \frac{a_2}{g_2} \in K[x]_{<x>}$, such that $(\frac{a_1}{g_1}, \frac{a_2}{g_2})$ satisfies the condition $(\star)$. We conclude that $<x> \in D(g_2) \subseteq \{<x>, \{0\}\}$, hence $g_2 \notin \{0\}, g_2 \not\in <x>$, but $g_2 \in <x-1>$ and $g_2 \in <x-2>$. Consequently, g_2 is of the form $g_2 = (x-1)^{c_1}(x-2)^{c_2}$ with $c_1, c_2 \geq 1$. Similarly, we get $g_1 = (x-1)^{d_1}(x-2)^{d_2}$ with $d_1, d_2 \geq 1$. Moreover, we must have $a_1 g_2 = g_1 a_2$. We see that the derivation of all elements of $\mathcal{O}(U_6)$ is equal and we can choose one representative from $\mathcal{O}(U_6)$. (This observation will be further elaborated in the remark after Corollary 4). We choose $(\frac{1}{(x-1)(x-2)}, \frac{1}{(x-1)(x-2)})$. In the same way, we can choose one representative for each of the open sets:

$\mathcal{O}(\emptyset) \quad : 0$

$\mathcal{O}(U_2) : (1,1,1,1)$

$\mathcal{O}(U_3) : (\frac{1}{x}, \frac{1}{x}, \frac{1}{x})$

$\mathcal{O}(U_4) : (\frac{1}{x-2}, \frac{1}{x-2}, \frac{1}{x-2})$

$\mathcal{O}(U_5) : (\frac{1}{x-1}, \frac{1}{x-1}, \frac{1}{x-1})$

$\mathcal{O}(U_6) : (\frac{1}{(x-1)(x-2)}, \frac{1}{(x-1)(x-2)})$

$\mathcal{O}(U_7) : (\frac{1}{x(x-2)}, \frac{1}{x(x-2)})$

$\mathcal{O}(U_8) : (\frac{1}{x(x-1)}, \frac{1}{x(x-1)})$

Thus we obtain the following formal context:

$\ni$	$\mathcal{O}(\emptyset)$	$\mathcal{O}(U_2)$	$\mathcal{O}(U_3)$	$\mathcal{O}(U_4)$	$\mathcal{O}(U_5)$	$\mathcal{O}(U_6)$	$\mathcal{O}(U_7)$	$\mathcal{O}(U_8)$
$\{0\}$	×							
$<x>$	×		×				×	×
$<x-1>$	×				×	×		×
$<x-2>$	×			×		×	×	

Note that we have clarified the context $(Spec\,K[x], \overline{\mathcal{O}}, \triangle)$ only in order to be able to write it down and visualize it in the simplest possible way. We must keep in mind that the trick of the sheaf construction is to consider a formal context which is not only non-clarified but which is enlarged in such a way that it carries an additional mathematical structure.

The spectrum of a ring as it is described in Definition 14 is the prototype of an affine scheme. However, to give the exact definition we still need some preparations.

Definition 15 A **locally ringed space** is a topological space X together with a sheaf $\mathcal{O}_X$ of rings on X, such that for all $p \in X$ the stalk $\mathcal{O}_{X,p}$ of $\mathcal{O}_X$ at p is a local ring. (A ring is said to be local if it has only one maximal ideal). Two locally ringed spaces $(X, \mathcal{O}_X)$ and $(Y, \mathcal{O}_Y)$ are isomorphic if there is a pair of mappings $(f, f^\sharp)$ such that f is a homeomorphism from X to Y and $f^\#$ is an isomorphism between the sheaves $i_*\mathcal{O}_Y$ and $\mathcal{O}_X$. $\diamond$

We give an analogue to this definition from Algebraic Geometry in the language of contexts:

Definition 16 A **locally ringed context** is a context $(X, \overline{\mathcal{O}_X}, \triangle)$, where
(1) X is a topological space,
(2) $\mathcal{O}_X$ is a sheaf of rings on X and $\overline{\mathcal{O}_X}$ is defined as before,
(3) $\mathcal{O}_{X,p}$ is a local ring for each $p \in X$,
(4) the extents of $(X, \overline{\mathcal{O}_X}, \triangle)$ are exactly the closed subsets of the topological space X,
(5) $X_1^\triangle = \overline{\mathcal{O}(U)}$ with $U = X_1^c$ for any closed subset $X_1 \subseteq X$. $\diamond$

Corollary 4 *Let A be a ring and let $(Spec\,A, \mathcal{O}_{Spec\,A})$ be its spectrum. Then $(Spec\,A, \overline{\mathcal{O}_{Spec\,A}}, \triangle)$ is a locally ringed context.*

Proof Immediate from the Theorems 9 and 10.   □

Remark There is a standard way to define the incidence relation $\triangle$ on a locally ringed space $(X, \mathcal{O}_X)$ to obtain a locally ringed context. First, if $\mathcal{F}$ is an arbitrary sheaf, then $\mathcal{F}$ is canonically isomorphic to the following sheaf $\mathcal{F}^+$: for any open set U let $\mathcal{F}^+(U)$ be the set of functions s from U to the union $\bigcup_{P \in U} \mathcal{F}_P$ such that (1), for each $P \in U$, $s(P)$ is an element of $\mathcal{F}_P$ and such that (2), for each $P \in U$, there is a neighbourhood V of P contained in U and an element $t \in \mathcal{F}(V)$ such that, for all $Q \in V$, the germ t_Q of t at Q, (i.e., the equivalence class of t in $\mathcal{F}_Q$), is equal to $s(Q)$. (This method can also be used to construct a sheaf from a **presheaf**. A presheaf does not have to satisfy (4) and (5) of the definition of a sheaf).

Now using this isomorphism we can define $q \triangle s(P)_{P \in U} : \Longleftrightarrow q \notin U$ and we get a locally ringed context: obviously, $(X, \overline{\mathcal{O}_X}, \triangle)$ satisfies the conditions (1),(2), and (3). For $A \subseteq X$, we have $A^{\triangle\triangle} = \{q \mid q \notin U$ for all U with $x \notin U$ for all $x \in A\} = \{q \mid q \notin U$ for all $U \subseteq A^c\}$. To proof (4), we show $A^{\ni\ni} \subseteq A^{\triangle\triangle}$ and $A^{\triangle\triangle} \subseteq A^{\ni\ni}$ for all $A \subseteq X$. First of all, let A be closed, i.e. $A = A^{\ni\ni}$. Then we have $A^{\triangle\triangle} = \{q \mid q \notin A^c\} = A$, because A^c is open. "$\subseteq$:" Let $A \subseteq X$. $A \subseteq A^{\ni\ni}$ implies $A^{\triangle\triangle} \subseteq (A^{\ni\ni})^{\triangle\triangle}$ and the latter expression is equal to $A^{\ni\ni}$. "$\supseteq$:" For the other inclusion suppose that $x \in A^{\ni\ni}$, but $x \notin A^{\triangle\triangle}$, hence $x \in U$ for some $U \subseteq A^c$. Let us write for a moment $\overline{A}$ for the topological closure of A. We have $x \in \overline{A}$ and $x \in U \subseteq A^c$. $U \subseteq A^c$ implies that $U \subseteq \overset{\circ}{A^c}$, because U is open and $\overset{\circ}{A^c}$ is the largest open set contained in A^c. It follows that $U^c \supseteq (\overset{\circ}{A^c})^c$, and the latter expression is equal to $\overline{A}$ (cf.[Su64, 2.7]). Since $x \in \overline{A}$, we conclude $x \in U^c$, which is a contradiction to $x \in U$. This settles (4). It remains to show (5). If $s(P)_{P \in U} \in X_1{}^{\triangle}$, we have $q \notin U$ for all $q \in X_1$. Hence $X_1 \cap U = \emptyset$ and $U \subseteq X_1{}^c$, which shows $s(P)_{P \in U} \in \mathcal{O}(X_1{}^c)$. If on the other hand $s(P)_{P \in V} \in \mathcal{O}(X_1{}^c)$, then $V \subseteq X_1{}^c$ and $V \cap X_1 = \emptyset$. We conclude $s(P)_{P \in V} \in X_1{}^{\triangle}$.

The definition of $\triangle$, given here for arbitrary locally ringed spaces, coincides with the definition of $\triangle$ for the spectrum $(Spec\,A, \mathcal{O}_{Spec\,A})$ of a ring A. Indeed, we know that both definitions yield the same concept lattice, and according to the Basic Theorem on Concept Lattices ([GW99], p.20), we have for both versions of $\triangle$ the equivalence $q \triangle s(P)_P \Longleftrightarrow (q^{\triangle\triangle}, q^{\triangle}) \leq (\{s(P)_P\}^{\triangle}, \{s(P)_P\}^{\triangle\triangle})$. Since $q^{\triangle}$ and $\{s(P)_P\}^{\triangle}$ are equal for both versions we conclude that both definitions are equal. The reason why we use the more complicated version for spectra, is that we want to keep track of the original relation $\ni$ (and $\perp$ for coordinate algebras).

Theorem 11 *For commutative rings A and B the following statements are equivalent:*

a) A and B are isomorphic.

b) $(Spec\,A, \mathcal{O}_{Spec\,A})$ and $(Spec\,B, \mathcal{O}_{Spec\,B})$ are isomorphic as locally ringed spaces.

Proof Let A and B be isomorphic via an isomorphism ϕ. Then the contexts $(Spec\,A, A, \ni)$ and $(SpecB, B, \ni)$ are isomorphic and we obtain a homeomorphism from $Spec\,A$ to $Spec\,B$ by sending $\mathfrak{p}$ to $\phi(\mathfrak{p})$. Additionally, we must define an isomorphism of sheaves from $\mathcal{O}_{Spec\,B}$ to $\phi_* \mathcal{O}_{Spec\,A}$. For $V \subseteq Spec\,B$ we define $\phi_V^{\sharp} : \mathcal{O}_{Spec\,B}(V) \longrightarrow \phi_* \mathcal{O}_{Spec\,A}(V) = \mathcal{O}_{Spec\,A}(\phi^{-1}(V))$. We send the tuple $(\frac{a_{\mathfrak{p}}}{f_{\mathfrak{p}}})_{\mathfrak{p} \in V}$ to $(\frac{\phi^{-1}(a_{\mathfrak{p}})}{\phi^{-1}(f_{\mathfrak{p}})})_{\phi^{-1}(\mathfrak{p}) \in \phi^{-1}(V)}$. Analogous to this, we define a morphism $(\phi^{\sharp})^{-1}$ from $\mathcal{O}_{Spec\,A}$ to $\phi_* \mathcal{O}_{Spec\,B}$ by sending $(\frac{a_{\mathfrak{p}}}{f_{\mathfrak{p}}})_{\mathfrak{p} \in U}$ to $(\frac{\phi(a_{\mathfrak{p}})}{\phi(f_{\mathfrak{p}})})_{\phi(\mathfrak{p}) \in \phi(U)}$. Now it is clear that $(\phi^{\sharp})^{-1}$ and $\phi^{\sharp}$ are inverse morphisms, because ϕ^{-1} induces a homeomorphism.

Conversely, if $\phi^{\sharp}$ is an isomorphism from $Spec\,A$ to $Spec\,B$, then $\phi^{\sharp}$ yields an isomorphism of rings from $\mathcal{O}_{Spec\,A}(Spec\,A)$ to $\mathcal{O}_{Spec\,B}(Spec\,B)$. But these rings are isomorphic to A and B, respectively, (cf. Theorem 9) and the proof is completed. □

Theorem 12 *Let A and B be isomorphic. Then the contexts $(Spec\,A, \overline{\mathcal{O}_{Spec\,A}}, \triangle)$ and $(Spec\,B, \overline{\mathcal{O}_{Spec\,B}}, \triangle)$ are isomorphic.*

Proof Let ϕ be an isomorphism from A to B. Define a mapping $\tilde{\phi} : Spec\,A \longrightarrow Spec\,B$ by sending $\mathfrak{p}$ to $\phi(\mathfrak{p})$. Define $\phi^{\sharp} : \overline{\mathcal{O}_{Spec\,A}} \longrightarrow \overline{\mathcal{O}_{Spec\,B}}$ by sending $(\frac{a_{\mathfrak{p}}}{f_{\mathfrak{p}}})_{\mathfrak{p} \in U}$ to $(\frac{\phi(a_{\mathfrak{p}})}{\phi(f_{\mathfrak{p}})})_{\phi(\mathfrak{p}) \in \phi(U)}$. Then $(\tilde{\phi}, \phi^{\sharp})$ is an isomorphism of contexts since we have

$$\mathfrak{q} \triangle (\tfrac{a_{\mathfrak{p}}}{f_{\mathfrak{p}}})_{\mathfrak{p} \in U} \iff \mathfrak{q} \notin U \iff \phi(\mathfrak{q}) \notin \phi(U) \iff \phi(\mathfrak{q}) \triangle (\tfrac{\phi(a_{\mathfrak{p}})}{\phi(f_{\mathfrak{p}})})_{\phi(\mathfrak{p}) \in \phi(U)}. \qquad \square$$

We may ask under which circumstances the existence of an isomorphism of contexts between $(Spec\,A, \overline{\mathcal{O}_{Spec\,A}}, \triangle)$ and $(Spec\,B, \overline{\mathcal{O}_{Spec\,B}}, \triangle)$ does imply that A and B are isomorphic? Let $(\alpha, \beta) : (Spec\,A, A, \ni) \longrightarrow (Spec\,B, B, \ni)$ be an isomorphism of contexts. Then by the same reasoning as above, we obtain a pair of mappings $(\alpha, \beta^{\sharp})$ from $(Spec\,A, \overline{\mathcal{O}_{Spec\,A}}, \triangle)$ to $(Spec\,B, \overline{\mathcal{O}_{Spec\,B}}, \triangle)$ which is compatible with the incidence relations, if we derive $\beta^{\sharp}$ from β in the same way as we derived $\phi^{\sharp}$ from ϕ. Yet it is not clear if $\beta^{\sharp}$ is bijective because we do not know if β is an isomorphism.

If on the other hand $(Spec\,A, \overline{\mathcal{O}_{Spec\,A}}, \triangle) \cong (Spec\,B, \overline{\mathcal{O}_{Spec\,B}}, \triangle)$ via (α, β), we can define a mapping $(\alpha, \tilde{\beta})$ which keeps the incidence relation, but in general will also fail to be bijective. Let $g \in A$. Consider $U := D(g)$. We have $\beta(\mathcal{O}(D(g))) = \mathcal{O}(V)$, where $V = D(h)$ for some $h \in B$, because we have an isomorphism of contexts. Define a mapping $\tilde{\beta} : A \longrightarrow B$ by sending g to h. We claim that $(\alpha, \tilde{\beta})$ is compatible with the incidence relations. Let $\tilde{\beta}(f) := h$. We conclude $\alpha(\mathfrak{q}) \in \tilde{\beta}(f) \iff \forall (r_{\mathfrak{p}})_{\mathfrak{p} \in D(h)} : \alpha(\mathfrak{q}) \ni D(h) \iff \alpha(\mathfrak{q}) \triangle (r_{\mathfrak{p}})_{\mathfrak{p} \in D(h)} \iff \forall (s_{\mathfrak{p}})_{\mathfrak{p} \in D(f)} : \mathfrak{q} \triangle (s_{\mathfrak{p}})_{\mathfrak{p} \in D(f)} \iff \mathfrak{q} \notin D(f) \iff f \in \mathfrak{q}$.

The example $(Spec\,\mathbb{R}, \mathbb{R}, \ni) \cong (Spec\,\mathbb{C}, \mathbb{C}, \ni)$ shows that an isomorphism of two contexts $(Spec\,A, A, \ni)$ and $(Spec\,B, B, \ni)$ is not sufficient for the underlying rings to be isomorphic. At least we still see that for an isomorphism (α, β) the mapping β must be multiplicative on the clarified contexts. Since we have $fg \in \mathfrak{p}$ if and only if $f \in \mathfrak{p}$ or $g \in \mathfrak{p}$, we conclude $(\beta(fg))^{\ni} = \{\alpha(\mathfrak{p}) \mid \mathfrak{p} \ni fg\} = \{\alpha(\mathfrak{p}) \mid \mathfrak{p} \ni f\} \cup \{\alpha(\mathfrak{p}) \mid \mathfrak{p} \ni g\}$. And in the same way $(\beta(f)\beta(g))^{\ni} = \{\alpha(\mathfrak{p}) \mid \alpha(\mathfrak{p}) \ni \beta(f)\beta(g)\} = \{\alpha(\mathfrak{p}) \mid \alpha(\mathfrak{p}) \ni \beta(f)\} \cup \{\alpha(\mathfrak{p}) \mid \alpha(\mathfrak{p}) \ni \beta(g)\} = \{\alpha(\mathfrak{p}) \mid \mathfrak{p} \ni f\} \cup \{\alpha(\mathfrak{p}) \mid \mathfrak{p} \ni g\}$. However, we can neither say anything about units, because they belong to no prime ideal, nor about nilpotent elements, because they belong to each prime ideal. This means that we must at least presuppose the existence of multiplicative and additive bijections from the set A^{*} of units of A to the set B^{*} of units of B and from $Nil(A)$ to $Nil(B)$. It is not clear for which classes of rings these conditions guarantee that A and B are isomorphic.

Definition 17 An **affine scheme** is a locally ringed space $(X, \mathcal{O}_X)$ which is isomorphic as a locally ringed space to the spectrum of some ring. For an open set $U \subseteq X$, the pair $(U, \mathcal{O}_X|_U)$ is called an **open subscheme**. $\diamondsuit$

Proposition 3 *Let $(X, \mathcal{O}_X)$ be a locally ringed space and let $(U, \mathcal{O}_X|_U)$ be an* **open subscheme**. *The locally ringed context $(U, \overline{\mathcal{O}_X|_U}, \triangle)$ is isomorphic to a partially clarified version of the subcontext $(U, \overline{\mathcal{O}_X}, \triangle)$ of $(X, \overline{\mathcal{O}_X}, \triangle)$.*

Proof First note that $(U, \mathcal{O}_X|_U)$ is a locally ringed space since $(\mathcal{O}_X|_U)_p \cong$ $(\mathcal{O}_X)_p$ for all $p \in U$. W is an open subset of the topological space U if and only if there is an open set $W_1 \subseteq X$ such that $W = W_1 \cap U$. So $W \subseteq U$ is open in U if and only if W is open in X. Therefore we have $\mathcal{O}_X|_U(W) = \mathcal{O}_X(W)$ if $W \subseteq U$. In the subcontext $(U, \overline{\mathcal{O}_X}, \triangle)$ we have $\mathcal{O}_X(W)^\triangle = \mathcal{O}_X(W \cap U)^\triangle$, so we can (partially) clarify $(U, \overline{\mathcal{O}_X}, \triangle)$ by removing all $\mathcal{O}_X(W)$, where W is not a subset of U, and obtain $(U, \overline{\mathcal{O}_X|_U}, \triangle)$. $\qquad\qquad\square$

Definition 18 A **scheme** is a locally ringed space $(X, \mathcal{O}_X)$ in which every point has a neighborhood U such that $(U, \mathcal{O}_X|_U)$ is an affine scheme. $\qquad\qquad\Diamond$

We will encounter a standard example of a scheme which is not affine in the following chapter, where the projective spectrum of a graded ring is introduced. It plays the same role for projective varieties as the spectrum does for affine varieties. For this section we consider an example which stems from a construction principle for schemes found in Algebraic Geometry. We also describe a corresponding construction principle for locally ringed contexts. First we state the original definition.

Definition 19 Let X_1, X_2 be schemes, $U_1 \subseteq X_1$ and $U_2 \subseteq X_2$ be open subsets and let $\phi : (U_1, \mathcal{O}_{X_1}|_{U_1}) \longrightarrow (U_2, \mathcal{O}_{X_2}|_{U_2})$ be an isomorphism of locally ringed spaces. Define the scheme X obtained by **glueing** X_1 and X_2 along U_1 and U_2 via ϕ. The topological space of X is the quotient of the disjoint union $X_1 \cup X_2$ by the equivalence relation $x_1 \sim \phi(x_1)$ for each $x_1 \in U_1$, with the quotient topology. Thus there are mappings $i_1 : X_1 \longrightarrow X$ and $i_2 : X_2 \longrightarrow X$ and $V \subseteq X$ is open if and only if $i_1^{-1}(V) \subseteq X_1$ and $i_2^{-1}(V) \subseteq X_2$ are open. For $V \subseteq X$ we define $\mathcal{O}_X(V) := \{< s_1, s_2 > \mid s_1 \in \mathcal{O}_{X_1}(i_1^{-1}(V)), s_2 \in \mathcal{O}_{X_2}(i_2^{-1}(V)), \phi(s_1|_{i_1^{-1}(V) \cap U_1}) = s_2|_{i_2^{-1}(V) \cap U_2})\}$. Then X is a scheme (cf. [Ha77], II.2.3.5). $\qquad\qquad\Diamond$

The glueing of affine schemes is the standard tool to construct schemes from affine schemes. So far we have described affine schemes as formal contexts with certain additional properties. Also schemes can be understood as formal contexts since they are locally ringed spaces. It is natural to look for a construction principle for the respective contexts that models the glueing of the affine schemes. We define a procedure that can also be applied to extent-topological contexts. Its role concerning arbitrary formal contexts is discussed at the end of this section.

Definition 20 a) An **extent-topological context** is a formal context (X, M, I) where X is a topological space and where $A \subseteq X$ is an extent if and only if it is closed in the topology on X.
b) Let $\mathbb{K}_1 := (X_1, M_1, I_1)$ and $\mathbb{K}_2 := (X_2, M_2, I_2)$ be extent-topological contexts with $X_k^{I_k} \neq \emptyset$, $k = 1, 2$. Let $A_k \subseteq X_k$ be any extent and let $U_k := A_k^c$ be the complementary open set. Let $N_k := \{m_k \in M_k \mid m^{I_k} \cap U \neq \emptyset\}$, $k = 1, 2$. Let $(\alpha, \beta) : (U_1, N_1, I_1 \cap (U_1 \times N_1)) \longrightarrow (U_2, N_2, I_2 \cap (U_2 \times N_2))$ be an isomorphism of subcontexts. Let $X = X_1 \cup X_2 / \sim$ be the set of equivalence classes on

the disjoint union of X_1 and X_2, where $\sim$ is the equivalence relation given by $u_1 \sim \alpha(u_1)$ for $u_1 \in U_1$. Furthermore, let $i_k : X_k \longrightarrow X$ be the map that sends an element $x_k \in X_k$ to its equivalence class in X, $k = 1, 2$. Consider the context $(X, M_1 \times M_2, J)$ with

$$xJ(m_1, m_2) :$$
$$\Longleftrightarrow$$
$$1)\ i_k^{-1}(x)I_k m_k \text{ if } i_k^{-1}(x) \in A_k, (k = 1, 2),$$
$$2)\ i_1^{-1}(x)I_1 m_1 \wedge i_2^{-1}(x)I_2 m_2 \text{ if } i_1^{-1}(x) \in U_1.$$

(Recall that $i_1^{-1}(x) \in U_1$ is equivalent to $i_2^{-1}(x) \in U_2$.)
We will denote this context by $\mathbb{K}_1(U_1)\mathbb{K}_2$.   ◇

Note that in the case $U_1 := \emptyset$ we get the dual semi-product of $\mathbb{K}_1$ and $\mathbb{K}_2$, a construction principle which is known from Formal Concept Analysis. For reasons of completeness we insert the following observation:

Lemma 7 *If* $(\alpha, \beta) : (U_1, N_1, I_1 \cap (U_1 \times N_1)) \longrightarrow (U_2, N_2, I_2 \cap (U_2 \times N_2))$ *is as above, then* α *is a homeomorphism of the topological spaces* U_1 *and* U_2.

Proof Let $k = 1, 2$. α carries extents to extents because it belongs to an isomorphism of contexts. Thus it is enough to show that $(U_k, N_k, I_k \cap (U_k \times N_k))$ is an extent-topological context where U_k carries the topology induced by X_k. This means that we must show that the extents of $(U_k, N_k, I_k \cap (U_k \times N_k))$ are exactly the sets of the form $U_k \cap A_k$ where $A_k \subseteq X_k$ is closed. According to Lemma 1 each intent of $(U_k, N_k, I_k \cap (U_k \times N_k))$ is an intent of (X_k, M_k, I_k). Let $J_k := I_k \cap (U_k \times N_k)$. Hence, if $D^{J_k J_k} = D$, then $D^{I_k I_k} = D$ and $D^{J_k} = D^{I_k I_k J_k} = D^{I_k I_k I_k} \cap U_k = D^{I_k} \cap U_k$, and every intent is of the desired form. On the other hand, let $A_k \subseteq X_k$ be closed. We have $(U_k \cap A_k)^{J_k J_k} = (U_k \cap A_k)^{I_k J_k} = U_k \cap (U_k \cap A_k)^{I_k I_k} \subseteq U_k \cap A_k^{I_k I_k} = U_k \cap A_k$ which shows that $U_k \cap A_k$ is an extent of the subcontext.   □

Theorem 13 *Let* $\mathbb{K} := \mathbb{K}_1(U_1)\mathbb{K}_2$ *be as above and let* $C \subseteq X$. $C \subseteq X$ *is an extent if and only if* $i_1^{-1}(C) \subseteq X_1$ *and* $i_2^{-1}(C) \subseteq X_2$ *are extents. Hence* $\mathbb{K}_1(U_1)\mathbb{K}_2$ *is an extent-topological context whose underlying set* X *of objects is homeomorphic to the topological space* $X_1 \cup X_2 / \sim$ *obtained by glueing* X_1 *and* X_2 *via* α.

Proof (i) We start with the first assertion.
$'\Longleftarrow'$
Let $i_1^{-1}(C) \subseteq X_1$ and let $i_2^{-1}(C) \subseteq X_2$ be extents of $\mathbb{K}_1$ and $\mathbb{K}_2$, respectively. It is clear from the definition of J that we have $C^J = (i_1^{-1}(C))^{I_1} \times (i_2^{-1}(C))^{I_2}$. Let $x \in C^{JJ}$, without loss of generality $x \in C^{JJ} \cap i_1(X_1)$. We show $x \in C$. First of all, $(i_2^{-1}(C))^{I_2}$ is not empty since $X_2^{I_2}$ is not empty by definition. Hence there is a pair $(m_1, m_2) \in C^J$ for all $m_1 \in (i_1^{-1}(C))^{I_1}$. We conclude $x \in C^{JJ} \cap i_1(X_1) \Longrightarrow x \in$

$i_1(X_1) \wedge \forall (m_1, m_2) \in C^J : xJ(m_1, m_2) \implies \forall\, m_1 \in (i_1^{-1}(C))^{I_1} : i_1^{-1}(x)I_1 m_1$
$\implies i_1^{-1}(x) \in (i_1^{-1}(C))^{I_1 I_1} = i_1^{-1}(C)$. Hence $x \in C$ and C is an extent, which settles the first implication. (Note that we did not exploit the fact that the underlying contexts are topological).

$' \implies '$

Let $C^{JJ} = C$. We show $(i_1^{-1}(C))^{I_1 I_1} = i_1^{-1}(C)$. We first prove the following statement: $(+)$ $i_1^{-1}(x) \in (i_1^{-1}(C))^{I_1 I_1} \wedge \forall (m_1, m_2) \in C^J : i_1^{-1}(x)I_1 m_1 \implies$ $\forall (m_1, m_2) \in C^J : xJ(m_1, m_2)$. In order to prove this it remains to show that, if $i_1^{-1}(x) \in U_1$, the element $\alpha(i_1^{-1}(x)) = i_2^{-1}(x)$ is in relation with all $m_2 \in (i_2^{-1}(C))^{I_2}$. Suppose the latter condition is violated. Hence $i_2^{-1}(x) \notin (i_2^{-1}(C))^{I_2 I_2}$ and so $i_2^{-1}(x) \notin (i_2^{-1}(C) \cap U_2)^{I_2 I_2}$. By means of the isomorphism (α, β) we conclude $i_1^{-1}(x) \notin (i_1^{-1}(C) \cap U_1)^{I_1 I_1}$. Let us write $D := i_1^{-1}(C)$ and $y := i_1^{-1}(x)$. Then, in terms of topology, we have the following situation: y is in the closure $\overline{D}$ of a set D, y belongs to an open set U_1, but y does not belong to the closure of $D \cap U_1$. But this is impossible: in order to show that y must be in $\overline{D \cap U_1}$, we have to show that, if V is an open neighbourhood of y, then $U_1 \cap V \cap D$ is not empty. By definition, $y \in \overline{D}$ and $y \in U_1$ imply that $U_1 \cap D$ is not empty, because U_1 is an open neighbourhood of y. Yet, if V is any open neighbourhood of y, then so is $U_1 \cap V$ and we can apply the above reasoning on $U_1 \cap V$ and conclude that $U_1 \cap V \cap C$ is not empty. Hence, $y \in \overline{D}$ and $y \in U_1$ always imply $y \in \overline{D \cap U_1}$ and $(+)$ must be valid. Now we conclude $i_1^{-1}(x) \in (i_1^{-1}(C))^{I_1 I_1} \implies \forall m_1 \in (i_1^{-1}(C))^{I_1} : i_1^{-1}(x)I_1 m_1 \implies \forall (m_1, m_2) \in C^J : i_1^{-1}(x)I_1 m_1 \implies \forall (m_1, m_2) \in C^J : xJ(m_1, m_2) \implies x \in C^{JJ} = C$. Hence, $i_1^{-1}(x) \in i_1^{-1}(C)$ and $i_1^{-1}(C)$ is an extent.

(ii) Let C and D be extents. It is sufficient to show that $C \cup D$ is an extent, because in any formal context the intersection of extents is again an extent. By (i), $i_1^{-1}(C)$ and $i_1^{-1}(D)$ are extents of $\mathbb{K}_1$, i.e., they are closed in the topology on X_1. Since $\mathbb{K}_1$ is a topological context we conclude that $i_1^{-1}(C) \cup i_1^{-1}(D) = i_1^{-1}(C \cup D)$ is an extent of $\mathbb{K}_1$. In the same way $i_2^{-1}(C \cup D)$ is an extent of $\mathbb{K}_2$. Applying again (i), we see that $\mathbb{K}_1(U_1)\mathbb{K}_2$ is an extent-topological context.

(iii) It remains to show that $V \subseteq X$ is open if and only if $i_1^{-1}(V) \subseteq X_1$ and $i_2^{-1}(V) \subseteq X_2$ are open. $V \subseteq X$ is open if and only if $X \setminus V$ is closed and $X \setminus V$ is closed if and only if $i_1^{-1}(X \setminus V)$ is closed in X_1 and $i_2^{-1}(X \setminus V)$ is closed in X_2. $i_k^{-1}(X \setminus V)$ is closed in X_k if and only if $X_k \setminus [i_k^{-1}(X \setminus V)]$ is open, $k = 1, 2$. Since the latter set is equal to $i_k^{-1}(V)$, $k = 1, 2$, we have the desired equivalence. $\square$

(Note that because of part (ii) we have also proven that definition 7.16 of the quotient topology actually characterizes a topological space).

Corollary 5 *Let $(X, \mathcal{O}_{X_1})$ and $(X, \mathcal{O}_{X_2})$ be schemes and let $\phi : (U_1, \mathcal{O}_{X_1}|_{U_1}) \longrightarrow (U_2, \mathcal{O}_{X_2}|_{U_2})$ be an isomorphism of open subschemes (i.e. an isomorphism of locally ringed spaces). Let $\mathbb{K}_i := (X_i, \overline{\mathcal{O}_{X_i}}, \Delta)$, $i = 1, 2$. Then, after partial clar-*

ification, the context $\mathbb{K}_1(U_1)\mathbb{K}_2$ *is isomorphic to the context* $(X, \overline{\mathcal{O}_X}, \triangle)$, *where* X *is obtained by glueing* X_1 *and* X_2 *along* U_1 *and* U_2. *Moreover, the underlying locally ringed spaces are isomorphic.*

Proof We have $\mathbb{K}_1(U_1)\mathbb{K}_2 = (X, \overline{\mathcal{O}_{X_1}} \times \mathcal{O}_{X_2}, J)$. For a closed set $B \subseteq X$ we have $B^J = (i_1^{-1}(B))^\triangle \times (i_2^{-1}(B))^\triangle$. Since $(i_1^{-1}(B))^c = i_1^{-1}(B^c)$ we have $B^J = \{(s_1, s_2) \mid s_1 \in \mathcal{O}_{X_1}(i_1^{-1}(B^c)), s_2 \in \mathcal{O}_{X_1}(i_2^{-1}(B^c)),$ if $s_1 \in \overline{\mathcal{O}_{X_1}(i_1^{-1}(B^c) \cap U_1)}$ then $s_2 \in \overline{\mathcal{O}_{X_1}(i_2^{-1}(B^c) \cap U_2)}\}$. Hence we can clarify our context if we keep only those pairs (s_1, s_2) which satisfy $\phi(s_1|_{i_1^{-1}(V) \cap U_1}) = s_2|_{i_2^{-1}(V) \cap U_2}$. $\square$

We have already seen how Formal Concept Analysis can benefit from ideas from Algebraic Geometry. Another example is the glueing of arbitrary formal contexts.

Let $\mathbb{K}_1 := (G_1, M_1, I_1)$ and $\mathbb{K}_2 := (G_2, M_2, I_2)$ be formal contexts with $G_k^{I_k} \neq \emptyset$, $k = 1, 2$. Let $A_k \subseteq X_k$ be subsets and let $U_k := A_k^c$, $k = 1, 2$, be the complementary sets. Let $N_k := \{m_k \in M_k \mid m^{I_k} \cap U \neq \emptyset\}$, $k = 1, 2$. Let $(\alpha, \beta) : (U_1, N_1, I_1 \cap (U_1 \times N_1)) \longrightarrow (U_2, N_2, I_2 \cap (U_2 \times N_2))$ be an isomorphism of subcontexts. Let $G = G_1 \cup G_2 / \sim$ be the set of equivalence classes on the disjoint union of G_1 and G_2, where $\sim$ is the equivalence relation given by $u_1 \sim \alpha(u_1)$ for $u_1 \in U_1$. Furthermore, let $i_k : G_k \longrightarrow G$ be the map that sends an element $x_k \in X_k$ to its equivalence class in G, $k = 1, 2$. We consider again the context $(G, M_1 \times M_2, J)$ with

$$xJ(m_1, m_2) : \\ \Longleftrightarrow$$

$$1)\ i_k^{-1}(x)I_k m_k \text{ if } i_k^{-1}(x) \in A_k, (k = 1, 2),$$
$$2)\ i_1^{-1}(x)I_1 m_1 \wedge i_2^{-1}(x)I_2 m_2 \text{ if } i_1^{-1}(x) \in U_1.$$

Just as in the case of extent-topological contexts, we denote this context by $\mathbb{K}_1(U_1)\mathbb{K}_2$.

Theorem 14 *Let* $\mathbb{K}_1(U_1)\mathbb{K}_2$ *be as above and let* $C \subseteq G$.
1) If $i_1^{-1}(C) \subseteq G_1$ *is an extent of* $\mathbb{K}_1$ *and if* $i_2^{-1}(C) \subseteq G_2$ *is an extent of* $\mathbb{K}_2$, *then* C *is an extent of* $\mathbb{K}_1(U_1)\mathbb{K}_2$.
2) If $C = i_1(A_1) \cup (i_1(U_1) \cap C)$ *and if* $i_1^{-1}(i_1(U_1) \cap C)$ *is an extent of* $\mathbb{K}_1$, *then* C *is an extent of* $\mathbb{K}_1(U_1)\mathbb{K}_2$. *(The same statement is, of course, true if* $C = i_2(A_2) \cup (i_2(U_2) \cap C))$.

Proof 1) was already shown in Lemma 7.
2) We have $C^J = (i_1^{-1}(C))^{I_1} \times (i_2^{-1}(C))^{I_2} = (i_1^{-1}(C))^{I_1} \times (i_2^{-1}(i_2(U_2) \cap C))^{I_2}$, since $i_2^{-1}(C) \cap A_2$ is empty. We can write $C^{JJ} = F \cup T \cup H$, where
$F = \{i_1(g) \in i_1(U_1) = i_2(U_2) \mid gI_1m_1 \text{ for all } m_1 \in (i_1^{-1}(C))^{I_1},$
$gI_2m_2 \text{ for all } m_2 \in (i_2^{-1}(i_2(U_2) \cap C))^{I_2}\}$,
$T = \{i_1(g) \in i_1(A_1) \mid gI_1m_1 \text{ for all } m_1 \in (i_1^{-1}(C))^{I_1}\}$,
$H = \{i_2(g) \in i_1(A_2) \mid gI_1m_2 \text{ for all} m_2 \in (i_2^{-1}(C))^{I_2}\}$. We observe

$F = i_1(U_1) \cap i_1((i_1^{-1}(C))^{I_1 I_1}) \cap i_2((i_2^{-1}(C))^{I_2 I_2}) =$
$i_1(U_1) \cap i_1((i_1^{-1}(C))^{I_1 I_1}) \cap i_2((i_2^{-1}(i_1(U_2) \cap C))^{I_2 I_2})$, since $i_2^{-1}(C) \cap A_2 = \emptyset$.
Using the isomorphism of subcontexts we conclude $F = i_1(U_1) \cap i_1((i_1^{-1}(i_1(U_1) \cap C)^{I_1 I_1}) = i_1(U_1) \cap C$, since $i_1^{-1}(i_1(U_1) \cap C) = U_1 \cap i_1^{-1}(C)$ is an extent of $\mathbb{K}_1$.
Since $i_1(A_1) \subseteq C$, we have $T = i_1(A_1)$. Moreover, H is empty because $i_2(A_2) \cap C$
is empty. Therefore $C^{JJ} = (i_1(U_1) \cap C) \cup i_1(A_1) = C$ and C is an extent   □

Example 7.23   a) Consider the formal context $\mathbb{K}$:

I	1	2	3
a	×		×
b		×	×

Its concept lattice has four elements. Let $\overline{\mathbb{K}} := (\overline{H}, \overline{M}, \overline{I})$ be an isomorphic copy
of $\mathbb{K} := (H, M, I)$ and let $\mathbb{K}(U_1)\overline{\mathbb{K}} = (G, M \times \overline{M}, J)$ be obtained by glueing along
$U_1 := \{b\}$. If we write $G := \{a, \overline{a}, b\}$ and m instead of $\overline{m}$ for $\overline{m} \in M$ we obtain
the following context:

I	$(1,1)$	$(1,2)$	$(1,3)$	$(2,1)$	$(2,2)$	$(2,3)$	$(3,1)$	$(3,2)$	$(3,3)$
a	×	×	×				×	×	×
$\overline{a}$	×		×	×		×	×		×
b					×	×		×	×

We see that the concept lattice of $\mathbb{K}(U_1)\overline{\mathbb{K}}$ is isomorphic to the lattice of all
subsets of a set with three elements. When we continue glueing $\mathbb{K}(U_1)\overline{\mathbb{K}}$ and
an isomorphic copy along, say $U_2 := \{b\}^c$, we obtain again a boolean concept
lattice whose extents are exactly the subsets of a set with four elements. (This
follows immediately from the characterization of extents given in Lemma 7). In
this way, we get an iteration to construct all finite power set lattices.

b) The next example shows that there may be extents in a formal context obtained by glueing, which are neither of the form described in Theorem 14 1), nor
of the form described in Theorem 14 2). Consider the context $\mathbb{K}$

I	1	2	3	4	5	6
a	×	×				×
b	×	×		×	×	×
c		×	×	×		×
d		×		×		×
e			×		×	×
f					×	×

Let $A := \{a, b\}$ and $U_1 := \{c, d, e, f\}$ and consider $\mathbb{K}(U_1)\overline{\mathbb{K}}$, where $\overline{\mathbb{K}}$ is an
isomorphic copy of $\mathbb{K}$. Let us write $G := \{a, \overline{a}, b, \overline{b}, c, d, e, f\}$. One computes
$(5,3)^J = \{b, e\}$, which means that $C := \{b, e\}$ is an extent of $\mathbb{K}(U_1)\overline{\mathbb{K}}$. However,

$i_1^{-1}(C)$ is not an extent of $\mathbb{K}$, since $\{b, e\}^{II} = \{b, e, f\}$, and C is not of the form 2) either, because $C \cap i_1(A) = \{b\} \neq A$.

One could object that C is of the form $C = (i_1(U) \cap C) \cup (i_1(A) \cap C)$, where $i_1^{-1}(i_1(U) \cap C)$ and $i_1^{-1}(i_1(A) \cap C)$ are closed. However, the conjecture that sets of this form are always extents is not true. Consider the following context $\mathbb{K}$:

I	1	2	3	4	5	6
a				×	×	×
b				×		×
c	×		×			×
d		×	×			×

Let $A := \{c, d\}$ and $U_1 := \{a, b\}$ and consider again $\mathbb{K}(U_1)\overline{\mathbb{K}}$. The set $C := \{a, b, c\}$ is of the same form as above, but C is not an extent of $\mathbb{K}(U_1)\overline{\mathbb{K}}$. Indeed, we have $C^J = \{a, b, c\}^I \times \{\overline{c}\}^{\overline{I}} = \{6\} \times \{\overline{1}, \overline{3}, \overline{6}\}$ and so $C^{JJ} = \{a, b, c, d\}$.

The last two examples show that a complete characerization of the extents of a formal context obtained by glueing seems to be rather difficult. The topological case is especially "nice".

8 Projective Schemes

Finally, we wish to apply the terminology of locally ringed contexts to the projective case as well. Just as in the affine case, $\mathbb{P}^n(K)$ is a topological space when we take the projective varieties as closed sets. Every projective variety has a decomposition into irreducible projective varieties. We recall that an ideal I of the n-dimensional projective context is closed if and only if I is a reduced homogenous prime ideal not containing a power of the irrelevant ideal $< x_0, \ldots, x_n >$. A projective variety V is irreducible if and only if its ideal is prime. Therefore we can define $Proj\, K[x_0, \ldots, x_n] := \{\mathfrak{p} \,|\, \mathfrak{p}$ is a homogenous prime ideal not containing $< x_0, \ldots, x_n >\}$.

Theorem 15 *Let K be an algebraically closed field. $(Proj\, K[x_0, \ldots, x_n], K[x_0, \ldots, x_n], \ni)$ and $(\mathbb{P}^n(K), K[x_0, \ldots, x_n], \ni)$ have isomorphic concept lattices and the systems of intents are identical. In particular, $\sqrt{I} = \bigcap \{\mathfrak{p} \,|\, \mathfrak{p} \in Proj\, K[x_0, \ldots, x_n]\}$ holds for any homogenous ideal I not containing a power of the irrelevant ideal. $Proj\, K[x_0, \ldots, x_n]$ is a topological space.*

Proof Immediate from Lemma 6.   □

We construct a corresponding scheme. We generalize the situation to arbitrary **graded rings** S, where closed sets are defined using the relation $\ni$ as above. A graded ring is a ring S together with a decomposition $S = \sum_{d \geq 0} S_d$ of S into a direct sum of abelian groups, such that $S_d S_e \subseteq S_{d+e}$ for all $d, e \geq 0$. Elements of S_d are called **homogenous** of degree d. Thus, each element of S

can be written uniquely as a finite sum of homogenous elements. An ideal $\mathfrak{a}$ is called **homogenous** if $\mathfrak{a} = \sum_{d \geq 0}(\mathfrak{a} \cap S_d)$. The ideal $\sum_{d \geq 1} S_d$ is denoted by S_+. For example, $S := K[x_0, \ldots, x_n]$ is a graded ring if we define S_d to be the set of all monomials of total degree d. Here we have $S_0 = K$ and $S_+ = <x_0, \ldots, x_n>$ is the unique maximal homogenous ideal of S.

Let $Proj\, S$ be the set of all homogenous prime ideals that do not contain all of S_+. If $\mathfrak{p} \in Proj\, S$ let T be the multiplicative set of all homogenous elements not contained in $\mathfrak{p}$. Then $T^{-1}S$ is a graded ring via $deg\,(f/g) := deg\, f - deg\, g$. Now, let $S_{(\mathfrak{p})}$ be the subring of elements of degree zero of $T^{-1}S$. $S_{(\mathfrak{p})}$ is a local ring with maximal ideal $(\mathfrak{p}T^{-1}S) \cap S_{(\mathfrak{p})}$. Similarly, if f is a homogenous polynomial, we denote by $S_{(f)}$ the subring of elements of degree zero in the localized ring S_f.

Now, let $U \subseteq Proj\, S$ be open. We define $\mathcal{O}(U)$ as the set of tupels $(r_\mathfrak{p})_{\mathfrak{p} \in U}$, $r_\mathfrak{p} \in S_{(\mathfrak{p})}$, satisfying $(\star)$ $\forall \mathfrak{p} \in U \exists V : \mathfrak{p} \in V \subseteq U$ and elements $a, f \in S$ homogenous of the same degree, such that for all $\mathfrak{q} \in V$ the equality $s_\mathfrak{q} = a/f$ holds in $S_{(\mathfrak{q})}$.

Theorem 16 *Let S be a graded ring and let $(Proj\, S, \mathcal{O})$ be its **projective spectrum**.*
(a) For all $\mathfrak{p} \in S$ the stalk $\mathcal{O}_\mathfrak{p}$ is isomorphic to $S_{(\mathfrak{p})}$.
(b) For any $f \in S_+$ let $D_+(f)$ be the open complement of $f \ni$. These open sets cover $Proj\, S$ and for each f, we have an isomorphism of locally ringed spaces $Spec\, S_{(f)} \cong (D_+(f), \mathcal{O}|_{D_+(f)})$.
(c) The statements (a) and (b) imply that $(Proj\, S, \mathcal{O})$ is a scheme.

For a **proof** see ([Ha77], II.2.5).

Corollary 6 *Let S be a graded ring. Then $(Proj\, S, \overline{\mathcal{O}}, \triangle)$ is a locally ringed context if we use the standard definition of $\triangle$. $\triangle$ can also be defined analogously to the affine case using the relation $\ni$. We define $\mathfrak{q}\triangle(r_\mathfrak{p})_{\mathfrak{p} \in U}$ if and only if, for all $\mathfrak{p} \in U$, the denominator of a representation of $r_\mathfrak{p}$, that satisfies $(\star)$, is in $\mathfrak{q}$.*

Example   We wish to illustrate 8.2.(b) by constructing the corresponding isomorphism from $(D_+(f), \overline{\mathcal{O}|_{D_+(f)}}, \triangle)$ to $(Spec\, S_{(f)}, \overline{\mathcal{O}}, \triangle)$. First note that the sets of the form $D_+(f)$ cover $Proj\, S$, because we have $\mathfrak{p} \in Proj\, S$ if and only if $\mathfrak{p}$ does not contain all of S_+, which means that there is an element $f \in S_+$, such that $\mathfrak{p} \in D_+(f)$. For a homogenous prime ideal $\mathfrak{a} \subseteq S$ we define $\phi(\mathfrak{a}) = (\mathfrak{a}S_f) \cap S_{(f)}$. Here S_f is the ring of fractions arising from $T := \{1, f, f^2, \ldots\}$ when we regard S as a ring in the usual sense, ignoring the grading. By the properties of localization we know that, for a prime ideal $\mathfrak{p} \in D_+(f)$, we have $\phi(\mathfrak{p}) \in Spec\, S_{(f)}$ and that the induced mapping $\tilde{\phi}$ from $D_+(f)$ to $Spec\, S_{(f)}$ is a homeomorphism. Furthermore, we know that for any $\mathfrak{p} \in D_+(f)$ we have an isomorphism of rings $S_{(\mathfrak{p})} \cong (S_{(f)})_{\phi(\mathfrak{p})}$. Let $\psi_\mathfrak{p}$ be this isomorphism. Now, we can define a map $\tilde{\psi}$ from $\overline{\mathcal{O}|_{D_+(f)}}$ to $\overline{\mathcal{O}}$ as follows: if $(r_\mathfrak{p})_{\mathfrak{p} \in D_+(f) \cap U}$ is an element from $\mathcal{O}|_{D_+(f)}(U)$ its image under $\tilde{\psi}$ is defined to be $(\psi_\mathfrak{p}(r_\mathfrak{p}))_{\phi(\mathfrak{p}) \in Spec\, S_{(f)} \cap \tilde{\phi}(U)}$. Here $\psi_\mathfrak{p}(\frac{f}{g})$ is defined to be $\frac{\psi_\mathfrak{p}(f)}{\psi_\mathfrak{p}(g)}$. Then it is easily computed that $(\tilde{\phi}, \tilde{\psi})$ is an isomorphism of contexts.

References

[Be99]   T. Becker: *Formal Concept Analysis and Algebraic Geometry*. Dissertation, TU Darmstadt. Shaker Verlag, Aachen 1999.

[BW91]   T. Becker, V. Weispfennig: *Groebner Basis. A computational approach to Commutative Algebra*. Springer, New York 1991.

[Br89]   M. Brodmann: *Algebraische Geometrie. Eine Einführung*. Birkhäuser, Basel 1989.

[CLO92] D. Cox, J. Little, D. O'Shea: *Ideals, varieties and algorithms*. Springer, New York 1992.

[Fi94]   G. Fischer: *Ebene Algebraische Kurven*. Vieweg, Braunschweig 1994.

[GW99]   B. Ganter, R. Wille: *Formal Concept Analysis: mathematical foundations*. Springer, Heidelberg 1999.

[GD60]   A. Grothendieck, J.Dieudonne: *Elements de geometrie algebrique*. Publ. Math., No.4, Institut des Hautes Etudes Scientifiques, Paris 1960.

[Ha77]   R. Hartshorne: *Algebraic Geometry*. Springer, Heidelberg 1977.

[HKS99] G. Hartung, M. Kamara, C. Sacarea: A topological representation of polarity lattices. In: *Acta Math. Univ. Comenianae*, LXVIII, 49–70.

[Ku80]   E. Kunz: *Einführung in die Kommutative Algebra und algebraische Geometrie*. Vieweg, Braunschweig 1980.

[Ku91]   E. Kunz: Ebene algebraische Kurven. In: *Regensburger Trichter* 23, 1991.

[Ku97]   E. Kunz: *Einführung in die Algebraische Geometrie*. Vieweg, Braunschweig 1997.

[Sch64] H. Schubert: *Topologie*. B. G. Teubner, Stuttgart 1964.

[Xi93]   W. Xia: *Morphismen als formale Begriffe. Darstellung und Erzeugung*. Dissertation, TU Darmstadt. Verlag Shaker, Aachen 1993.

From Formal Concept Analysis
to Contextual Logic

Frithjof Dau and Julia Klinger

Technische Universität Darmstadt, Fachbereich Mathematik
Schloßgartenstr. 7, D-64289 Darmstadt
{dau,jklinger}@mathematik.tu-darmstadt.de

Abstract. A main goal of Formal Concept Analysis from its very beginning has been the support of rational communication. The source of this goal lies in the understanding of mathematics as a science which should encompass both its philosophical basis and its social consequences. This can be achieved by a process named 'restructuring'. This approach shall be extended to logic, which is based on the doctrines of concepts, judgments, and conclusions. The program of restructuring logic is named *Contextual Logic* (CL). A main idea of CL is to combine Formal Concept Analysis and *Concept Graphs* (which are mathematical structures derived from conceptual graphs). Concept graphs mathematize *judgments* which combine concepts, and conclusions can be drawn by inferring concept graphs from others. So we see that concept graphs can be understood as a crucial part of the mathematical implementation of CL, based on Formal Concept Analysis as the mathematization of the doctrine of concepts.

1 Overview

Formal Concept Analysis (FCA) is a mathematical theory applied successfully in a wide range (there have been more than 200 projects in various academic and commercial fields). The impact and success of FCA and the large number of applications in the real world cannot be explained solely with the mathematical results and the mathematical power of FCA. The driving force behind FCA lies in our understanding of mathematics as a science which encompasses the philosophical basis and the social consequences of this discipline as well.

A main goal of Formal Concept Analysis from its very beginning has been the support of rational communication and the representation and processing of knowledge. This goal is based on and carried out by a process termed 'restructuring'. In the first section, we will describe the ideas and purposes of this restructuring process as well as further philosophical foundations of FCA. Moreover, we will argue why FCA fulfils the purposes of restructuring to a large extent.

In the next section, we will report how this restructuring approach is extended to logic. We will point out that the purely extensional and mechanistic attempt of contemporary formal mathematical logic is too narrow for our purposes. For

B. Ganter et al. (Eds.): Formal Concept Analysis, LNAI 3626, pp. 81–100, 2005.

this reason, we revitalize the traditional philosophical understanding of logic. This understanding tries to capture and to investigate the laws of thinking and is based on the doctrines of concepts, judgments, and conclusions.

The outcome of restructuring our understanding of logic will be called *Contextual Logic* (CL). As this logic starts with the doctrine of concepts, we see that CL builds upon FCA. However, further formalizations of the doctrines of judgments and conclusions are needed.

A promising approach is to use the system of conceptual graphs (CGs) of John Sowa, which will be introduced in Section 4. The philosophical background of this system is similar to that of FCA (cf. [MSW99]), and the system allows us to formalize judgments and conclusions in a way which is much nearer to human reasoning than predicate logic. But the system of conceptual graphs is not elaborated mathematically. One reason for this is that this system is very open-minded and extended in various ways which are often not clearly defined. Thus, it is huge, without sharp borders, and contains several ambiguities, gaps and even flaws and mistakes.

For this reason, in Section 5, a mathematical formalization for a core of CGs is provided. The mathematical structures modelling CGs are called *concept graphs*. As said above, it is impossible to find a definition for concept graphs which covers all aspects and features of CGs at once. Instead of that, different versions of concept graphs which correspond to different fragments of CGs are elaborated. An overview over the different kinds of concept graphs is presented in Section 6.

2   Formal Concept Analysis

From the main theorem of Formal Concept Analysis we know that the concept lattices of FCA are -up to isomorphism- exactly the complete lattices as defined in lattice theory. Thus, from a purely mathematical point of view, FCA can be seen as the theory of complete lattices, presented in an unfamiliar way. The broad results of lattice theory have been applied in various fields of mathematics, but only little outside mathematics. In contrast to that, FCA has been successfully applied in various real-world projects. Thus, in order to explain the power and success of FCA, it is obviously not sufficient to look at FCA as a solely mathematical structure theory. The question is: What is the unique peculiarity that makes FCA that usable? What is the advantage of FCA compared to the usual form of lattice theory? The answer lies in the underlying philosophy of FCA. The main idea is a program termed 'restructuring (lattice theory)', relying on the concept of *Wissenschaftsdidaktik* by Hartmut von Hentig. In his book '*Magier oder Magister? Über die Einheit der Wissenschaft im Verständigungsprozess.*' ('Magician or Magister? On the Unity of Science in the Process of Understanding', cf. [He74], p. 136f), the restructuring of scientific disciplines is explained as follows:

Sciences have to examine their disciplinary, and this means: To uncover the unconscious purposes, to declare their conscious purposes, to select and to adjust their means according to those purposes, to explain possible consequences comprehensible and publicly, and to make accessible their ways of scientific finding and their results by the every-day language.

The program of restructuring is based on a philosophical background which goes back to the *Pragmatism* of Charles Sanders Peirce (see [Pe35]) and which is adopted and continued in the *Discourse Philosophy* of Karl-Otto Apel (cf. [Ap89]) and Jürgen Habermas (see [Ha81]). The main idea of Pragmatism is that the significance of any conception consists exclusively in its effects. In particular, each scientific concept and theory has to be judged by all the effects it may produce. This establishes a tight connection between theory and practice.

Another crucial point is that in Pragmatism and discourse philosophy, the basis and origin of reasoning lies within intersubjective communication and argumentation. It is important to note that intersubjective communication takes place not only between members of a specific scientific community, but between members of different communities and even between scientists and non-scientists. Thus, a transdisciplinary communication (see [Wi02c])has to be enabled and established. For this reason, Hentig demands the use of every-day language, so that a scientific theory, including its results and effects, can be understood, applied and critizised by people standing outside that specific scientific community. In another place in [He74], p. 33f, he says (italics by Dau/Klinger):

> The restructuring of scientific disciplines within themselves becomes more and more necessary to make them more learnable, mutually available, and criticizable in more general surroundings, *also beyond disciplinary competence.* This restructuring may and must be performed by general patterns of perception, thought, and action of our civilization.

The development of FCA is inspired by Hentigs restructuring program. Lattice theory is reworked in order to integrate and to rationalize origins, connections to and interpretations in the real world. One main goal of FCA is to support a rational discourse, not only between mathematicians, but between mathematicians and non-mathematicians, even non-scientists, as well. Thus the results of lattice theory have to be presented in a way which makes them understandable, learnable, available and criticizable, particularly for non-mathematicians. As Wille says in [Wi96]:

> The aim is to reach a structured theory which unfolds the formal thoughts according to meaningful interpretations allowing a broad communication and critical discussion of the content.

We have to discuss why and how FCA achieves the requirements of the restructuring program. The starting point of FCA is the philosophical understanding of *concepts* as the basic units of thought. A concept is constituted by two counterparts: its *extension* which consists of all objects belonging to the concept, and its *intension* which contains all attributes shared by all objects of the extension. Due to Peirce, in any reasoning or argumentation process, we can only grasp a limited part of the reality. Our universe of discourse is always a re-stricted context. These considerations lead to the well-known basic definitions of FCA which formalize these ideas (see [Wi82]). FCA starts with the notion of a *formal context* (G, M, I) consisting of a set G of (formal) objects (in German: 'Gegenstände'), a set M of (formal) attributes (in German: 'Merkmale'), and an incidence-relation $I \subseteq G \times M$. The relationship gIm (with $g \in G$ and $m \in M$) indicates that the object g has the attribute m. A *formal concept* is a pair (A, B) with $A \subseteq G$ and $B \subseteq M$, which satisfies $B = \{m \in M \,|\, gIm$ for all $g \in A\}$ and $A = \{g \in G \,|\, gIm$ for all $m \in B\}$. This is clearly a mathematical formalization of the philosophical concepts.

From a mathematical point of view, formal contexts and formal concepts could be reduced to classical relational structures (which are purely extensional) resp. to unary predicates. But humans structure the world conceptually and meaningfully, and the meaning of concepts cannot be explained solely by their extensions. On the contrary: The meaning of concepts is heavily constituted by their intensions and by the intermediate relationships between the concepts (cf. [Se01] and [Br94]). Moreover, a formal context can be represented by crossta-bles which are very common in our daily-life culture and therefore easy to com-prehend. Thus, from a human point of view, formal contexts are easier to un-derstand and much more meaningful than relational structures.

The most important relation between concepts is given by the relation-ships *subconcept* and *superconcept*. For formal concepts, they are defined as follows: Given two formal concepts (A_1, B_1) and (A_2, B_2), we set $(A_1, B_1) \leq (A_2, B_2) :\Longleftrightarrow A_1 \subseteq A_2 (\Longleftrightarrow B_2 \subseteq B_1)$ and say that (A_1, B_1) is a *subconcept* of (A_2, B_2) resp. (A_2, B_2) is a *superconcept* of (A_1, B_1). The set of all formal concepts of a given formal context, together with the relation $\leq$, is a complete lattice, the *concept lattice* of the formal context.

The concept lattice of a formal context can be represented as a *labelled line diagram*. This visualization of the underlying formal context is the next advan-tage of FCA. With a small amount of experience, these diagrams are easy to understand and comprehend. Our experience with projects shows that when a specific domain is formalized by a formal context, the examination of the line diagram of the corresponding concept lattice often leads to unexpected insights by the domain experts. For this, no mathematical knowledge of FCA is needed.

Still more effects come into play when so-called many-valued contexts are considered. Many-valued contexts are transformed into ordinary (one-valued) formal contexts by the process of *conceptual scaling*. During this process, each attribute $m \in M$, together with its values, is interpreted by a formal context itself. There is no standard or even neccessary interpretation of a many-valued attribute: It has to be *decided* by the domain expert which scale is appropriate for a given attribute with respect to a given purpose. The fact that the process of scaling cannot be automated should be understood not as a drawback, but a great advantage: In our projects it turned out, when a TOSCANA-System[1] was

---

[1] TOSCANA is a computer program which allows to explore relational databases with FCA-methods, see for instance [BH03]

implemented by TOSCANA-consultants (which are experts on FCA, but which usually have no or only few experience in the specific domain), that the right choice of a scale is usually far from being trivial and often raises discussions among the field experts as well as between the field experts and the TOSCANA-consultants. Thus, already the process of scaling supports rational communication, even between members of different research fields. Moreover, as the choice of the scales is left to the experts and not done automatically by the machine, the cognitive autonomy and the responsibility of the experts is preserved.

Finally, many-valued contexts can be understood as mathematical versions of tables in relational databases. As most information is stored as data in relational databases, FCA turns out to be an adequate instrument for a meaningful and conceptual exploration of the stored data.

3 Logic

The aim of restructuring logic seems to be self-evident, as logic, understood as the investigation of the laws of thinking, is another fundamental source for reasoning. Due to the pragmatic paradigm, the first step in the restructuring process has to make clear the purposes and effects of logic.

In the previous section, it was already argued that the purely extensional approach to predicate logic and their models (relational structures) is too narrow to be used in rational and meaningful discourses. As another point, the mechanistic approach of predicate logic shall be mentioned, particulary the calculi which act on formulas, understood as a priori meaningless sequences of signs. The reason for this is clear: The purpose of predicate logic has never been to model or support human reasoning, but to provide an instrument which shall explain and contribute to the structure of mathematical argumentations only. Thus, for supporting reasoning, we have to find a broader understanding of logic which goes beyond classical predicate logic. As Apel says in [Ap89] (italics by Dau/Klinger):

> In view of this problematic situation [of rational argumentation] it is more obvious not to give up reasoning at all, but to break with the concept of reasoning which is orientated by the pattern of *logic-mathematical proofs*.

A convincing approach is to revitalize the traditional, philosophical understanding of logic, given by 'the three essential main functions of thinking – concepts, judgments, and conclusions' [Ka88]. Concepts, the basic units of thought, are already formalized in FCA. If we combine concepts to meaningful statements, we obtain propositions; and judgments are propositions which are asserted, i. e. valid propositions (due to Peirces pragmatism and the discourse philosphy, the validity of judgments has to be confirmed by a rational discourse in the intersubjective community of communication). With conclusions, new judgments are obtained from already existing ones.

The process of restructuring this understanding of logic, i. e., the mathematical formalization of philosophical logic shall lead to a theory called *Contextual*

Logic (CL) (see [Wi00b]). In CL, we have to formalize concepts, judgments, and conclusions. Particularly, logical junctors like conjunction oder negation have to be incorporated into CL, as well as the possibility to draw judgments which range over objects (of a given universe of discourse), i.e., quantification, has to be enabled. Hence, as Wille says in [Wi00a]: 'Contextual Logic may reach at least the expressibility of first order predicate logic.'

As the mathematization of concepts has already been elaborated in FCA, the question of how to proceed with judgements and conclusion arises.

4   Conceptual Graphs

For the formalization of judgments, we use the theory of *Conceptual Graphs (CG)* of John Sowa (cf. [So84]). In Figure 1, we provide two well-known examples of CGs.

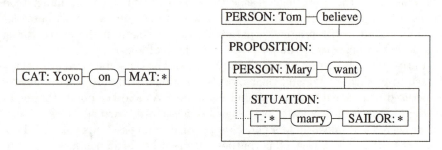

Fig. 1. Two conceptual graphs

Conceptual graphs are assembled of concept boxes, each of them containing a type and a reference belonging to the type. The star in the right concept box of the left graph is the so-called *generic marker* and has to be understood as an unqualified object; i.e. the generic marker can be read as an existential quantifier. Concept boxes may be nested (see the right graph for an example). In this case, a box stands for a context, and the reference of the box is a graph itself which describes this context. Concept boxes may be connected by relation ovals or by a dotted line (so-called *coreference-links*). These connections describe relationships between the references of the boxes. In particular, a coreference-link between two concept boxes means that their references are identical.

The meaning of the left graph is therefore 'The cat Yoyo is on a mat'. The right graph contains two contexts, namely the concept boxes of type PROPOSITION and SITUATION (which are common types of contexts). The graph can be read as follows: The person Tom believes a proposition, which is described by a graph itself. The proposition says that the person Mary wants a situation, which again is described by a graph. In this situation we have a concept box ⊤ : * connected with a coreference link to the concept box PERSON:Mary in the context above. So the situation is that Mary marries a sailor. The understanding of the whole graph is now: The person Tom believes the proposition that the person Mary wants the situation that Mary marries a sailor. In short: Tom believes that Mary wants to marry a sailor.

In [So92], Sowa explains the foundations and the purpose of conceptual graphs as follows:

> Conceptual graphs are a system of logic based on the existential graphs of Charles Sanders Peirce and the semantic networks of artificial intelligence. The purpose of the system is to express meaning in a form that is logically precise, humanly readable, and computationally tractable. With their direct mapping to language, conceptual graphs can serve as an intermediate language for translating computer-oriented formalisms to and from natural languages. With their graphic representation, they can serve as a readable, but formal design and specification language.

As this quotation, shows the philosophy behind conceptual graphs is based on Peirce's pragmatism, and it is very close to Hentig's restructuring program. Thus the spirit of CGs can directly by adcopted by CL, which makes a mathematical elabarotion of CGs a promising approach for formalizing judgments (and conclusions, as argued below) in CL.

In philosophy, the considered judgments are often elementary, i. e., judgments of the form 'an object belongs to the extension of a concept' or 'one concept is a subconcept of another concept'. The first kind of judgments corresponds to the boxes in conceptual graphs, the second kind is coded in the so-called *type hierarchy* of conceptual graphs. But as we can see in Figure 1, with conceptual graphs, much more complex judgments can be formulated. This higher expressiveness is clearly a further advantage.

Moreover, CGs can be *rhetorically structured*. That is, they bear the possibility to represent information in different ways. Assume a person working with graphs wants to represent some amount of information with a CG. First of all, this amount of information can be captured by mathematically different, but semantically equivalent graphs, and he can choose one graph among these. For example, he could choose a graph which does not contain any redundancies. This possibility of choice is usually given in other (e.g., linear) representations of logic as well. Moreover, the chosen graph can be graphically represented in different ways (i.e., how its concept boxes and relation ovals are arranged on the plane). A 'good' graphical representation can improve the readability of the graph to a large extent. Note that this freedom in the representation is not given in linear representations of logic.

Finally, Sowa provides rules for formal deduction procedures on conceptual graphs. Thus the system of conceptual graphs can offer a formalization of conclusions as well.

In some sense the system of conceptual graphs is not fixed, but open-minded. It is designed to be used in fields like software specification and modelling, knowledge representation, natural language generation and information extraction, and these fields have to cope with problems of implementational, mathematical, linguistic or even philosophical nature. In order to deal with such problems, different modifications and extensions of conceptual graphs are suggested. For this reason it is impossible (and perhaps not even desirable) to provide a definition which covers all possible aspects and features of conceptual graphs at once. On the other hand a closer observation shows that this leads to a lack of preciseness which causes several difficulties and fallacies, ranging from ambiguities over minor gaps to major flaws.

As Sowa's system is not mathematically elaborated, we have to provide mathematical definitions for conceptual graphs, and we have to formalize how reasoning can be carried out with them. As argued above, it cannot be expected to find a mathematical definition which covers all aspects of conceptual graphs at once. Instead of this, different forms of concept graphs (as mathematization of CGs) with different levels of expressiveness (and further differences) have been developed during the last years. An overview of the several forms of concept graphs will be provided in the next sections.

When the meanings of CGs are explained, the references of concept boxes are often interpreted by objects of the real world, and the types and relations are interpreted by concepts of the natural language. In order to formalize some aspects of reasoning with graphs, this informal world semantics of CGs has to be replaced by a mathematically elaborated semantics. In FCA, the step of concepts (as basic units of thoughs) to *formal* concepts has already be done. Thus it is an evident approach to interpret the references of concept boxes by objects and the types of conceptual graphs by formal concepts in a formal context. This idea yields a mathematically defined semantics which is based on formal contexts and therefore called *contextual semantics*. In contrast to the purely extensional relational models of FOL, contextual models bear intensional information, too. They are conceptually structured and represent information in a formalized way. The same holds for concept graphs: They are formalized judgments, particularly they represent information. Hence, when a concept graph is interpreted in a contextual model, it is checked whether the information which is encoded by the graph is a part of the information of the contextual model. This tight connection between graphs and models, based on the *informational contents* of graphs and models, yields a crucial difference to the common understanding (e.g. in FOL) of interpreting formulas in models by means of *truth values*.

5   Concept Graphs and Contextual Logic

As explained in the previous sections, Contextual Logic can be understood as a formalization of the traditional philosophical logic with its doctrines of concepts, judgments, and conclusions. In this section we explain briefly how the theory of Conceptual Graphs (CGs) combined with FCA led to the theory of so-called *concept graphs* in its several distinct formings (this new term was introduced in order to distinguish Sowa's approach from the FCA-inspired mathematical theory).

Concept graphs are abstracted from CGs. However, as elaborated in the previous section, CGs were designed to be of use in a wide variety of different fields, resulting in difficulties in the mathematization. Hence, concept graphs as mathematization of CGs only cover restricted parts of Sowa's Theory. In general, two accesses can be distinguished, namely semantical approaches and those based on a separation in syntax and semantics.

Semantical theories deal with the elaboration of a mathematical structure theory of concept graphs of a given power context family in an algebraic manner (where a power context family (PCF) is a family of contexts which are connected via their object sets; the mathematical definition can be found below). In particular, the forms and relations of those concept graphs are investigated. This includes operations on those graphs (see for instance [Wi01]) and a thorough study of the properties of the corresponding algebra of concept graphs of a given power context family. Since semantical approaches to the theory of concept graphs are concerned with all 'valid propositions' of a power context family, they are understood as a formalization of the doctrine of judgments.

The other approaches are logical ones, using a separation of syntax and semantics as it is common in mathematical logic: Concept graphs as syntactical constructs are defined over an alphabet consisting of object-, concept- and relation-names. They are then equipped with an explicit contextual semantics based on power context families (instead of the traditional implicit semantics of CGs via a translation into predicate logic). In many of the approaches, the results of the corresponding semantical access are used. Since the different specificities of logic systems of concept graphs include an adequate calculus (i. e. a set of inference rules which are sound and complete) or (for the theories of Prediger and Klinger) an effective method to do reasoning via standard models/PCFs, these theories are considered to be a formalization of the doctrine of conclusions.

We will now briefly explain the basic notions of those two approaches, starting with concept graphs as semantical structures.

The first approach to Contextual Judgment Logic was proposed by Wille in [Wi97] where he connected FCA with the theory of CGs. Wille defined concept graphs of power context families as semantical structures.

Since this first access was further developed and specified, we will use a slightly more recent version of concept graphs. The following definitions are taken from [Wi00b]: A *power context family* $\overrightarrow{\mathbb{K}} := (\mathbb{K}_0, \mathbb{K}_1, \mathbb{K}_2, \ldots, \mathbb{K}_n)$ is a family of contexts $\mathbb{K}_k := (G_k, M_k, I_k)$ with $G_0 \neq \emptyset$ and $G_k \subseteq (G_0)^k$ for each $k = 1, \ldots, n$. The formal concepts of $\mathbb{K}_k$ with $k = 1, \ldots, n$ are called *relation concepts* because they represent $k-$ary relations on the object set G_0 by their extent. A *concept graph* is defined as a structure $\mathfrak{G} := (V, E, \nu, \kappa, \rho)$ consisting of two sets V and E and three functions $\nu, \kappa,$ and ρ such that the following conditions are satisfied: Firstly, (V, E, ν) has to be a *relational graph* consisting of two disjoint sets V and E whose elements are called vertices and edges, respectively, and a function $\nu \colon E \to \bigcup_{k=1,\ldots,n} V^k$ which maps each edge to the ordered tuple of its adjacent vertices. For $u \in V \cup E$ we set $|u| = 0$ if $u \in V$ and $|u| = k$ if $\nu(u) \in V^k$. Secondly, $\kappa \colon V \cup E \to \bigcup_{k=0,\ldots,n} \mathfrak{B}(\mathbb{K}_k)$ is a mapping such that $\kappa(u) \in \mathfrak{B}(\mathbb{K}_k)$ for all $u \in V \cup E$ with $|u| = k$. Finally, $\rho \colon V \cup E \to \bigcup_{k=0,\ldots,n} \mathfrak{P}(G_k) \setminus \{\emptyset\}$

is a mapping such that $\rho(u) \subseteq \mathrm{Ext}(\kappa(u))$ for all $u \in V \cup E$ and, furthermore, $\rho(u) = \rho(v_1) \times \cdots \times \rho(v_k)$ if $|u| = k > 0$ and $\nu(u) = (v_1, \ldots, v_k)$.

A partial order on the set of all concept graphs of a power context family is then introduced via the so-called conceptual content, which is studied extensively in [Wi03]. It is even shown that the conceptual contents of concept graphs of a power context family can be described as extents of concepts of a suitably constructed power context family again.

As an example, we consider the different theories of concept graphs with papers they were published in, the kind of approach which was taken and the relation 'is an extension of' between these theories. The first graph of Figure 4, for instance, is a semantical concept graph of the PCF shown in Figure 2. It is read as follows: Triadic Concept Graphs in [Wi98] have a semantic approach, Concept Graphs in [Wi97] as well and the theory presented in [Wi98] extends [Wi97].

$\mathbb{K}_0$:

	Approach: Semantic	Approach: Sep. Syntax/Semantics
Concept Graphs [Wi97]	×	
Triadic Concept Graphs [Wi98]	×	
Concept Graphs [Pr00]		×
Concept Graphs [DH03]		×

$\mathbb{K}_2$:

	Extension
(Concept Graphs [Wi97], Concept Graphs [Wi97])	
(Triadic Concept Graphs [Wi98], Concept Graphs [Wi97])	×
(Concept Graphs [Wi97], Concept Graphs [Pr00])	
(Concept Graphs [DH03], Concept Graphs [Pr00])	

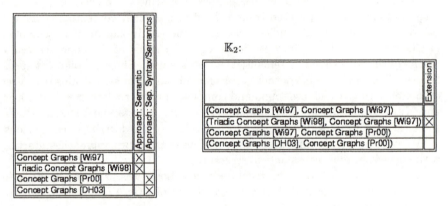

Fig. 2. The power context family $(\mathbb{K}_0, \mathbb{K}_2)$

$\mathcal{G}$:= {[Wi97], [Wi98], [Pr00], [DH03]}

$\mathcal{C}$:= {$\bot$, SEMANTIC APPROACH, LOGIC APPROACH, $\top$}

$\mathcal{R}$:= {EXTENSION}

$\leq_{\mathcal{C}}$:=

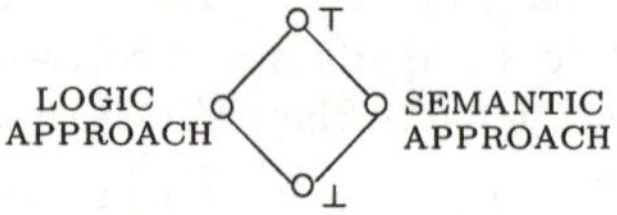

$\leq_{\mathcal{R}} := \mathrm{id}_{\mathcal{R}}$

Fig. 3. Example for an Alphabet

1. Semantic Concept Graph:

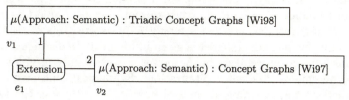

2. Syntactic Concept Graph:

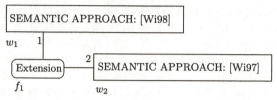

3. Syntactic Semiconcept Graph:

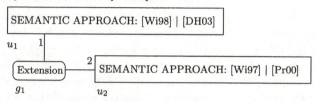

4. Syntactic Concept Graph with Cuts:

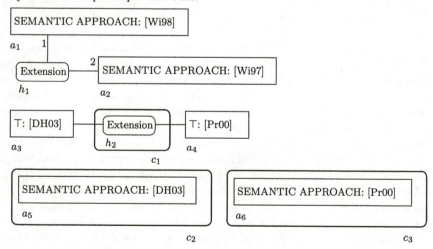

Fig. 4. Examples for Concept Graphs

As an approach based on the separation of syntax and semantics, we describe how concept graphs were introduced in [Pr98b], where the approach of [Wi97] was adopted and modified in order to obtain a logical theory:

The first step towards a syntactical implementation of concept graphs is the definition of an alphabet $\mathcal{A} := (\mathcal{G}, \mathcal{C}, \mathcal{R})$ consisting of a set $\mathcal{G}$ of object names, an ordered set $\mathcal{C}$ of concept names and a family $\mathcal{R}$ of ordered sets of relation names. An example for such an alphabet is given in Figure 3. To distinguish the syntactical names from the elements of the power context family used for the interpretation, we employ different capitalizations. Syntactical concept graphs over an alphabet are then introduced as mathematical structures of the form $\mathfrak{G} := (V, E, \nu, \kappa, \rho)$, consisting of a relational graph, a function κ assigning concept names to vertices and relation names to edges, and a function ρ which assigns non-empty sets of object names to the vertices (as references). As an example, consider the second graph in Figure 4 which shows a syntactical concept graph over the alphabet depicted in Figure 3.

For the semantics, the names of the alphabet are interpreted in a given power context family $\overrightarrow{\mathbb{K}} := (\mathbb{K}_0, \mathbb{K}_1, \ldots, \mathbb{K}_n)$ via a so-called interpretation λ: This interpretation specifies how the syntactical elements of the alphabet are related to elements of $\overrightarrow{\mathbb{K}}$ such that object names are mapped to objects of $\mathbb{K}_0$, concept names to concepts of $\mathbb{K}_0$ and relation names to elements of $\mathfrak{B}(\mathbb{K}_k)$. Moreover, the orders specified in the alphabet are preserved. The resulting structures $(\overrightarrow{\mathbb{K}}, \lambda)$ are called *context interpretations*. Now we say that a concept graph is *valid* in a context interpretation $(\overrightarrow{\mathbb{K}}, \lambda)$ (and call $(\overrightarrow{\mathbb{K}}, \lambda)$ a *model* for the graph) if the so-called vertex- and edge condition for the vertices respectively edges are both satisfied. The vertex condition for a vertex v is fulfilled if the interpreted object names of v belong to the extent of the interpreted concept name of that vertex. Similarly, the edge condition for an edge e holds if the objects along e are in the relation concept assigned to that edge. It is easy to see that the second graph in Figure 4 is (using a suitable interpretation) indeed a concept graph of the PCF in Figure 2.

Why does the theory of concept graphs meet the claims for Contextual Logic? The aim of Contextual Logic is not to enter into competition with First Order Predicate Logic (FOPL), in particular since the expressiveness of most of the theories corresponds only to restricted parts of FOPL. The goal was rather to find an approach which may support rational communication and argumentation. In contrast to FOPL, CGs have been developed and used as a language for knowledge representation with its focus on connections of concepts. They aim at capturing the rhetoric structure of common language and are graphically representable. The contextual foundation of concept graphs as the mathematization of CGs enables us to make the restrictions which occur during the transition from real to formal data explicit and hence discussable. The various extensions for the basic theory of concept graphs as proposed by Wille in [Wi97] and developed by Prediger in [Pr98a], [Pr98b], yield a broad expressiveness for Contextual Logic. These extensions include negation on the level of propositions ([Da01], [Da03a]), negation on the level of concepts and relations ([Wi01],[Wi02a], [Kl01a],[Kl01b]), existential ([Da01],[Wi02a], [Kl02]) and universal quantifiers ([Ta00]), and Nestings ([Wi98], [Pr00], [SW03], [DH03]). Moreover, the Contextual Logic of Relations has been developed ([PoW00], [Wi00c], [Ar02]) as a Contextual Attribute

Logic ([GW99b]) on the relational contexts of a power context family and so-called *relation graphs* have been introduced as algebraic structures for represent-ing relations and operations on relations ([Po01],[Po02]). In the next section we will discuss these extensions in more detail.

6   Different Forms of Concept Graphs

In this section, our aim is to give an overview over the characteristics and the diverse states of development of the different theories of concept graphs. Please note that the notion of 'concept graphs' is used as a generic term for all these approaches.

In Figure 5–7, several concept lattices are shown, which all have the same object set consisting of theories of concept graphs along with a reference for the corresponding paper. In each lattice, the attributes are chosen with respect to a certain focus (e. g. 'logical properties'). We will explain the attributes and give examples for the objects. Since the most fundamental differentiation for theories of concept graphs is that in semantic approaches and those established by a separation in syntax and semantics, this attribute occurs in each of the concept lattices.

The concept lattice in Figure 5 is about models and the kind of concepts which are considered (e. g. concepts, semiconcepts or protoconcepts for PCFs) for each semantical and syntactical theory in question. First, we explain the attributes: All approaches have a so-called contextual semantics, which is based on FCA and describes concepts as the constituents of concept graphs in a formal and comprehensive way. The components of such a contextual model are specific

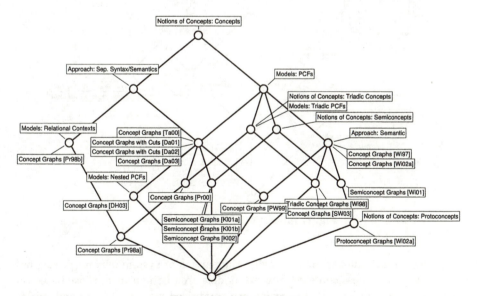

Fig. 5. Models and Units

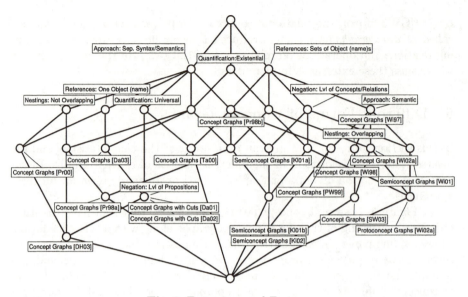

Fig. 6. Extensions and Features

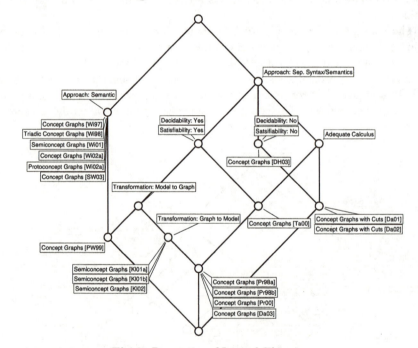

Fig. 7. Properties of Logical Theories

kinds of formal contexts (depending on the particular system of concept graphs) and concepts of these contexts. Several models are distinguished: those based on *power context families* (PCFs) which are comprised by both *triadic* and *nested*

PCFs, and *relational contexts*. For PCFs, there are three notions of concepts: The set of all *protoconcepts* of a context (see [Wi00a]) contains all *semiconcepts* which in turn include all *concepts*. Obviously, triadic PCFs lead to *triadic concepts*, and for nested PCFs and relational contexts, the usual dyadic formal concepts are considered. In order to include, for instance, a negation on the level of concepts and relations, power context families with semiconcepts (or protoconcepts) can be employed as model since they include a negation on concepts.

The concept lattice depicts which papers employ which kind of semantics. For instance, we find that that relational contexts as models were only utilized by Prediger in [Pr98a] and [Pr98b]. They were then substituted by PCFs which yield a richer description of relations.

Figure 6 is concerned with extensions to the basic theory of concept graphs. There are several extensions and features, including the number of references, negation, nestings and quantification. We will now briefly summarize the meaning of the different attributes.

The first two attributes we consider are concerned with negation, which can be introduced on different levels. If the traditional philosophical understanding is taken into account, then (interpreted) concept graphs as judgments are regarded as valid propositions, hence as meaningful combinations of concepts. In particular, this means that the definition should not allow the construction of a self-contradictory graph. In order to still be in the position to employ a (though restricted) negation, in [Wi01] a *negation on the level of concepts and relations* was introduced. For the theory of concept graphs with cuts ([Da01]), the focus is different: Here the aim is to reach a logical equivalence to FOPL, thus the introduction of a *negation on the level of propositions* is necessary. Moreover, as we have self-contradictory formulas in FOPL, it is allowed to construct self-contradictory graphs. The two attributes can be found on the upper right hand side and the lower left hand side of the diagram, respectively.

Two different types of *references* are considered, which are the objects (or, in logic approaches, the object names) assigned to the vertices of the relational graph. In some approaches there may be *sets of objects* assigned to each vertex, while other approaches only allow *single* objects. These first approaches therefore allow a more condensed representation of knowledge than the latter. These attributes are depicted in the upper half of the diagram.

As for quantifiers, we find that there are two kinds of quantification for concept graphs (the corresponding attributes are in the upper half, right in the middle of the diagram). With *existential quantification*, existentially quantified variables are introduced as references, whereas *universal quantification* allows the assertion of propositions about *all* objects satisfying certain conditions. There is only one paper ([Ta00]) where universal quantifiers are introduced directly. In all other cases the topic of the corresponding paper is a system of concept graphs with existential quantification and global negation.

Finally we consider *nestings* (or *subdivisions*). Nestings form an extension to the language of concept graphs, which allows the bundling of information and the assertion of propositions which refer to different contexts (thus the coding of 'modal information'). There are theories which only consider *disjoint* nestings and those in which parts of the subdivision may *overlap*.

The concept lattice shows that all semantic theories have sets of objects as references, and, conversely, that only in the papers of Dau and one paper of Prediger the references are restricted to single objects. Moreover, by combining Figure 5 and 6 we can observe that all theories including nestings/subdivisions (which would comprise [Wi98] and [SW03] as semantic approaches and [Pr98a], [Pr00] and [DH03] as logical ones) require PCFs with additional structures, such as triadic or nested PCFs.

Finally, Figure 7 addresses properties of logical theories, i.e. of approaches based on a separation in syntax and semantics. Again, we first outline the definitions of the attributes.

A logic system of concept graphs is called *satisfiable* if each syntactical concept graph is valid in at least one model, thus if the construction of self-contradictory concept graphs is not possible. Moreover, in accordance with [Ba77], we say that a theory based on the separation of syntax and semantics is *decidable*, if there is an algorithm which determines for each pair $\mathfrak{G}_1, \mathfrak{G}_2$ of concept graphs over the same alphabet whether $\mathfrak{G}_1$ entails $\mathfrak{G}_2$ (hence if $\mathfrak{G}_2$ is valid in every model for $\mathfrak{G}_1$) or not.

Since the previous two attributes do not make sense for semantical approaches, we chose to state explicitly whether a logic approach satisfies them or not.

As for the remaining attributes: A logic system is said to have an *adequate calculus*, if there exists a sound and complete set of inference rules. Furthermore, we say that it is possible to *transform a model into a concept graph*, if for a given model (in particular, for a given PCF), a concept graph can be constructed which codes the same information as the PCF. This so-called standard graph then entails all other (valid) concept graphs of the corresponding model. Finally, the attribute *Transformation: Graph to Model* denotes that for each syntactical concept graph of the logic system in question it is possible to construct a model (or simply a power context family) coding the same information. Via this so-called standard model, entailment can be characterized.

In the concept lattice we can now see that the attribute 'Transformation: Graph to Model' implies both decidability and satisfiability. Satisfiability follows since in each of the corresponding papers the 'standard model' is indeed a model for the concept graph. Furthermore, in every paper having this attribute the entailment relation is characterized via a finite standard model or a finite standard PCF, hence the attribute 'Decidability: Yes' is implied, too.

After having discussed the properties of several theories, we again turn to the examples for the different systems of concept graphs provided in Figure 4. The first two graphs were already explained in the previous section in order to exemplify semantic approaches and those based on a separation in syntax and semantics. The third graph is a (syntactical) semiconcept graph over the alphabet in Figure 3 and includes a negation on the concept and the relation level. It represents that the semantic approach in [Wi98] is an extension of the semantic approach described in [Wi97], that both [DH03] and [Pr00] are **not** semantic approaches, and that [DH03] is not a generalization of [Pr00] (which is interesting since both approaches deal with nestings, and [DH03] was published three years after [Pr00]). Hence, the references on the right side of the stroke are negative references with respect to both the corresponding concepts and relations. The last graph of Figure 4 is a concept graph with cuts (assembled of three contiguous subgraphs) representing the same information as the third graph. As already mentioned, these graphs include a negation on the level of propositions. Informally speaking, a part of the graph is negated if it is nested in a so-called cut, which is represented in the Figure by a bold oval. The subgraphs not containing cuts are read in the same way as concept graphs. However, each cut negates everything within it, so the two bottommost parts of the graph are read: [DH03] and [Pr00] are 'T', [DH03] is not an extension of [Pr00], and neither [DH03] nor [Pr00] are semantic approaches. Concept Graphs with nestings are omitted in this example, since a triadic (or nested) PCF would be required to provide a reasonable explanation.

In this section we have given a brief overview over the most prominent distinguishing attributes of the different theories of concept graphs. As we have shown, the set of all these theories is rather large, each theory incorporating different features (some of which suggested in Sowa's theory of Conceptual Graphs) and thereby broadening the program of Contextual Logic. For an overview of the theories of Prediger, Dau and Klinger with a more formal focus we refer to [KV03].

7   Outlook

The restructuring process of logic is still in its early stages. Some syntactical approaches still lack adequate calculi. Much effort has to be spent on further extensions of concept graphs in order to model further aspects of reasoning and communication. This will include various forms of background knowledge, like material implications as they are described in the book of Brandom ([Br01]). Furthermore, modal and contextual reasoning or different kinds of quantification have to be incorporated.

On the other hand, a restructured logic has to prove itself in practice. Thus, it is desirable to implement concept graphs, for example as computer programs, and to test out their usability in real life projects. Only projects can show whether our formalizations of logic are adequate for our purpose.

It will certainly be a long way to carry out all these steps. But as our experience with FCA shows, we can be quite optimistic that these goals can be reached in the long run.

References

[Ap89]   K.-O. Apel: Begründung. In H. Seifert, G. Radnitzky (Eds.): Handlexikon der Wissenschaftstheorie. Ehrenwirth, München 1989, 14–19.

[Ar02]   M. Arnold: Einführung in die Relationenlogik. Diplomarbeit. FB Mathematik, TU Darmstadt 2002.

[Ba77]   J. Barwise (Ed.): Handbook of Mathematical Logic. North–Holland Publishing Company, Amsterdam–New York–Oxford 1977.

[BH03]   P. Becker, J. Hereth Correia: The ToscanaJ Suite for Implementing Conceptual Information Systems. This volume.

[Br94]   R.B. Brandom: Making it explicit. Reasoning, Representing, and Discursive Commitment. Harvard University Press, Cambridge 1994.

[Br01]   R.B. Brandom: Begründen und Begreifen. Eine Einführung in den Inferentialismus. Suhrkamp 2001.

[Da01]   F. Dau: Concept Graphs and Predicate Logic. In: H. S. Delugach, G. Stumme (Eds.): Conceptual Structures: Broadening the Base, Springer Verlag, Berlin–New York 2001, 72–86.

[Da03a]   F. Dau: The Logic System of Concept Graphs with Negations (and its Relationship to Predicate Logic). Springer Verlag, Berlin–Heidelberg, 2003.

[Da03b]   F. Dau: Concept Graphs without Negations: Standardmodels and Standardgraphs. In: A. de Moor, W. Lex, B. Ganter (Eds.): Conceptual Structures for Knowledge Creation and Communication. Springer Verlag, Berlin–New York 2003, 243-256.

[DH03]   F. Dau, J. Hereth Correia: Nested Concept Graphs: Mathematical Foundations and Applications in Databases. In: B. Ganter, A. de Moor (Eds.): Using Conceptual Structures: Contributions to ICCS 2003. Shaker Verlag, Aachen 2003, 125-141.

[GW99a]   B. Ganter, R. Wille: Formal Concept Analysis: Mathematical Foundations. Springer, Berlin–New York 1999.

[GW99b]   B. Ganter, R. Wille: Contextual Attribute Logic. In: W. Tepfenhart, W. Cyre (Eds.): Conceptual Structures: Standards and Practices, Springer, Berlin Heidelberg New York 1999, 401–414.

[Ha81]   J. Habermas: Theorie kommunikativen Handelns. 2 Bände. Suhrkamp, Frankfurt 1981.

[He74]   H. von Hentig: Magier oder Magister? Über die Einheit der Wissenschaft im Verständigungsprozess. Suhrkamp Verlag, Frankfurt 1974.

[HLSW00]   C. Herrmann, P. Luksch, M. Skorsky, R. Wille: Algebras of Semiconcepts and Double Boolean Algebras. Contributions to General Algebra 13, 2000.

[Ka88]   I. Kant: Logic. Dover, New York 1988.

[Kl01a]   J. Klinger: Simple Semiconcept Graphs: a Boolean Logic Approach. In: H. S. Delugach, G. Stumme (Eds.): Conceptual Structures: Broadening the Base, Springer Verlag, Berlin–New York 2001, 115–128.

[Kl01b]   J. Klinger: Semiconcept Graphs: Syntax and Semantics, Diplomarbeit, FB Mathematik, TU Darmstadt 2001.

[Kl02]   J. Klinger: Semiconcept Graphs with Variables. In: U. Priss, D. Corbett, G. Angelova (Eds.): Conceptual Structures: Integration and Interfaces, Springer Verlag, Berlin–New York 2002, 382–396.

[KV03]   J. Klinger, B. Vormbrock: Contextual Boolean Logic: How did it develop? In: B. Ganter, A. de Moor (Eds.): Using Conceptual Structures: Contributions to ICCS 2003. Shaker Verlag, Aachen 2003, 143-156.

[MSW99]   G. Mineau, G. Stumme, R. Wille: Conceptual Structures Represented by Conceptual Graphs and Formal Concept Analysis. In; W. Tepfenhart. W, Cyre (Eds.): Conceptual Structures: Standards and Practices, Springer Verlag, Berlin–New York 1999, 423-441.

[Pe35]   Ch. S. Peirce: Collected Papers. Harvard Uni. Press, Cambridge 1931–35.

[Po01]   S. Pollandt: Relational Constructions on Semiconcept Graphs. In: G.
            Mineau (Ed.): Conceptual Structures: Extracting and Representing Seman-
            tics. Dept. of Computer Science. University Laval, Quebec, Canada, 2001,
            171–185.

[Po02]   S. Pollandt: Relation Graphs - A Structure for Representing Relations in
            Contextual Logic of Relations. In: U. Priss, D. Corbett, G. Angelova (Eds.):
            Conceptual Structures: Integration and Interfaces, Springer Verlag, Berlin–
            New York 2002, 382–396.

[PoW00]   S. Pollandt, R. Wille: On the Contextual Logic of Ordinal Data. In: B.
            Ganter, G. W. Mineau (Eds.): Conceptual Structures: Logical, Linguistic,
            and Computational Issues, Springer Verlag, Berlin–New York 2000, 249–
            262.

[Pr98a]   S. Prediger: Kontextuelle Urteilslogik mit Begriffsgraphen, Shaker Verlag,
            Aachen 1998.

[Pr98b]   S. Prediger: Simple Concept Graphs: A Logic Approach. In: M.-L. Mugnier,
            M. Chein (Eds): Conceptual Structures: Theory, Tools and Application,
            Springer Verlag, Berlin–Heidelberg 1998, 225–239.

[Pr00]   S. Prediger: Nested Concept Graphs and Triadic Power Context Families:
            A Situation-Based Contextual Approach. In: B. Ganter, G. W. Mineau
            (Eds.): Conceptual Structures: Logical, Linguistic, and Computational Is-
            sues, Springer Verlag, Berlin–New York 2000, 249–262.

[PW99]   S. Prediger, R. Wille: The lattice of concept graphs of a relationally scaled
            context. In: W. Tepfenhart, W. Cyre (Eds.): Conceptual Structures: Stan-
            dards and Practices, Springer, Berlin Heidelberg New York 1999, 401–414.

[Se01]   T. B. Seiler: Begreifen und Verstehen: Ein Buch über Begriffe und Bedeu-
            tungen. Verlag Allgemeine Wissenschaft, Mühltal 2001.

[SW03]   L. Schoolmann, R. Wille: Concept Graphs with Subdivision: a Semantic
            Approach. In: U. Priss, D. Corbett, G. Angelova (Eds.): Conceptual Struc-
            tures: Integration and Interfaces, Springer Verlag, Berlin–New York 2002,
            271-281.

[So84]   J. F. Sowa: Conceptual Structures: Information Processing in Mind and
            Machine. Adison-Wesley, Reading 1984.

[So92]   J. F. Sowa: Conceptual Graphs Summary. In: T. E. Nagle, J. A. Nagle, L.
            L. Gerholz, P. W. Eklund (Eds.): Conceptual Structures: Current Research
            and Practice. Ellis Horwood, 1992, 3–51.

[Ta00]   J. Tappe: Simple Concept Graphs with Universal Quantifiers. In: G.
            Stumme (Ed.): Working with Conceptual Structures. Contributions to
            ICCS 2000. Shaker, Achen 2000, 94–108.

[Wi82]   R. Wille: Restructuring Lattice Theory: An Approach Based on Hierarchies
            of Concepts. In: I. Rival (Ed.): Ordered Sets. Reiderl, Dordrecht–Boston,
            445-470.

[Wi96]   R. Wille: Restructuring Mathematical Logic: An Approach based on
            Peirce's Pragmatism. In: A. Ursini, P. Agliano (Eds.): Logic and Algebra.
            Marcel Dekker, New York 1996, 267–281.

[Wi97]   R. Wille: Conceptual Graphs and Formal Concept Analysis. In: D. Lukose,
            H. Delugach, M. Keeler, L. Searle, J. Sowa (eds.): Conceptual Structures:
            Fullfilling Peirce's Dream. Springer, Berlin–New York 1997, 290–303.

[Wi98]   R. Wille: Triadic Concept Graphs. In: M.-L. Mugnier, M. Chein (Eds):
            Conceptual Structures: Theory, Tools and Application, Springer Verlag,
            Berlin–New York 1998, 194–208.

[Wi00a] R. Wille: Boolean Concept Logic. In: B. Ganter, G.W. Mineau (Eds.): Conceptual Structures: Logical, Linguistic, and Computational Issues, Springer Verlag, Berlin–New York 2000, 317–331.

[Wi00b] R. Wille: Contextual Logic Summary. In: G. Stumme (Ed.): Working with Conceptual Structures. Contributions to ICCS 2000. Shaker, Aachen 2000, 256–276.

[Wi00c] R. Wille: Lecture Notes on Contextual Logic of Relations. FB4-Preprint, TU Darmstadt 2000.

[Wi01] R. Wille: Boolean Judgment Logic. In: H. S. Delugach, G. Stumme (Eds.): Conceptual Structures: Broadening the Base, Springer Verlag, Berlin–New York 2001, 115–128.

[Wi02a] R. Wille: Existential Concept Graphs of Power Context Families. In: U. Priss, D. Corbett, G. Angelova (Eds.): Conceptual Structures: Integration and Interfaces, Springer Verlag, Berlin–New York 2002, 382–396.

[Wi02b] R. Wille: The Contextual-Logic Structure of Distinctive Judgments. In: U. Priss, D. Corbett, G. Angelova (Eds.): Foundations and Applications of Conceptual Structures - Contributions to ICCS 2002, Bulgarian Academiy of Sciences, 92–101.

[Wi02c] R. Wille: Transdisziplinarität und Allgemeine Wissenschaft. FB4-Preprint No. 2200, TU Darmstadt 2002.

[Wi03] R. Wille: Conceptual Content as Information - Basics for Contetxual Judgment Logic. In: A. de Moor, W. Lex, B. Ganter (Eds.): Conceptual Structures for Knowledge Creation and Communication. Springer Verlag, Berlin–New York 2003, 1-15.

Contextual Attribute Logic
of Many-Valued Attributes

Bernhard Ganter

Institut für Algebra
Technische Universität Dresden

1   Contextual Attribute Logic

Sometimes even the most elementary data type of Formal Concept Analysis, that of a *formal context*, can be difficult to handle. This is typically the case when the context under consideration is not fully available, because e.g. it is too large to be completely recorded. Then even the question "Which attribute combinations are possible?" cannot simply be answered by giving all concept intents, because such a list may be huge and therefore of little insight. In such a situation, the weaker information that certain attribute combinations are possible and others are not, may be of interest. A language to systematically address such information was introduced in [8] under the name of "Contextual Attribute Logic". It activates (with an entirely different semantic in mind) basic notions of mathematical Propositional Logic for the investigations of Formal Concept Analysis. Instead of "propositions" and "formulae" we prefer to speak of "attributes" and "compound attributes", because this better fits our intended interpretation. But the formulas we use are essentially the same. For example, the compound attribute

$$(m_1 \vee m_2) \wedge (\neg m_3)$$

can be interpreted in the obvious way in any formal context having m_1, m_2, m_3 in its attribute set. The extent of this compound attribute is

$$((m_1 \vee m_2) \wedge (\neg m_3))' = (m_1' \cup m_2') \setminus m_3'.$$

It is evident from this example how the extent of an arbitrary compound attribute is defined. The extent of a set A of compound attributes is the intersection of their individual extents:

$$A' := \bigcap_{a \in A} a'.$$

If a compound attribute has all objects in its extent, then it is *all–extensional* (in the context under consideration). Two compound attributes are *extensionally equivalent* in (G, M, I) if they have the same extent, and *globally equivalent* if they are extensionally equivalent in every context. There is one context that suffices for testing global equivalence, as the following proposition shows:

B. Ganter et al. (Eds.): Formal Concept Analysis, LNAI 3626, pp. 101–113, 2005.

Proposition 1 *Two compound attributes of an attribute set M are globally equivalent iff they are extensionally equivalent in the* test context $(\mathfrak{P}(M), M, \ni)$.

Global equivalence of course corresponds to the logical equivalence of the propositional formulae.

Let us recall from [8] that two types of compound attributes are especially important: *implications*,

$$A \rightarrow S \quad A \subseteq M, S \subseteq M,$$

because they are easy to handle, and *sequents* (also called *clauses*, and also denoted by $A \multimap S$),

$$(A, S) \quad A \subseteq M, S \subseteq M,$$

because of their expressive power.

A *sequent* (A, S) over M is a compound attribute that is globally equivalent to

$$\bigvee (S \cup \{\neg m \mid m \in A\}).$$

Every compound attribute is globally equivalent to some conjunction of sequents.

An *implication* $A \rightarrow B$ is globally equivalent to

$$\bigvee \{\neg m \mid m \in A\} \vee \bigwedge B.$$

A perhaps more intuitive equivalence is that of

$$A \rightarrow B \quad \text{with} \quad \bigwedge A \rightarrow \bigwedge B$$

$$\text{and} \quad A \multimap B \quad \text{with} \quad \bigwedge A \rightarrow \bigvee B.$$

The implications that are all-extensional in (G, M, I) form the *implication logic* (or *implicational theory*) of (G, M, I). The all-extensional sequents form the *sequent logic* of (G, M, I).

The expressivity of implications is strictly weaker than that of sequents. The implication logic characterizes the system of all *intents* and thereby the concept lattice up to isomorphism. The sequent logic characterizes the system of all *object intents*. Assuming finiteness, we may say that the implication logic determines the context up to object reduction, while the sequent logic determines it up to object clarification.

Although sequents are more expressive than implications, it seems that in everyday logic implications are preferred over sequents. One of the reasons is that of *inference*. We say that a compound attribute m can be *inferred* from a set A of compound attributes if

$$A' \subseteq m'$$

holds in the test context. It is well known that it is easy ("linear time") to decide if a given implication can be inferred from other given implications, and that it is difficult ("$\mathcal{NP}$-complete") to decide inference for sequents. This is of importance for *attribute exploration*, an interactive procedure to determine a *base* for the implication logic. We come back to this later.

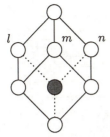

The implication $\{l, n\} \rightarrow \{m\}$ forces the absence of the shaded concept.

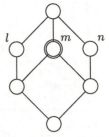

The sequent $\{m\} \multimap \{l, n\}$ expresses that the marked concept is not an object concept.

Fig. 1. The different effects of implications and sequents

2   Many-Valued Attributes

How Formal Concept Analysistreats many-valued attributes is well known and well documented. Instead of repeating the definitions (for which we refer to [7]), we simply discuss an example. Figure 2 shows some data about turtles on the Galapagos islands. The elephant turtles on these islands have developed different shapes of their shells (saddle, dome, intermediate). This may be connected to the island size and the type of cacti (opuntia) that grow on it. The data set allows a first discussion of such a potential interplay, and the reader is referred to the biological literature for details. Here the data only serve as an example for (plain) conceptual scaling: The data set can be interpreted as a many-valued context. To give it a conceptual structure, we use some scaling that interprets the many-valued attributes as one-valued ones. Using the scales that are shown

Galapagos island	Island size	Opuntia		Turtle type		
		bushy	treelike	dome	intermediate	saddle
Albemarle	4278 km²	+	-	+	-	-
Indefatigable	1025 km²	+	-	+	-	-
Narborough	650 km²	+	-	+	-	-
James	574 km²	+	-	-	+	-
Chatham	about 500 km²	+	-	-	+	-
Charles	about 200 km²	+	-	-	+	+
Hood	<100 km²	+	-	-	-	+
Bindloe	<100 km²	-	+	-	-	-
Abingdon	<100 km²	+	-	-	-	+
Barringdon	<100 km²	+	-	-	+	+
Tower	<100 km²	-	+	-	-	-
Wenman	<100 km²	-	+	-	-	-
Culpepper	<100 km²	-	+	-	-	-
Jervis	<100 km²	+	-	-	+	+

Fig. 2. Turtles on the Galapagos islands

Island size	small	not large	not small	large
< 100 km²	×	×		
100–1000 km²		×	×	
> 1000 km²			×	×

Opuntia bushy	treelike	bushy	treelike
+	−	×	
−	+		×

Turtle type dome	intermediate	saddle	dome	intermediate or saddle	saddle	intermediate
−	−	−				
+	−	−	×			
−	+	−		×	×	
−	−	+		×		×
−	+	+		×	×	×

Fig. 3. Scales for the turtles context

in Figure 3, we obtain the derived context in Figure 4 and the concept lattice in Figure 5.

The structure of the derived context in Figure 4 comes from two sources: One is the original data (Figure 2) and the other is the scaling. Even without knowing the original data, we know from the scaling that the derived context will be of a very special form: Each row is composed from rows of the scale contexts in Figure 3. More formally: The derived context necessarily is a subcontext of the semiproduct of the scales. In our example, the scales have 3, 2, and 5 objects (scale values), respectively. Their semiproduct therefore has $3 \cdot 2 \cdot 5 = 30$ objects.

This gives an idea of what the Contextual Attribute Logic of a many valued context *with respect to a fixed scaling* might be: It is the attribute logic of the derived attributes, but with a modified notion of global equivalence and of inference. Two compound attributes will be globally equivalent iff they are extensionally equivalent in *every derived context for this scaling*. As in the general case, this can be tested in a single context, but this is no longer the test context $(\mathfrak{P}(M), M, \ni)$ that was introduced above. Instead, we get a smaller *relative test context*:

Definition 1 The relative test context for a fixed scaling scheme is the semiproduct of the scales.   ◊

In our example, the relative test context has 30 objects, while the test context for the 10-element attribute set has $2^{10} = 1024$ objects.

	island size: large	island size: not small	island size: not large	island size: small	opuntia: bushy	opuntia: treelike	turtles: dome	turtles: intermediate or saddle	turtles: saddle	turtles: intermediate
Albemarle	×	×			×		×			
Indefatigable	×	×			×		×			
Narborough		×	×		×		×			
James		×	×		×			×		×
Chatham		×	×		×			×		×
Charles		×	×		×			×	×	×
Hood			×	×	×			×	×	
Bindloe			×	×		×				
Abingdon			×	×	×			×	×	
Barringdon			×	×	×			×	×	×
Tower			×	×		×				
Wenman			×	×	×					
Culpepper			×	×	×					
Jervis			×	×	×			×	×	×

Fig. 4. The derived context

This approach does not only work for derived contexts. More generally, we can introduce Contextual Attribute Logic *with respect to a relative test context*. This relative test context can be an arbitrary subcontext

$$(\mathcal{R}, M, \ni), \qquad \mathcal{R} \subseteq \mathfrak{P}(M),$$

of the test context $(\mathfrak{P}(M), M, \ni)$. Each such family $\mathcal{R} \subseteq \mathfrak{P}(M)$ is the extent of a set of sequents, because for each $X \subseteq M$ we find that the extent of the sequent $X \multimap (M \setminus X)$ is $\mathfrak{P}(M) \setminus \{X\}$ and thus

$$\{X \multimap (M \setminus X) \mid X \subseteq M, X \notin \mathcal{R}\}' = \mathcal{R}$$

in the test context. Rather than of "Contextual Attribute Logic with respect to a relative test context" we can therefore speak of "Contextual Attribute Logic with respect to a set of *background sequents*".

The case of plain scaling is subsumed, because we can describe each scale context (up to object clarification) by its sequent logic; taking these sequents as background sequents will result in a relative test context which is isomorphic to the semiproduct of the scales. For example, the sequent logic of the "Island size"

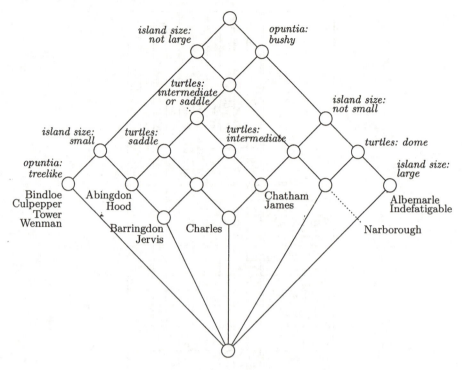

Fig. 5. The concept lattice of the scaled context

scale (in Figure 3) contains the following:

$$\{ \text{ small, not small } \} \multimap \emptyset$$
$$\{ \text{ large, not large } \} \multimap \emptyset$$
$$\{ \text{ small } \} \multimap \{ \text{ not large } \}$$
$$\{ \text{ large } \} \multimap \{ \text{ not small } \}$$
$$\emptyset \multimap \{ \text{ small, not small } \}$$
$$\emptyset \multimap \{ \text{ large, not large } \}.$$

The first four lines contain implications. These form, in fact, the *stem base* of the "Island size" scale. One of the nice features of the stem base is its uniqueness. One can easily give examples that this is not preserved in our more general setting here. If we loosen, for example, the background information slightly and only assume that "small" and "not small" as well as "large" and "not large" are negations of each other, then the background sequents would be

$$\{ \text{ small, not small } \} \multimap \emptyset$$
$$\{ \text{ large, not large } \} \multimap \emptyset$$
$$\emptyset \multimap \{ \text{ small, not small } \}$$
$$\emptyset \multimap \{ \text{ large, not large } \}.$$

The relative test context for this sequent set is

	small	not small	large	not large
{small, large}	×		×	
{small, not large}	×			×
{not small, large}		×	×	
{not small, not large}		×		×

In this relative test context, the two implications

$$\{\text{ small }\} \rightarrow \{\text{ not large }\}$$
$$\{\text{ large }\} \rightarrow \{\text{ not small }\}$$

are extensionally equivalent. Thus each of the two can be inferred from the other, and therefore each of the two generates the implication logic of the "Island size" scale with respect to these background sequents. We see from the example that a uniqueness result as for the stem base cannot be expected.

Although there is no natural choice for a "base", a relative generating set of the implication logic can be very useful. Again, this can be seen from our example. The stem base of the formal context in Figure 4 consists of 14 implications. These 14 implications are necessary to exclude all object intents of the test context except for those which represent intents of the context in Figure 4. A *relative* generating system consists of implications that exclude all unwanted object intents of the relative test context. It is not difficult to see that this can be done with six implications, see Figure 6. The advantage of such a split into background sequents and relative implications is not that the system gets simplified. But it separates the interesting information from "what we knew anyway" (the scaling-related information).

Note however that an important information of the original data does not occur at all in Figure 6: Treelike opuntiae occur on all islands with no turtles (and only on those). This fact was hidden by our unskillful choice of the third scale. In fact, the implication

$$\neg \text{ dome } \wedge \neg \text{ intermediate\_or\_saddle } \rightarrow \text{ treelike}$$

does not follow from the information in Figure 6. On the first glance this is surprising, because this implication, interpreted as a compound attribute, is all-extensional in the turtles context (Figure 4). But the concept lattice in Figure 5 has an intent that does not respect this implication, namely

$$\{\text{not small, not large, bushy}\}.$$

The reader should therefore be warned that implications between compound attribute do not automatically carry over to the closure system of intents, unless these attributes are included in the set M of (atomic, non-compound) attributes.

Island size:

$$\text{small, not small } \multimap \emptyset$$
$$\emptyset \multimap \text{small, not small}$$
$$\text{large, not large } \multimap \emptyset$$
$$\emptyset \multimap \text{large, not large}$$
$$\text{small } \multimap \text{not large}$$

Opuntia:

$$\text{bushy, treelike } \multimap \emptyset$$
$$\emptyset \multimap \text{bushy, treelike}$$

Turtles:

$$\text{dome, intermediate\_or\_saddle } \multimap \emptyset$$
$$\text{intermediate\_or\_saddle } \multimap \text{intermediate, saddle}$$
$$\text{saddle } \multimap \text{intermediate\_or\_saddle}$$
$$\text{intermediate } \multimap \text{intermediate\_or\_saddle}$$

$$\text{island size: large } \rightarrow \text{turtles: dome}$$
$$\text{island size: not small } \rightarrow \text{opuntia: bushy}$$
$$\text{turtles: dome } \rightarrow \text{island size: not small}$$

$$\left.\begin{array}{l}\text{island size: not large,}\\ \text{turtles: intermediate\_or\_saddle}\end{array}\right\} \rightarrow \text{opuntia: bushy}$$

$$\left.\begin{array}{l}\text{island size: small,}\\ \text{opuntia: bushy}\end{array}\right\} \rightarrow \text{turtles: saddle}$$

$$\left.\begin{array}{l}\text{island size: not small, not large,}\\ \text{turtles: intermediate\_or\_saddle}\end{array}\right\} \rightarrow \text{turtles: intermediate.}$$

Fig. 6. The lower box shows a generating set for the implication logic of the derived context in Figure 4, relative to the scaling-induced logic. Generating sequents for the scaling-induced are listed in the upper box. Set brackets are omitted

Such considerations are, of course, only the beginning of *Contextual Attribute Logic with background knowledge*. One of the encouraging results of this theory is that implication inference remains easy:

Theorem 1 (Ganter, Krauße [6]) *Implication inference with a fixed size set of background sequents is of linear time complexity.*

Theorem 1 is of interest for *attribute exploration with background knowledge* (introduced in [5]), which will be discussed below. Before we do so, we consider a much more elementary problem.

3 Excursus: Exploring a Concept with Preconcepts and Convex Sets

A *preconcept* of (G, M, I) is a pair (A, B) satisfying

$$A \subseteq G, B \subseteq M, A \subseteq B', \text{ and } B \subseteq A'.$$

(A, B) is a *protoconcept* if it is a preconcept satisfying

$$A' = B'', A'' = B'.$$

Each preconcept (A, B) specifies an interval

$$[(A'', A'), (B', B'')]$$

of the concept lattice, and it is a protoconcept iff this interval consists of only one element.

Consider the very elementary task of specifying some formal concept (A, B) of a known context (G, M, I). The simplest solution would be to give one of the two sets A or B, since each of them specifies the concept uniquely. But if (G, M, I) is large (perhaps infinite), it may be difficult to note down A or B, and we may be forced to use smaller bits of information.

We could, for example, name *some* objects from A and some attributes from B, say $A_1 \subseteq A$ and $B_1 \subseteq B$. What we have given then is a preconcept (A_1, B_1) and thereby the information, that (A, B) lies in the interval $[(A_1'', A_1'), (B_1', B_1'')]$ of the concept lattice. Such an interval is itself a concept lattice. We therefore arrive at the same task, but for a smaller lattice. Iterating this procedure, we obtain a sequence

$$(A_1, B_1), (A_2, B_2), \ldots, \qquad A_1 \subseteq A_2 \subseteq \ldots, \qquad B_1 \subseteq B_2 \subseteq \ldots$$

of nested preconcepts that (hopefully) eventually converges to a protoconcept specifying (A, B).

In a computer-aided version of such an exploration process, the computer should make valid suggestions for extending the sets A_t and B_t of the t-th iteration step. It is quite clear which objects and attributes can be added in each step. Given (A_t, B_t), an object g can be added to A_t only if $g \in B_t'$, because each other choice would lead to a contradiction. But if $g \in A_t''$, then the preconcept $(A_t \cup \{g\}, B_t)$ defines the same interval as (A_t, B_t), and no new information about (A, B) is obtained. Therefore the objects that can be meaningfully added are those from $B_t' \setminus A_t''$, and the attributes are those from $A_t' \setminus B_t''$.

The situation gets slightly more involved if we allow for statements that certain objects or attributes do *not* belong to A or B, respectively. Then our information sequence would consist instead of preconcepts of expressions of the form

$$(A_t \mid C_t, B_t \mid D_t),$$

expressing that

$$A_t \subseteq A, \quad C_t \cap A = \emptyset, \quad B_t \subseteq B, \quad D_t \cap B = \emptyset.$$

It is not difficult to see that the set $C(A_t \mid C_t, \, B_t \mid D_t) :=$

$$\{(X,Y) \in \underline{\mathfrak{B}}(G,M,I) \mid A_t \subseteq X, \quad C_t \cap X = \emptyset, \quad B_t \subseteq Y, \quad D_t \cap Y = \emptyset\}$$

of those concepts that match this condition is a *convex* subset of the concept lattice, and the exploration procedure therefore results in a nested sequence of convex sets.

Again the question arises which objects and attributes can be added, giving more information. Clearly, adding an object g to $(A_t \mid C_t, \, B_t \mid D_t)$ makes the corresponding convex set smaller iff some, but not all concepts in $C(A_t \mid C_t, \, B_t \mid D_t)$ contain g, and dually for attributes. We define the *saturation* of $C(A_t \mid C_t, \, B_t \mid D_t)$ by

$$\bar{A}_t := \bigcap\{X \mid (X,Y) \in C(A_t \mid C_t, \, B_t \mid D_t)\},$$

$$\bar{B}_t := \bigcap\{Y \mid (X,Y) \in C(A_t \mid C_t, \, B_t \mid D_t)\},$$

$$\bar{C}_t := \bigcap\{G \setminus X \mid (X,Y) \in C(A_t \mid C_t, \, B_t \mid D_t)\},$$

$$\bar{D}_t := \bigcap\{M \setminus Y \mid (X,Y) \in C(A_t \mid C_t, \, B_t \mid D_t)\}.$$

Obviously

$$C(A_t \mid C_t, \, B_t \mid D_t) = C(\bar{A}_t \mid \bar{C}_t, \, \bar{B}_t \mid \bar{D}_t),$$

and the saturation procedure is a closure. The objects which give additional information thus are precisely those which do not belong to $\bar{A}_t \cup \bar{C}_t$, and the meaningful attributes are those not in $\bar{B}_t \cup \bar{D}_t$.

This shows that an exploration, even computer-aided, is possible for the more general setting allowing for negated objects and attributes, too.

4   The Contexts of Sequents and Implications

We come back to the theory of sequents and implications. A more formal way of understanding their rôle is the following. From the given set M we construct two formal contexts, where in both cases the set of objects is the power set $\mathfrak{P}(M)$ of the set M, and the set of attributes is the square $\mathfrak{P}(M)^2$ of this power set. The elements of $\mathfrak{P}(M)^2$ are pairs of subsets of M, and such a pair (A,B) may be interpreted as an implication $A \to B$ or a sequent $A \multimap B$. These two interpretations suggest different incidence relations:

– $(\mathfrak{P}(M), \mathfrak{P}(M)^2, \models_{\mathrm{imp}})$, where

$$X \models_{\mathrm{imp}} A \to S : \Longleftrightarrow A \not\subseteq X \text{ or } B \subseteq X,$$

– $(\mathfrak{P}(M), \mathfrak{P}(M)^2, \models_{\mathrm{seq}})$, where

$$X \models_{\mathrm{seq}} A \to S : \Longleftrightarrow A \not\subseteq X \text{ or } B \cap X \neq \emptyset.$$

The formal concepts of these two contexts can easily be described:

– The extents of $(\mathfrak{P}(M), \mathfrak{P}(M)^2, \models_{\mathrm{imp}})$ are precisely the closure systems on M. The intents are the respective implicational theories.

- The extents of $(\mathfrak{P}(M), \mathfrak{P}(M)^2, \models_{seq})$ are all subsets of $\mathfrak{P}(M)$. The intents are the respective sequent logics.

Every extent of the first context is also an extent of the second. The concepts of the first therefore can canonically be mapped to those of the second (mapping a concept of the first to the concept of the second with the same extent). This is in fact a $\bigwedge$-embedding (since any intersection of closure systems yields a closure system).

5   Attribute Exploration with Background Knowledge

Attribute exploration is a well-known interactive procedure for stepwise exploring a closure system of intents or, in other words, an implication logic on a given set M. The basic structure of the procedure is that in each step either an implication or an (object-)intent is added. Using the stem base theorem of Duquenne and Gigues [3] this can nicely be supported by a computer program that at each step suggests an undecided implication.

Attribute exploration can be generalized in several ways. One approach by Burmeister and Holzer [1] is presented in this volume. In earlier papers ([4], [5], [6]), we have discussed a twofold generalization:

- We allow many–valued contexts with a fixed scaling, or more generally an arbitrary set of background sequents, and
- we allow that object intents ("counter examples") can be partially given in form of two sets $A, S \subseteq M$ expressing that " there is an intent containing each attribute from A, but no attribute from S ".

Here we omit details of the algorithms. Instead, we give an abstract view on the procedure.

An attribute exploration can be seen as an instance of the exploration described above in Section 3. The aim of attribute exploration is to determine the implicational theory of some formal context and thereby a formal concept of the formal context

$$(\mathfrak{P}(M), \mathfrak{P}(M)^2, \models_{imp})$$

described in the previous section. Adding an implication means adding an attribute of $(\mathfrak{P}(M), \mathfrak{P}(M)^2, \models_{imp})$, and each counterexample is an object of this context. The intermediate results of an attribute exploration therefore give a sequence of preconcepts of this context, as described in section 3.

We might as well consider *sequent exploration*, which would amount to exploring a concept in

$$(\mathfrak{P}(M), \mathfrak{P}(M)^2, \models_{seq}).$$

The only difference is that instead of implications we may specify, more generally, sequents as attributes. This type of exploration is less popular, because of its algorithmic complexity and its less intuitive results. But we should keep in mind that attribute exploration is simply a special case of sequent exploration. In simple words: *attribute exploration is sequent exploration with the restriction that only implications are used as input (for attributes).*

When working with background knowledge, or, in other words, with a relative test context $(\mathcal{R}, M, \ni)$, then attribute exploration is performed in the subcontext

$$(\mathcal{R}, \mathfrak{P}(M)^2, \models_{imp}).$$

Each intent of this context is an intent of $(\mathfrak{P}(M), \mathfrak{P}(M)^2, \models_{imp})$ and therefore an implication logic. The closure systems described by these logics are precisely the ones in which the background sequents are all-extensional.

A different view of attribute exploration with background knowledge is obtained when working with the context

$$(\mathfrak{P}(M), \mathfrak{P}(M)^2, \models_{seq}).$$

It may be understood as follows: *attribute exploration with background knowledge is sequent exploration with the restriction that, except for the first step, only implications are used as input (for attributes).* However, we should keep in mind that the extent of the explored concept then no longer needs to be a closure system.

The considerations in Section 3 allow the use of negated objects and attributes. We find that not very intuitive in the case of $(\mathcal{R}, \mathfrak{P}(M)^2, \models_{imp})$, but quite clear for the context of sequents, $(\mathfrak{P}(M), \mathfrak{P}(M)^2, \models_{seq})$. The statement that a certain set $A \subseteq M$ is not an object intent can be encoded by the sequent

$$A \multimap M \setminus A.$$

Negated objects will therefore not be considered. Partially given counterexamples can be encoded by excluded attributes: A statement about a partially given counterexample " there is an intent containing each attribute from A, but no attribute from S " is expressed by the negated sequent

$$A \multimap S.$$

It now becomes apparent what the status of attribute exploration with partially given counter examples is: *sequent exploration with negated attributes, with the restriction that only implications are used as positive attributes.* An algorithmic realization and implementation of this method has yet to be developed.

References

1. P. Burmeister and R. Holzer. *Treating incomplete knowledge in Formal Concept Analysis.* This volume.
2. I. Eibl-Eibesfeldt. *Galapagos.* Piper Verlag, München 1962.
3. J.-L. Guigues and V. Duquenne. *Familles minimales d'implications informatives resultant d'un tableau de données binaires.* Math. Sci. Humaines, 95 (1986), 5–18.
4. B. Ganter. *Begriffe und Implikationen.*

5. B. Ganter. *Attribute exploration with background knowledge.* Theoretical Computer Science 217 (2), 1999.

6. B. Ganter, R. Krauße. *Pseudo-Models and propositional Horn inference.* Discrete Applied Mathematics 6133 (2004).

7. B. Ganter, R. Wille. *Formal Concept Analysis – Mathematical Foundations.* Springer Verlag 1999.

8. B. Ganter, R. Wille. *Contextual Attribute Logic.* in: W. Tepfenhart, W. Cyre (eds.), *Conceptual Structures: Standards and Practices.* Springer LNAI, 377–388, 1999.

9. Joachim Jaenicke (Ed.). *Materialienhandbuch Kursunterricht Biologie, Band 6: Evolution.* Aulis Verlag Köln 1997, ISBN 3-7614-1966-X. (Refers to [2] as scientific source.)

Treating Incomplete Knowledge
in Formal Concept Analysis

Peter Burmeister and Richard Holzer

Technische Universität Darmstadt

Introduction

Whenever human knowledge is considered, one has to take into account that such knowledge may be incomplete. Since Formal Concept Analysis (FCA for short) deals with the representation and investigation of such knowledge which can be connected to data tables – in FCA these are represented as formal one- or many-valued contexts –, it seems to be useful to have ways of representing (and dealing with) situations, where one knows about the incompleteness of parts of the represented knowledge. In this note we try to give a survey of what has been done so far in connection with the treatment of incomplete knowledge in FCA. In particular, we shall compare different algorithms developed so far in connection with the knowledge acquisition tool of "attribute exploration", which also treat incomplete knowledge – cf. in particular [B91], [G99] and [H01]. Moreover, at the end, different treatments of incomplete knowledge in many-valued contexts and databases are discussed.

1 Some Basic Concepts

We refer to [GW99] for most of the basic definitions needed here, however we repeat some of those concepts which we want to comment here in connection with our topic. E.g. *many-valued contexts* are quadruples (G, M, W, I), where G is a set the elements of which are called *objects*, M is a set the elements of which are called *attributes*, W is a set (or sometimes a family $(W_m)_{m \in M}$) the elements of which are called *values*, and I is a ternary relation $I \subseteq G \times M \times W$ with $(g, m, w_i) \in I$ for $i = 1, 2$ implying $w_1 = w_2$.[1] In connection with such many-valued contexts incomplete knowledge is usually easily recognizable from the so-called "missing values", i.e. from the pairs $(g, m) \in G \times M$ for which there is **no** $w \in W$ with $(g, m, w) \in I$. In this note we shall indicate among other things – along the lines of how it is done with relational databases[2] – how one can deal with such missing data. The basic structures in FCA, however, are the so-called one-valued contexts (G, M, I) – representing data tables, in which one has only two kinds of entries – with G and M sets of objects and attributes,

---

[1] Because of this requirement one often considers I as the graph of a (partial) mapping
$I : G \times M \to W$

[2] Cf. e.g. Maier [Ma83]

B. Ganter et al. (Eds.): Formal Concept Analysis, LNAI 3626, pp. 114–126, 2005.
© Springer-Verlag Berlin Heidelberg 2005

respectively, as in the many-valued case, and $I \subseteq G \times M$ just a binary relation $-(g, m) \in I$, or briefly gIm, meaning that "the object g has the attribute m". Here only the relation I is "meaningful"[3], and $(g, m) \notin I$

(i) could mean that the object g does not have the attribute m,
(ii) or it could mean that it is **unknown**, whether or not the object g has the attribute m, but this has not been made explicit,
(iii) or it could even have any other meaning not covered by "the object g has the attribute m".

I.e. in such contexts it is impossible to detect incompleteness of the represented knowledge without additional information. However, often one knows from the history of how a data table originated that some of the cases $(g, m) \notin I$ really mean that one has case (ii) above. Then the data table becomes more informative, if one uses an additional symbol like a question mark "?" for such a situation. The corresponding formal context then actually becomes "kind of a three-valued context"[4] $\mathbb{K}_i = (G, M, \{\times, ?, -\}, J)$. We call it an *incomplete context*. And in connection with incomplete contexts we shall always assume that, for any $g \in G$ and $m \in M$,

- $(g, m, \times) \in J$ shall mean that *it is known that the object g does have the attribute m*, and we shall indicate this sometimes by writing $gJ^\times m$,
- $(g, m, ?) \in J$ shall mean that *it is not known whether or not the object g has the attribute m*, and we shall indicate this sometimes by writing $gJ^? m$,
- $(g, m, -) \in J$ shall mean that *it is known that the object g does **not** have the attribute m*, and we shall indicate this sometimes by writing $gJ^- m$,
- and for each object g and each attribute m one has exactly one of the above cases.

Although this means that there will remain a difference between an ordinary formal context $\mathbb{K} = (G, M, I)$ and an incomplete context $\mathbb{K}_i = (G, M, \{\times, ?, -\}, J)$ even in the case that there is no instance $(g, m, ?)$ in J – i.e. when the binary "subrelation" $J^?$ is empty (in symbols: $J^? = \emptyset$) –, we shall not distinguish such an incomplete context with $J^? = \emptyset$ from the ordinary one $\mathbb{K}_i^\times := (G, M, J^\times)$ – sometimes it will also be useful to consider the further contexts $\mathbb{K}_i^- := (G, M, J^-)$ and $\mathbb{K}_i^? := (G, M, J^?)$ derived from $\mathbb{K}_i$ in order to have the corresponding derivation operators[5] around. Given an incomplete context $\mathbb{K}_i = (G, M, \{\times, ?, -\}, J)$, every ordinary context (G, M, I) with $J^\times \subseteq I$ – i.e. $(g, m, \times) \in J$ implying gIm

---

[3] In a table representing (G, M, I) the rows correspond to the objects, the columns to the attributes, gIm is represented by a $\times$ at the crossing of the corresponding line and column, while the other entries are left blank – or sometimes full stops "." are used instead of blanks for a better orientation in the table

[4] We call it "kind of a three-valued context", since it looks like a three-valued context, but while one usually "scales" (See [GW99].) many-valued contexts in order to transform them into ordinary ones, such scalings in most cases do not seem appropriate to us for incomplete contexts; see below for more details

[5] See below

$-$, and $J^- \subseteq (G \times M) \setminus I$ – i.e. $(g, m, -) \in J$ implying that (g, m) does not belong to I – is called a *completion* of $\mathbb{K}_i$.

In comparison to the derivation operators for ordinary contexts we have two basic kinds of derivation operators for incomplete contexts. Namely, for sets $A \subseteq G$ of objects and $B \subseteq M$ of attributes we define

$A^\square := \{m \in M \mid (g, m, \times) \in J \text{ for all } g \in A\}$ to be the set of all attributes in M applying to all objects in A – we call it the *certain intent* generated by A;

$A^\diamond := \{m \in M \mid ((g, m, \times) \in J \text{ or } (g, m, ?) \in J) \text{ for all } g \in A\}$ to be the set of all attributes in M possibly applying to all objects in A – we call it the *potential intent* generated by A;

$B^\square := \{g \in G \mid (g, m, \times) \in J \text{ for all } m \in B\}$ to be the set of all objects in G having all attributes from B – we call it the *certain extent* generated by B;

$B^\diamond := \{g \in G \mid ((g, m, \times) \in J \text{ or } (g, m, ?) \in J) \text{ for all } m \in B\}$ to be the set of all objects in G possibly having all attributes from B – we call it the *potential extent* generated by B.

Observe that for an ordinary context (G, M, I) one has e.g. for $B \subseteq M$ that

$$B^\square = B^\diamond =: B^I =: B'$$

is just the usual derivation operator. Morover, if $B = \{m\}$ consists of just a single attribute, then we write $m^\square$, $m^\diamond$ and m^I instead of $\{m\}^\square$, $\{m\}^\diamond$ and $\{m\}^I$ – and the same for objects.

2   Attribute Logic

We want to follow in this note as much as possible the line of [GW00] for a set theoretical representation of attribute logic. Since the attribute logic we use for incomplete contexts is based on the one for their completions, i.e. for ordinary contexts, using KRIPKE semantics (see below), we start with ordinary contexts: First we introduce as in [GW00] *compound attributes* as follows: Let (G, M, I) be any ordinary context, and let $A, B \subseteq M$, $m \in M$, then we define

- the *negation* $\neg m$ of m is a compound attribute with

$$(\neg m)^I := G \setminus m^I \tag{1}$$

- the *conjunction* $\bigwedge A := \bigwedge_{a \in A} a$ with

$$\left(\bigwedge A\right)^I := \bigcap_{a \in A} a^I \tag{2}$$

- the *disjunction* $\bigvee A := \bigvee_{a \in A} a$ with

$$\left(\bigvee A\right)^I := \bigcup_{a \in A} a^I \tag{3}$$

Iteration of these processes lead to more complicated compound attributes, among which *implications* $c \to d := \neg c \vee d$ are in this note of most importance for arbitrary compound attributes c and d. In particular, one considers, for arbitrary sets $A, B \subseteq M$ of attributes,

- *attribute implications* $A \to B := \bigwedge A \to \bigwedge B$, where its extent is computed as

$$(A \to B)^I := (\bigwedge A \to \bigwedge B)^I = \bigcup_{a \in A} (G \setminus a^I) \cup \bigcap b^I, \qquad (4)$$

- and *sequents*[6] $< A, B > := \bigwedge A \to \bigvee B$ (which is semantically equivalent to $\bigvee_{a \in A} \neg a \vee \bigvee_{b \in B} b$), where

$$< A, B >^I = \bigcup_{a \in A} (G \setminus a^I) \cup \bigcup_{b \in B} b^I \qquad (5)$$

An attribute or compound attribute c is called *universal*[7] w.r.t. an ordinary context $\mathbb{K} = (G, M, I)$, iff $c^I = G$. An attribute or compound attribute c is called *universal* w.r.t. an incomplete context $\mathbb{K}_i = (G, M, \{\times, ?, -\}, J)$, iff it is universal w.r.t. every completion of $\mathbb{K}_i$. And we want to call such a compound attribute c *potentially universal* w.r.t. $\mathbb{K}_i$, iff c is universal w.r.t. **at least one completion** of $\mathbb{K}_i$. We call this the KRIPKE-*semantics* of compound attributes for incomplete contexts.

If one wants to treat compound attributes of an incomplete context with a so-called three-valued logic, then the KLEENE-logic comes closest to it, but KLEENE-logic is not sufficient to characterize all cases, as we shall see below:

For any set G let us consider the KLEENE-algebra $\mathfrak{A}_G = (A_G; \wedge, \vee, \to, \neg, \bot, \top)$ as e.g. defined in [P97] or [H01], where[8]:

- $\mathfrak{A}_G := \{ (P, N) \mid P, N \subseteq G, \text{ and } P \cap N = \emptyset \}$;
- $\top := (G, \emptyset)$ and $\bot := (\emptyset, G)$;
- $\neg(P_1, N_1) := (N_1, P_1)$;
- $(P_1, N_1) \wedge (P_2, N_2) := (P_1 \cap P_2, N_1 \cup N_2)$;
- $(P_1, N_1) \vee (P_2, N_2) := (P_1 \cup P_2, N_1 \cap N_2)$;
- $(P_1, N_1) \to (P_2, N_2) := (N_1 \cup P_2, P_1 \cap N_2)$.

When $\mathbb{K}_i = (G, M, \{\times, ?, -\}, J)$ is any incomplete context, then we consider its set theoretical KLEENE-*algebra representation* by defining for $m \in M$ its corresponding element m_{KL} of the KLEENE-algebra $\mathfrak{A}_G$ as follows:

$$m_{\text{KL}} := (m^{J^\times}, m^{J^-}), \qquad (6)$$

---

[6] These correspond in Boolean logic to so-called *clauses*. Contrary to [GW00] we use here pointed brackets instead of parentheses, since those will already be used for the representation of (compound) attributes within KLEENE-algebras (see below)

[7] In [GW00] this property has been called "all-extensional"

[8] Since implications will play an important role in this note, we have included it among the basic operations, as has been done in [H01], while it is not contained in the signature in [P97] since it can also be defined as $c \to d := \neg c \vee d$ like in the Boolean case

and then taking the subalgebra generated by these elements[9]. This then yields the representations of the compound attributes w.r.t. $\mathbb{K}_i$. This way one can also extend the derivation operators $\square$ and $\diamond$ w.r.t. $\mathbb{K}_i$ to compound attributes c either directly by

$$\text{If } c_{\text{KL}} = (P,\, N), \text{ then } c^{\square} := P \text{ and } c^{\diamond} := G \setminus N\,;$$

or recursively

by defining for an arbitrary compound attribute c and an arbitrary set of compound attributes A (w.r.t. $\mathbb{K}_i$), for which the operators have already been computed:

- $(\neg c)^{\square} := G \setminus c^{\diamond}$ and $(\neg c)^{\diamond} := G \setminus c^{\square}$,
- $(\bigwedge A)^{\square} := \bigcap_{a \in A} a^{\square}$ and
  $(\bigwedge A)^{\diamond} := \bigcap_{a \in A} a^{\diamond}$,
- $(\bigvee A)^{\square} := \bigcup_{a \in A} a^{\square}$ and
  $(\bigvee A)^{\diamond} := \bigcup_{a \in A} a^{\diamond}$.

Unfortunately, we usually cannot express universality or potential universality by just using the corresponding derivations or the corresponding representation in the KLEENE-algebra. A typical example is the attribute implication $\{m\} \to \{m\}$ for some $m \in M$ and the following incomplete context:

$\mathbb{K}_0$	a	m	b
g	×	?	−

With the KRIPKE-semantics it easily follows that $\{m\} \to \{m\}$ is universal in every complete and incomplete context, however, w.r.t. $\mathbb{K}_0$ we obtain $(\{m\} \to \{m\})^{\square} = \emptyset \neq \{g\} = G$.

On the other hand it has been proved in [H01], Satz 2.25:

Theorem 1 *Let $\mathbb{K}_i = (G, M, \{\times, ?, -\}, J)$ be any incomplete context and c a compound attribute such that each "original attribute", i.e. each element from M, occurs at most once in c, then*

(i) c is universal w.r.t. $\mathbb{K}_i$ iff $c^{\square} = G$, i.e. iff $c_{\text{KL}} = \top$.
(ii) c is potentially universal w.r.t. $\mathbb{K}_i$ iff $c^{\diamond} = G$, i.e. iff $c_{\text{KL}} = (A,\, \emptyset)$ for some $A \subseteq G$.

This means that in such a case (and in general only then) the usual three-valued KLEENE-logic is sufficient to decide whether or not a compound attribute is universal or potentially universal.

As a consequence one gets for every attribute implication $A \to B$ that

$$A \to B \text{ is universal in } \mathbb{K}_i, \text{ iff } (A \to (B \setminus A))^{\square} = G, \tag{7}$$

---

[9] If $\mathbb{K}_i$ is complete then the concept lattice of $\mathbb{K}_i$ is canonically embedded as a $\bigwedge$-semilattice into this KLEENE-algebra representation of $\mathbb{K}_i$ (see [H01])

as has already been observed in other terms in [B91]. In that paper it has also been discussed that plain scalings of incomplete contexts are of relatively little interest in connection with attribute logic. And a *Boolean derivation* of an incomplete context has been introduced. In it every object "with n question marks in its row" is replaced by $n + 1$ new objects: One with each "?" replaced by "$\times$", and for each "?" an object, where exactly one "?" is replaced by "$-$", while all other "?"s are replaced by "$\times$". In the resulting context exactly those attribute implications are universal, which are universal in $\mathbb{K}_i$.

If one identifies a compound attribute with the corresponding propositional formula (using M as set of propositional variables), then one can use the classical calculus of propositional logic as a sound and complete rule system for universal compound attributes of (complete or incomplete) contexts[10]: A compound attribute c is derivable from a set A of compound attributes iff c is universal in every context $\mathbb{K}$ in which all compound attributes of A are universal. If one is only interested in attribute implications there are smaller rule systems, for example the Armstrong rules[11] or the rule system of [H01], which are sound and complete for universal compound attributes of (complete or incomplete) contexts. In [H01] and part I of [H04] there is also a sound and complete rule system for potentially universal attribute implications of incomplete contexts.

The use of the operators $^\square$ and $^\diamond$ suggests that one could also use modal logic. S.Obiedkov has done this in [O02]. He uses a three-valued modal logic for propositional formulas, and he characterises the universal and the potentially universal propositional formulas by applying this logic on incomplete contexts. Another characterisation in modal logic can be found in [H01].

3   Attribute Exploration

Attribute exploration is an interactive computer algorithm for knowledge acquisition. Its main intention is to obtain complete knowledge about the attribute implications of an (unknown) universe, i.e. a formal context $\mathbb{K}_\mathfrak{U} = (G_\mathfrak{U}, M_\mathfrak{U}, I_\mathfrak{U})$, where the set $G_\mathfrak{U}$ of objects is usually too large to allow the context to be represented or even to be known in all details. The (finite) set $M_\mathfrak{U}$ contains the attributes of most interest among all the attributes conceivable for $G_\mathfrak{U}$. There are many different algorithms for attribute explorations. The following three algorithms also consider incomplete knowledge:

(1) Algorithm by Peter Burmeister (1986) and Holzer (2001): This algorithm was implemented in the program "ConImp" by Peter Burmeister[12]. Background knowledge in form of implications known to be universal and some objects $g \in G_\mathfrak{U}$ together with their (possibly incomplete) context rows can be entered before the exploration starts. So the algorithm starts with a possibly incomplete context $\mathbb{K}_1$ such that a completion of $\mathbb{K}_1$ is a subcontext

---

[10] See [H01]

[11] See [G00]

[12] See [B86/01]

of the universe $\mathbb{K}_\mathfrak{U}$. During the exploration the computer program asks in each step $n = 1, 2, 3, \ldots$ for an attribute implication $A \to B$, where A is a minimal set such that A respects[13] all implications accepted as universal and one has $B = A^{\square\diamond} \neq A$ in the incomplete context $\mathbb{K}_n$. This property can also be expressed by so called pseudoclosed sets, where the pseudoclosed sets of [GW99] can be generalized to incomplete contexts[14]: A set $A \subseteq M$ is called pseudoclosed in an incomplete context if one has $A^{\square\diamond} \neq A$ and $C^{\square\diamond} \subseteq A$ for each pseudoclosed proper subset $C \subset A$. The exploration program choses a minimal pseudoclosed set A which is not a premise of an accepted implication and asks whether $A \to A^{\square\diamond}$ is universal. The expert can either give the answer "yes" or he gives a counterexample or he gives the answer "unknown". If the expert does not know whether the proposed implication is universal the program offers two possibilities: Either the program continues as if the unknown implication would have been accepted as universal or the program asks the expert for the implications $A \to b$ for each $b \in B$, and for each of these implications $A \to b$ which is unkown to the expert a fictitious counterexample is added to the current context. The context row of this fictitious counterexample is the smallest context row (with respect to the information order) such that $A \to b$ is not potentially universal: The object has all attributes of A but not the attribute b, and for all other attributes the entry in the table is "?". If the user decides to continue the exploration as if the implication would have been accepted as universal, then the program asks less questions, so it comes earlier to an end, but in this case some informations may be missing: At the end of the exploration he only gets an upper bound for the universal implications (every universal implication of $\mathbb{K}_\mathfrak{U}$ is potentially universal in the last context of the exploration) and a lower bound (every implication accepted as universal), but there are many implications between these bounds and the program does not help the expert to find an answer for these implications. So it is better to use the second possible continuation: The program creates fictitious counterexamples against the unknown implications. In this case the upper and lower bounds are better than in the first case. The expert gets maximal information with respect to his knowledge: If the user can finally decide for every fictitious counterexample (i.e. unknown implication) whether it corresponds to an implication universal in $\mathbb{K}_\mathfrak{U}$ or to an object of $\mathbb{K}_\mathfrak{U}$, then he has complete knowledge about the attribute implications of $\mathbb{K}_\mathfrak{U}$ and (up to isomorphy) about the concept lattice of $\mathbb{K}_\mathfrak{U}$.

(2) Algorithm by Richard Holzer: The algorithm in [H01] and part I of [H04][15] is similar to (1), but the background knowledge is more general: The expert can enter a frame context to restrict the possible object intents of the universe. Every set of compound attributes (implications, sequents, ...) which are known to be universal can be described by such a frame context. While

---

[13] A set A respects an implication $C \to D$ if $C \subseteq A$ implies $D \subseteq A$
[14] See [H01] and part I of [H04]
[15] This algorithm has not yet been implemented

the background implications of (1) are only used to answer some questions automatically, the algorithm in [H01] also offers the possibility to get a base with respect to the background knowledge, that means every universal implication is derivable by the base with respect to the Armstrong rules and the exhaustion rule[16]. So the base at the end of the exploration does not contain implications which already can be derived from the background knowledge.

(3) Algorithm by Bernhard Ganter: The algorithm of [G99] was implemented in the program "Impex" by Rüdiger Krauße. Background knowledge in form of universal and existential sequents can be entered before the exploration starts, where an existential sequent is a sequent that is known to be not universal in $\mathbb{K}_\mathfrak{U}$. During the exploration algorithm the program asks by and by whether some implications[17] $A \to B$ are universal and the expert can either answer by a universal sequent (X, N) with $X \subseteq A$ and $N \cap A = \emptyset$ or by an existential sequent (X, N) with $A \subseteq X$ and $N \cap B \neq \emptyset$. Note that an existential sequent (X, N) can be seen as an incomplete context row of a counterexample: The existential sequent (X, N) means, that there is an object having all attributes of X but none of N, so this object is a counterexample against $A \to B$ because of $A \subseteq X$ and $N \cap B \neq \emptyset$.

The main difference between these algorithms is, that in the third algorithm the expert can not answer with "unknown". So in this algorithm only incomplete context knowledge (in form of existential sequents describing incomplete context rows) is considered but not incomplete implication knowledge.

In all three algorithms the background knowledge and the answers of the expert can be used to reduce incomplete knowledge: If there is a question mark in a context row of an object g in the column of an attribute m and if it can be derived from the informations entered by the expert that the object must have the attribute m then the question mark can be replaced by "×". Analogously it can be replaced by "−" if it can be derived that the object does not have the attribute.

In the program "ConExp" another algorithm for attribute exploration was implemented by Sergey Yevtushenko. This implementation only deals with complete knowledge, so the user can not use incomplete context rows for counterexamples and he is not allowed to give the answer "unknown" if he does not know whether the proposed implication is universal. Other exploration algorithms in formal concept analysis can be found in [Z91] (rule exploration) and in [S95] and [S97] (concept exploration).

4   Attribute Logic for Many-Valued Contexts

For many-valued contexts $\mathbb{K} = (G, M, W, I)$ there are many possibilities to define the attribute logic of $\mathbb{K}$. In [G99] the many-valued context is scaled by a family

---

[16] The exhaustion rule contains the background knowledge, see [G00]
[17] For the details how these implications are computed, see [G99]

of scales $\mathbb{S} = (\mathbb{S}_m)_{m \in M} = (G_m, M_m, I_m)_{m \in M}$ and the attribute logic of the derived context $\mathbb{K}' = (G, \bigcup_{m \in M} \{m\} \times M_m, J)$ with $(g, (m, p)) \in J$ iff $(I(g, m), p) \in I_m$ is used[18]. But during this process of scaling the attribute set changes: The attributes of the derived context are the disjoint union of the attributes of the scales. So if the user is interested in the dependencies between the attributes of the many-valued context then he needs a different attribute logic. In [H01] the logic is defined with respect to a fixed relation $\rho = (\rho_m)_{m \in M} \subseteq W^\tau$ with $\tau \geq 0$. The relation describes the "dependencies of interest", for example functional dependencies ($\rho = id_W$) or ordinal dependencies[19] ($\rho = \leq_W$). Then for each compound attribute the extent (with respect to ρ) is a subset of G^τ and it is defined recursively:

For $m \in M$ the extent of m is defined by

$$m' = \{(g_1, g_2, \ldots, g_\tau) \in G^\tau \mid (I(g_1, m), I(g_2, m), \ldots, I(g_\tau, m)) \in \rho\}$$

For the compound attributes the extents are defined by

$$(\neg m)' = G^\tau \setminus m'$$
$$\left(\bigwedge A\right)' = \bigcap_{a \in A} a'$$
$$\left(\bigvee A\right)' = \bigcup_{a \in A} a'$$

The many-valued context $\mathbb{K}$ can be transformed into a one-valued context $\mathbb{K}_\rho := (G^\tau, M, I_\rho)$ with $((g_1, g_2, \ldots, g_\tau), m) \in I_\rho$ iff $(I(g_1, m), I(g_2, m), \ldots, I(g_\tau, m)) \in \rho$. This is called ρ-transformation. Note that the extent of each compound attribute m of $\mathbb{K}$ is exactly the extent of m in the transformed context $\mathbb{K}_\rho$. The difference to plain scaling is that here the attribute set remains the same, but the object set is a power of the original object set G. For plain scaling the object set remains the same but the attribute set is a direct sum of the attribute sets of the scales: $\bigcup_{m \in M} \{m\} \times M_m$. For $\rho = id_W$ an attribute implication $A \to B$ just means that the columns of B functionally depend on the columns of A: The implication $A \to B$ is universal iff for each $g, h \in G$ with $I(g, a) = I(h, a)$ for all $a \in A$ we get $I(g, b) = I(h, b)$ for all $b \in B$. For an order relation $\rho = \leq_W$ universality of an attribute implication $A \to B$ means ordinal dependency (greater values in the columns of A imply greater values in the columns of B).

For incomplete many-valued contexts $\mathbb{K} = (G, M, W, I)$ we again have different possibilities to define the attribute logic: We can use a family $\mathbb{S} = (\mathbb{S}_m)_{m \in M} = (G_m, M_m, I_m)_{m \in M}$ of scales to derive an incomplete context:

---

[18] Usually one only requires that the relation $I \subseteq G \times M \times W$ of a many-valued context is the graph of a partial map, but with a new value "?" such a partial map can always be completed, so we can use the relation I as a map $I : G \times M \to W$

[19] See [W88]

$$\mathbb{K}_i = (G, \bigcup_{m \in M} \{m\} \times M_m, \{\times, -, ?\}, J)$$

$$J(g, (m, p)) = ? \text{ iff } I(g, m) = ?$$
$$J(g, (m, p)) = \times \text{ iff } I(g, m) \neq ? \text{ and } (I(g, m), p) \in I_m$$
$$J(g, (m, p)) = - \text{ iff } I(g, m) \neq ? \text{ and } (I(g, m), p) \notin I_m$$

Then we can use the attribute logic of the derived incomplete context. Again we have the problem that the set of attributes changes during the scaling. Like for complete many-valued contexts it is also possible to fix a relation $\rho \subseteq (W \setminus \{?\})^\tau$ for the dependencies of interest and transform $\mathbb{K}$ with respect to the relation ρ to keep the set of attributes[20]:

$$\mathbb{K}_\rho = (G^\tau, M, \{\times, -, ?\}, I_\rho)$$
$$I_\rho((g_1, g_2, \ldots, g_\tau), m) = ? \quad \text{iff} \quad I(g_j, m) = ? \text{ for some } j \leq \tau$$
$$I_\rho((g_1, g_2, \ldots, g_\tau), m) = \times \quad \text{iff} \quad I(g_j, m) \neq ? \text{ for } j \leq \tau \text{ and}$$
$$(I(g_1, m), I(g_2, m), \ldots I(g_\tau, m)) \in \rho$$
$$I_\rho((g_1, g_2, \ldots, g_\tau), m) = - \quad \text{iff} \quad I(g_j, m) \neq ? \text{ for } j \leq \tau \text{ and}$$
$$(I(g_1, m), I(g_2, m), \ldots I(g_\tau, m)) \notin \rho$$

In this case we can use the attribute logic of the incomplete context $\mathbb{K}_\rho$. For both cases of transforming $\mathbb{K}$ into an incomplete context it is possible to improve the transformation by reducing the number of question marks in the derived context: For the plain scaling if there exists an attribute $p \in M_m$ of a scale $\mathbb{S}_m$ such that the column of p is empty in $\mathbb{S}_m$ then it is not possible that an object $g \in G$ has the attribute (m, p) in the derived context, so if a question mark appears in the column of (m, p) in the derived context then it can be replaced by "$-$". On the other side if every object $x \in G_m$ has the attribute p in the scale $\mathbb{S}_m$ then of course also in the derived context every object $g \in G$ has the attribute (m, p) and the question marks in this column can be replaced by "$\times$". In a similar way some question marks can be reduced from the derived context for the ρ-transformation: If for every completion $\mathbb{K}_c = (G, M, W, I_c)$ of $\mathbb{K}$ we have $(I_c(g_1, m), I_c(g_2, m), \ldots, I_c(g_\tau, m)) \in \rho$ then we know that in the derived context $\mathbb{K}_\rho$ the object $(g_1, g_2, \ldots g_\tau)$ must have the attribute m, so a question mark can be replaced by "$\times$". Analogously some question marks can be replaced by "$-$".

If we would like to explore the dependencies between the columns of a many-valued context $\mathbb{K}_\mathfrak{U} = (G_\mathfrak{U}, M_\mathfrak{U}, W_\mathfrak{U}, I_\mathfrak{U})$ with an attribute exploration algorithm then we can either scale $\mathbb{K}_\mathfrak{U}$ with a family of scales $\mathbb{S} = (\mathbb{S}_m)_{m \in M}$ and use an exploration algorithm of one-valued universes described above[21], or we can explore the dependencies with respect to a fixed relation $\rho \subseteq W_\mathfrak{U}^\tau$:[22] The program proposes some implications $A \to B$ and the expert can either accept the implication to be universal in $\mathbb{K}_\mathfrak{U}$ or he gives a counterexample (which is a tuple $(g_1, g_2, \ldots g_\tau)$ of objects which is not in the extent of the compound attribute

---

[20] See [H01]
[21] See [G99]
[22] See [H01]

$A \to B$). It is also possible to give the answer "unknown". In this case the program asks for which attributes $b \in B$ the implication $A \to b$ is unknown, and for each of these attributes a fictitious object is added to the current context as a counterexample against $A \to b$. At the end of the algorithm the expert has a list of implications which are known to be universal, a list of implications accepted as unknown, a list of normal objects which are counterexamples against all implications known to be not valid and a list of fictitious objects which are counterexamples against the implications accepted as unknown. If the expert can decide which of the unknown implications are universal in $\mathbb{K}_\mathfrak{U}$ then he has complete knowledge about the attribute implications of $\mathbb{K}_\mathfrak{U}$.

5   Incomplete Databases

Let W be a set and $k > 0$. An incomplete database[23] is a finite subset of $(W \cup \{?\})^k$. For each incomplete database $D = \{d_1, d_2, \ldots, d_{|D|}\}$ there exists a canonical incomplete many-valued context corresponding to D:

$$\mathbb{K} = (G, M, W \cup \{?\}, I)$$
$$G = \{1, 2, 3, \ldots, |D|\}$$
$$M = \{1, 2, 3, \ldots, k\}$$
$$I(g, m) = d_g(m)$$

On the other hand for each incomplete many-valued context $\mathbb{K}$ there exists a canonical incomplete database:

$$D = \{(I(g, m_1), I(g, m_2), \ldots, I(g, m_k)) \mid g \in G\}$$

where $M = \{m_1, m_2, \ldots m_k\}$ with $k = |M|$. A question mark in a database may have different meanings:

1. the value is unknown
2. the value does not exist
3. the value is undefined
4. the value is invalid
5. the person who created the database did not fill in a value

To define a completion of an incomplete database is more difficult than for incomplete many-valued contexts because in databases the completion may have a different number of elements: It might happen, that two different incomplete tuples of D are completed in such a way, that both tuples coincide, so in the completion some tuples are identified. On the other hand, if one creates an incomplete database D to describe an (unknown) database $D_\mathfrak{U}$, then it might happen that two different tuples of $D_\mathfrak{U}$ are represented by the same incomplete tuple in D. So a completion of an incomplete database does not mean "replace

---

[23] See [K99]

each question mark by one value" but it means "replace each question mark by a set of values"[24].

For the dependencies between the columns of a database many logics have been analysed, for example the Kleene-Logic, the logic of Lukasiewicz, a five-valued logic[25], and Kripke-like logics in form of possible answers and certain answers[26]. Further information about incomplete databases can be found in [C79], [C86], [D86], [D89], [K99].

References

[B86/01] P.Burmeister (assisted by A.Rust, P.Scheich, N.Newrly and C.Bang). *ConImp – A program on formal concept analysis of one-valued contexts* (running under MS-DOS or Linux resp.). Darmstadt University of Technology, 1986/01.

[B91]   P.Burmeister. *Merkmalimplikationen bei unvollständigem Wissen.* In: W.Lex (Ed.): *Proceedings: Arbeitstagung Begriffsanalyse und Künstliche Intelligenz (Clausthal-Zellerfeld 6. - 8. 10. 1988)*; Technische Universität Clausthal, 1991, pp. 15–46.

[B00]   P.Burmeister. *ConImp – Ein Programm zur Formalen Begriffsanalyse.* In: G.Stumme, R.Wille (Eds.) *Begriffliche Wissensverarbeitung: Methoden und Anwendungen*, Springer, 2000.   See also
          Formal Concept Analysis with ConImp: Introduction to the Basic Features;
          (TUD 2003) at
          http://www.mathematik.tu-darmstadt.de/~burmeister
          which is more than a translation of the above paper.

[BH00]   P.Burmeister, R.Holzer. *On the treatment of incomplete knowledge in Formal Concept Analysis.* TU Darmstadt, Preprint No. 2063, Jan. 2000. Appeared in: B.Ganter, G.W.Mineau (Eds.): *Conceptual Structures: Logical, Linguistic, and Computational Issues* (Proceedings of the ICCS 2000 at Darmstadt, August 2000), LNAI 1867, Springer 2000, pp. 385–398.

[C79]   E.F.Codd. *Extending the database relational model to capture more meaning.* ACM Transactions on Database Systems 4 (4), 397-434, 1979

[C86]   E.F.Codd. *Missing information (applicable and inapplicable) in relational databases.* Sigmod, 15(4), 53-78, 1986

[D86]   C.J.Date. *Null Values in Database Management.* Chapter 15, Addison-Wesley, Reading, MA, 313-334, 1986

[D89]   C.J.Date. *NOT is not 'Not' (Notes on three-valued logic and related matters).* Chapter 8, Addison-Wesley Reading, MA, 217-248, 1989

[G99]   B.Ganter. *Attribute exploration with background knowledge.* Theoretical Computer Science 217, page 215-233. 10, 1999

[G00]   B.Ganter. *Begriffe und Implikationen.* In: G.Stumme, R.Wille (Eds.) *Begriffliche Wissensverarbeitung: Methoden und Anwendungen*, Springer, 2000.

[GW99]   B.Ganter, R.Wille. *Formal Concept Analysis – Mathematical Foundations.* Springer, 1999.

---

[24] See [K97]
[25] See [K97] and [K99]
[26] See [L76], [L79], [L81], [L84]

[GW00]   B.Ganter, R.Wille. *Contextual Attribute Logic*. In: W.Töpferhart, W.Cyre (eds.): Conceptual Structures: Standards and Practices. Springer Berlin-Heidelberg-New York, 2000, pp. 377-388.

[H01]   R.Holzer. *Methoden der formalen Begriffsanalyse bei der Behandlung unvollständigen Wissens*. Dissertation, Shaker Verlag, 2001.

[H04]   R.Holzer. *Knowledge acquisition under incomplete knowledge using methods from formal concept analysis Parts I and II*. Fundamenta Informaticae, Volume 63, No. 1, p. 17-39 and p. 41-63, 2004

[K97]   H.-J.Klein. *Gesicherte und mögliche Antworten auf Anfragen an relationale Datenbanken mit partiellen Relationen*. 1997

[K98]   H.-J.Klein. *Model Theoretic and Proof Theoretic View of Relational Databases with Null Values: A Comparison*. 1998

[K99]   H.-J.Klein. *Efficient Algorithms for Approximating Answers to Queries Against Incomplete Relational Databases*. Proceedings of the 6th International Workshop on Knowledge Representation meets Databases KRDB'99, Sweden, 1999

[L76]   W.Lipski. *Informational systems with incomplete information*. Proc. 3rd Int. Symp. on Automata, Languages and Programming (Hrg.: S.Michaelson, R.Milner), Edinburgh, 120-130, 1976

[L79]   W.Lipski. *On semantic issues connected with incomplete information databases*. ACM Trans. on Database Systems 4 (3), 262-296, 1979

[L81]   W.Lipski. *On databases with incomplete information*. J. of the ACM 18(1), 41-70, 1981

[L84]   W.Lipski. *On relational algebra with marked nulls*. Proc. 3rd ACM Symp. on Principles of Database Systems, 201-203, 1984

[Ma83]   D.Maier. The Theory of Relational Databases. Computer Science Press, 1983.

[O02]   S.Obiedkov. *Modal Logic for Evaluating Formulas in Incomplete Contexts*. In: U.Priss, D.Corbett, G.Angelova (Eds.): Conceptual Structures - Integration and Interfaces. LNAI 2393, Springer, Heidelberg, 314-325, ICCS 2002

[P97]   P.Pagliani. *Information Gaps as Communication Needs: A New Semantic Foundation for Some Non-Classical Logics*. Journal of Logic, Language and Information 6, 1997, pp. .

[S95]   G.Stumme. *Knowledge acquisition by distributive concept exploration*. In: G. Ellis, R.A. Levinson, W.Rich, J.F. Sowa (eds.), Supplementary proceedings of the third international conference on conceptual structures, Santa Cruz, CA, USA, 98-111, 1995

[S97]   G.Stumme. *Concept exploration – A tool for creating and exploring conceptual hierarchies*. In: D.Lukose, H.Delugach, M.Keeler, L.Searle, J.F.Sowa (eds.): Conceptual Structures: Fulfilling Peirce's Dream, LNAI 1257, Springer, Berlin, 318-331, 1997

[W88]   R.Wille. *Dependencies between many-valued attributes*. In: Classification and Related Methods of Data Analysis, H.H.Bock (Editor), 1988

[W89]   R.Wille. *Knowledge acquisition by methods of formal concept analysis*. In: E.Diday (ed.): *Data analysis, learning symbolic and numeric knowledge*. Nova Science Publishers, New York – Budapest, pp. 365–380, 1989

[Wo93]   K.E.Wolff. *A first course in formal concept analysis*. In: F.Faulbaum (ed.): *SoftStat'93, Advances in Statistical Software 4*, pp. 429–438, 1993

[Z91]   M.Zickwolff. *Rule exploration: first order logic in formal concept analysis*. Dissertation, TH Darmstadt, 1991

States, Transitions, and Life Tracks in Temporal Concept Analysis

Karl Erich Wolff

Mathematics and Science Faculty
Darmstadt University of Applied Sciences
Schoefferstr. 3, D-64295 Darmstadt, Germany
karl.erich.wolff@t-online.de
http://www.fbmn.fh-darmstadt.de/home/wolff

Abstract. Based on Formal Concept Analysis, we introduce Temporal Concept Analysis as a temporal conceptual granularity theory for movements of general objects in abstract or "real" space and time such that the notions of states, situations, transitions and life tracks of objects in conceptual time systems are defined mathematically. The life track lemma is a first approach to granularity reasoning. Applications of Temporal Concept Analysis in medicine and in chemical industry are demonstrated as well as recent developments of computer programs for graphical representations of temporal systems. Basic relations between Temporal Concept Analysis and other temporal theories, namely theoretical physics, mathematical system theory, automata theory, and temporal logic are discussed.

1 Introduction

The purpose of this paper is to present the actual state of Temporal Concept Analysis (TCA), a conceptual granularity theory for the treatment of temporal phenomena. In TCA, not only space and time but also objects and their movements are represented conceptually, including a granularity description based on the notion of formal concepts and conceptual scales. The classical point of view on temporal phenomena is dominated by classical mechanics describing space and time using the continuum of real numbers and by automata theory using an abstract notion of discrete states and transitions without an explicit time description. Therefore, it is desirable to develop a general temporal theory covering continuous as well as discrete temporal systems.

Clearly, such a unification demands a background theory based on general basic concepts. Such a theory emerged from lattice theory, introduced by Garrett Birkhoff [Bir67] as a common generalization of ordered structures in geometry and logic. Rudolf Wille [Wil82] brought a vivid real world relevance into the theory of abstract lattices by his introduction of *formal contexts* and their *concept lattices*. His purpose was to restructure lattice theory in the sense of Hartmut von Hentig's claim to restructure sciences [vHe72]. Concept lattices are used to

B. Ganter et al. (Eds.): Formal Concept Analysis, LNAI 3626, pp. 127–148, 2005.

describe the conceptual structures inherent in data tables without loss of information by means of *line diagrams* yielding valuable visualizations of real data.

The conceptual representation of temporal phenomena started with the usual order representation of time as a chain; later on *interordinal scales* proved extremely useful for working with temporal (or spatial) intervals. Based on the idea of an infinite interordinal scale Rudolf Wille introduced *linear continuum structures* "making mathematically explicit the Aristotelian conception of a time continuum" [Wil04].

Clearly, time has to be investigated in connection with a notion of space to represent movements of objects. Based on experiences with real data from psychological and industrial applications the author [Wol00a] combined the idea of a *time granule* like "this morning" with the idea of a *state* by introducing the mathematical notion of a *Conceptual Time System*. That led to a conceptual investigation of the notion of an *object* in the sense of a *spatio-temporal object*. Such an object is given by its *actual objects* which are connected by a *time relation* yielding a *life track* which represents the object [Wol02a, Wol02b]. That led to a purely conceptual understanding of movements of objects in continuous or discrete space and time – without employing the classical algebraic, metric and analytic structures. In the following sections we give a short overview over the main ideas in Temporal Concept Analysis as it is developed now. For that purpose we start with a simple example of a journey.

2   Contextual Description of a Journey

In this section we discuss some basic contextual descriptions of temporal and spatial aspects of a journey. In the following we assume that the reader is familiar with the basic definitions in Formal Concept Analysis, in particular with its Conceptual Scaling Theory [GaWi89, GaWi99]. For a short introduction we refer to [Wil97a, Wol94].

2.1   John's Journey

In this subsection we start with an example of a typical spatio-temporal description, namely a story about a journey. This example will be used throughout the paper to introduce the main ideas in Temporal Concept Analysis.

> **The Story of John's Journey:** John flies from Frankfurt to Napoli leaving Frankfurt on Thursday, returning on Sunday. John takes a flight on Thursday morning, arriving at Napoli in the afternoon. He visits a conference on Friday morning and the conference dinner on Saturday evening; he leaves Napoli on Sunday afternoon arriving at Frankfurt in the evening.

The following representation of this story does not represent its full linguistic structure. We only try to grasp the spatio-temporal structure and the granularity of the story. First we describe some basic temporal aspects. For that purpose we focus on the days from Thursday to Sunday. To represent the natural ordering of these days we employ a chain with four elements. The contextual representation of such a chain is given in Table 1:

Table 1. A formal context for a chain with four elements

greater or equal	1	2	3	4
1	×			
2	×	×		
3	×	×	×	
4	×	×	×	×

Replacing the four numbers by the four days of interest we get from the *abstract ordinal scale* in Table 1 the *concrete scale* for a chain of the four days. The line diagrams in Figure 1 represent the corresponding concept lattices.

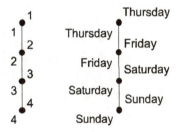

Fig. 1. Concept lattices of an "abstract" and a corresponding "concrete" scale

Similarly, the day times "morning", "afternoon", "evening" are described by a concept lattice which is a chain with three elements. The direct product of these chains represents the "time schedule of John's journey" (in Figure 2) which is again described as a concept lattice.

We consider the corresponding formal context since it gives us a first hint towards an understanding of the notion of *time granules*. The chosen granularity of the temporal description yields $4 \times 3 = 12$ "possible time granules", for example (Saturday, afternoon), which are the formal objects in Figure 2. Table 2 shows three of the twelve rows of the mentioned context, namely the *time granules* of Saturday:

The complete formal context of this simple and important combination of two scales has as objects all pairs of objects of the two given formal contexts; it has as attribute set the (disjoint) union of the two given attribute sets; and its incidence relation is constructed by copying the given incidence relations, for example: (Saturday, afternoon) gets a cross at all those attributes of the first context where "Saturday" has a cross there, and a cross at all those attributes of the second context where "afternoon" has a cross there. (The formal definition of a semiproduct of two contexts is given in [GaWi99], p.46.)

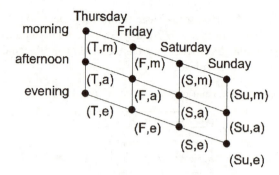

Fig. 2. A concept lattice representing the "time schedule of John's journey"

Table 2. Three of twelve rows of the so-called semiproduct of the two time scales

	Thursday	Friday	Saturday	Sunday	morning	afternoon	evening
(Saturday, morning)	×	×	×		×		
(Saturday, afternoon)	×	×	×		×	×	
(Saturday, evening)	×	×	×		×	×	×

To represent the spatial descriptions of the journey we construct a scale for the mentioned places "Frankfurt" and "Napoli" . We want to say that the town Napoli is situated south of Frankfurt. That is done in the following formal context:

Table 3. An ordinal conceptual scale for the places

southern of or equal to	Frankfurt	Napoli
Frankfurt	×	
Napoli	×	×

Clearly, the concept lattice of this context is, as a chain with two elements, a very simple map; a typical plane map with many towns can be represented in the same way by a direct product of two chains with many elements. The metric embedding into the usual plane can be made as fine as necessary; that is not discussed here. In the next section, we continue our example, describing the introduction of *conceptual time systems*, *time granules*, *situations*, and *states*.

3   Basic Notions in Temporal Concept Analysis

The author started the conceptual investigation of temporal phenomena with the key-idea that the *states* of a temporal system should be described as the object concepts of a suitable formal context. Since his search for useful descriptions of *states* in Mathematical System Theory, in physics, in Automata Theory, and several other domains did not yield, up to now, a conceptually satisfactory result [Zad64, Arb70, Eil74, Cast98, But99], the notion of a *state* in a *conceptual time system* has been introduced [Wol00a, Wol00b, Wol02b].

3.1   Time Granules as Formal Objects, States as Object Concepts

Searching for a general notion of a *state of a system*, we introduce first the definition of a *conceptual time system*. A general system description has to contain the elementary system descriptions that occur when we observe a real system and write down a finite protocol, usually represented as a data table. Therefore, we develop the main ideas in that framework. Indeed, we shall see that even infinite temporal systems can be described in the same way.

Let us imagine that we observe a real system. For a single observation we need some time, may be one minute or only one millisecond. Often we abstract from the duration of an observation and use the notion of a *point of time*, usually represented as a real number.

In the following, we do not assume any internal structure of such a *point of time*, as for example the assumption that it is a real interval or a real number. We just start from a set G; the elements of G are called *time granules*. For describing the observations, we introduce a many-valued context with G as its set of formal objects. In a data table of this many-valued context the row of a time granule g shows in column m the value m(g) of the *measurement* m *at time granule* g.

For clarifying our idea of a *conceptual time system*, we first consider the data table for John's journey in Table 4 where the integers 0,1,...,5 represent the six *time granules* which "occurred in the story of John's journey". Their meaning is described by the values in the two columns of the *time part* of the data table. The place of John at each of these time granules is described in the *event part* (or *space part*) of the data table. Together with the previously mentioned scales for the time part and the scale for the event part, we obtain an initial example of a *conceptual time system*.

Table 4. A data table of a conceptual time system

time granules	time part		event part
	day	day time	place
0	Thursday	morning	Frankfurt
1	Thursday	afternoon	Napoli
2	Friday	morning	Napoli
3	Saturday	evening	Napoli
4	Sunday	afternoon	Napoli
5	Sunday	evening	Frankfurt

Definition [Wol00a]: "conceptual time system, situations, states"
Let $\mathbf{T} := ((G, M, W, I_T), (\mathbf{S}_m \mid m \in M))$ and $\mathbf{C} := ((G, E, V, I), (\mathbf{S}_e \mid e \in E))$ be scaled many-valued contexts on the same object set G. Then the pair $(\mathbf{T}, \mathbf{C})$

is called a *conceptual time system on the set G of time granules.* **T** is called the *time part* and **C** the *event part* or the *space part* of (**T**, **C**). The derived context of **T** ([GaWi89, GaWi99]) is denoted by $\mathbb{K}_T$, the derived context of **C** is denoted by $\mathbb{K}_C$, and the apposition of $\mathbb{K}_T$ and $\mathbb{K}_C$ is denoted by $\mathbb{K}_{TC} := \mathbb{K}_T|\mathbb{K}_C$. It is called the derived context of the conceptual time system (**T**, **C**).

The object concepts of $\mathbb{K}_{TC}$ are called *situations*, the object concepts of $\mathbb{K}_C$ are called *states*, and the object concepts of $\mathbb{K}_T$ are called *time states*. The sets of situations, states, and time states are called *the situation space, the state space,* and *the time state space* of (**T**, **C**), respectively. The object concept mappings of $\mathbb{K}_{TC}$, $\mathbb{K}_T$, and $\mathbb{K}_C$ are denoted by γ, γ_T, and γ_C, respectively.

For the conceptual time system of John's journey the derived context $\mathbb{K}_{TC}$ is represented in the next table.

Table 5. The derived context of John's journey

$\mathbb{K}_{TC}$	time part $\mathbb{K}_T$							event part $\mathbb{K}_C$	
time gran.	Thursday	Friday	Saturday	Sunday	morn.	aftern.	evening	Frankfurt	Napoli
0	×				×			×	
1	×				×	×		×	×
2	×	×			×			×	×
3	×	×	×		×	×	×	×	×
4	×	×	×	×	×	×		×	×
5	×	×	×	×	×	×	×	×	

The subcontext $\mathbb{K}_C$ given by the first column and the two last columns of Table 5 is called the "the event part of $\mathbb{K}_{TC}$". The concept lattice of $\mathbb{K}_C$ is drawn in Figure 3. Its object concepts represent quite well our usual understanding of *states*, namely that each system is at each time granule in exactly one state.

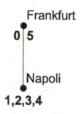

Frankfurt

0 5

Napoli

1,2,3,4

Fig. 3. The concept lattice of the event part $\mathbb{K}_C$ for John's journey

To visualize the time states we embed the concept lattice of the time part $\mathbb{K}_T$ into the lattice in Figure 2. The black circles in Figure 4 represent the concepts of the time part; the black ones which are numbered represent the six time states. In the right part of Figure 4, we have drawn some arrows indicating the temporal sequence in which John's journey happens. That will be discussed more extensively in the next subsection.

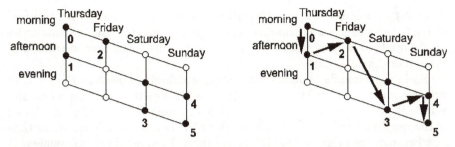

Fig. 4. Embedding the concept lattice of the time part into the "time schedule"

3.2 The Time Relation, Transitions, and Life Tracks

For our approach it is important that the notion of a *state* is introduced in a meaningful way without using an ordering of *the time*. What is *the time* in a conceptual time system? We have introduced several temporal notions, namely *time granules* and a *time part*, whose attributes are interpreted as *time measurements* and their scales as *time scales*. Thence we have mathematically defined *situations*, *states*, and *time states*. But for all that we did not need any notion of an ordering. In the example, we have used the integers $0,1,\ldots,5$ to represent time granules. We have written them down in the first column of Table 5 in their natural ordering. But since the sequence of the names of the (formal) objects in a data table of a (many-valued) context is not represented in the mathematical definition of a (many-valued) context we have to make it explicit formally.

3.3 The Time Relation

In many temporal systems we wish to express the "natural temporal ordering". To investigate carefully the conceptual role of temporal orderings we have to decide where we should introduce some ordinal structure; there are three main possibilities: in the time scales, on the time values, or on the time granules. In the following we describe a simple way to represent "the temporal ordering" by introducing a relation R, called the *time relation*, on the set G of time granules of a conceptual time system. Then we speak of a *conceptual time system with a time relation* (**CTST**).

Definition [Wol02a]: "conceptual time system with a time relation"
Let $(\mathbf{T}, \mathbf{C})$ be a conceptual time system on G and $R \subseteq G \times G$. Then the triple $(\mathbf{T}, \mathbf{C}, R)$ is called a *conceptual time system (on G) with a time relation*.

To distinguish clearly between some order theoretic and graph theoretic notions we again look at the conceptual time system of John's journey. On the set G $:= \{0,1,2,3,4,5\}$ of its time granules we introduce the relation $R := \{(0,1),(1,2),(2,3),(3,4),(4,5)\}$, shortly described as $0 \to 1 \to 2 \to 3 \to 4 \to 5$. Clearly, in that example the directed graph (G,R) is a directed path. It is not yet an ordered set since it is neither reflexive nor transitive (but antisymmetric). The reflexive and transitive closure of it is just the usual natural order on the set G. The chosen time relation R is just the neighborhood relation of that ordered set.

As in the example of John's journey, in standard applications the set G of time granules will be finite, say $G := \{0,\ldots,n-1\}$; then the time relation usually will be chosen as the neighborhood relation on these integers. If G is an interval of the usual real order, we emphasize taking the real order relation as the time relation since the neighborhood relation of the ordered set of the real numbers is empty. Now we are ready to introduce *transitions* and *life tracks* in conceptual time systems with a time relation.

3.4   Transitions

The basic idea of a *transition* is a "step from one point to another". We shall use *transitions* in several *spaces*, mainly in the *situation space*, the *state space*, and the *time state space*. The idea is to generate these transitions by the *R-transitions* (g,h) which are by definition the elements of the time relation R.

That is demonstrated for John's journey in the right part of Figure 4. Each arrow in Figure 4 represents a "transition of John" and is described by the R-transition (g,h) and by the pair of points say (f(g), f(h)) to which g and h are mapped. In this example the mapping f is the object concept mapping of the time part, which maps a time granule onto its object concept in the time state space.

In general, for any **CTST** and any mapping f from the set G of time granules into some other set X we define an *f-transition of the CTST in the set X* as a pair ((g,h), (f(g), f(h))) of two pairs, namely an R-transition and its image under f. That allows for describing "multiple transitions" between two given states (or situations or time states) at different time granules.

3.5   Life Tracks

The transitions in Figure 4 form a *life track* of John. To introduce life tracks mathematically we shall define a life track as a set which is structured by the induced time relation. In the three diagrams of Figure 5 John's life track is represented by the bold arrows. The thin arrows show the not yet told journey of John's wife, Mary. The formal representation of persons like John and Mary as subsystems will be discussed in the next section.

Figure 5 shows three related diagrams labelled by the names "states", "time states", and "situations". The time granules of John are represented in bold; those of Mary are thin; they are drawn only in the situation space; they can be reconstructed in the state space and in the time state space by projection from the situation space – which will be discussed later. The "state space" in the upper left of Figure 5 tells us that John and Mary make a journey from Frankfurt to Napoli and back. The time state space in the form of the schedule in Figure 2 tells us when they make their transitions. In the direct product of these two

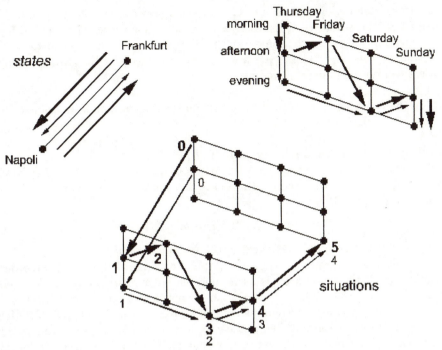

Fig. 5. John's and Mary's journey

spaces we see all "situations" as object concepts in the "situation space" of the journey. To be clear, we just tell the story of Mary's journey:

The Story of Mary's Journey: Mary takes a flight from Frankfurt to Napoli on Thursday afternoon arriving at Napoli in the evening. She visits the conference dinner on Saturday evening and leaves Napoli on Sunday afternoon arriving at Frankfurt in the evening.

To prepare the definition of a "life track of a **CTST**" we assume that we are interested in some mapping f (for example the object concept mapping) from the set G of time granules into some other set X (for example the set of states or the set of situations).

Definition: "transitions and life tracks"
Let $(\mathbf{T}, \mathbf{C}, R)$ be a conceptual time system on G with a time relation. Then any pair $(g, h) \in R$ is called an *R-transition on G*. Let X be a set and f: $G \to X$, then f induces the mapping $f_R : R \to \{\ (f(g), f(h))\ |\ (g,h) \in R\ \}$ where $f_R((g,h)) := (f(g), f(h))$. The element $(\ (g,h),\ (f(g),f(h))\ ) \in f_R$ is called the *f-induced R-transition on X leading from the start point* $(g, f(g))$ *to the endpoint* $(h, f(h))$. The set $f = \{(g, f(g))|\ g \in G\}$ is called the *life track of f in X*.

Now we are interested in some special choices of f. Let $\mathbb{K}_{TC} := \mathbb{K}_T | \mathbb{K}_C$ be the derived context of the CTS $(\mathbf{T}, \mathbf{C})$. For the object concept mapping γ: G $\rightarrow$ γG of $\mathbb{K}_{TC}$ the γ-induced R-transitions on the situation space γG are called the *R-transitions on γG*. In the same way the R-transitions on the *state space* γ_C G and on the *time state space* γ_T G are defined as induced by the corresponding object concept mappings γ_C and γ_T.

In the following definition we introduce on the life track an isomorphic copy of the time relation R:

Definition: "the life track digraph $(\mathbf{f}, R_f)$"
Let $(\mathbf{T}, \mathbf{C}, \mathbf{R})$ be a conceptual time system on G with a time relation. Let X be a set and f: G $\rightarrow$ X, then the relation R_f is defined on the life track $\mathbf{f} = \{(g, f(g)) | \ g \in G\}$ by

$$(g, f(g)) \ R_f \ (h, f(h)) :\Leftrightarrow g \ R \ h.$$

The directed graph $(\mathbf{f}, R_f)$ is called the *life track digraph of R*.

The life track digraph $(\mathbf{f}, R_f)$ is isomorphic to (G, R). Hence, if R is an order relation on G, the relation R_f is an isomorphic order relation on the life track f. If (G, R) is a chain, then $(\mathbf{f}, R_f)$ is an isomorphic chain yielding the usual trajectories in dynamical systems as defined for example in [Kr98], p.8. If (G, R) represents a directed graph-theoretic path, then $(\mathbf{f}, R_f)$ is an isomorphic path; representing that path on the set X (for example the state space) using labels (as in Figure 3) we get a directed graph with point labels and usually with loops (x, x). In Figure 5 we have omitted the loops (in the state diagram).

In the next section, we introduce "objects" or "persons", like John and Mary, as subsystems.

4   Objects as Subsystems

In Figure 5 we have visualized the life tracks of two persons. Since we represented John's journey as a CTST we would like to do the same for Mary. Hence the question arises of how to combine two CTST's in a meaningful way; for example, in such a way that the life tracks of these two systems appear in the same space; then the systems should have the same many-valued attributes and the same scales. In this case the tables are arranged in subposition, for example, the table of Mary is just written under the table of John. The formal definition of subposition of formal contexts can be found in [GaWi99], p.40. The subposition of many-valued contexts is defined analogously.

The following Table 6 shows the many-valued context of "John's and Mary's journey" where we introduced "actual objects"; for example, (John,5) describes "John at time granule 5". To obtain the life tracks of John and Mary as drawn in Figure 5 we introduce the time relation on the set of actual objects by:

$$(J,0) \rightarrow (J,1) \rightarrow (J,2) \rightarrow (J,3) \rightarrow (J,4) \rightarrow (J,5)$$
$$(M,0) \rightarrow (M,1) \rightarrow (M,2) \rightarrow (M,3) \rightarrow (M,4).$$

Table 6. The data table of John's and Mary's journey

time granules	time part		event part
	day	day time	place
(J,0)	Thursday	morning	Frankfurt
(J,1)	Thursday	afternoon	Napoli
(J,2)	Friday	morning	Napoli
(J,3)	Saturday	evening	Napoli
(J,4)	Sunday	afternoon	Napoli
(J,5)	Sunday	evening	Frankfurt
(M,0)	Thursday	afternoon	Frankfurt
(M,1)	Thursday	evening	Napoli
(M,2)	Saturday	evening	Napoli
(M,3)	Sunday	afternoon	Napoli
(M,4)	Sunday	evening	Frankfurt

Together with the above mentioned scales and the previously mentioned time relation Table 6 shows a first example of a "conceptual time system with actual objects and a time relation" (CTSOT). Its derived context yields the concept lattice indicated in Figure 5 with the two life tracks of John and Mary.

The following definition of a CTSOT was introduced by the author in [Wol02a].

Definition: "CTSOT"
"conceptual time systems with actual objects and a time relation"
Let P be a set (of "persons", or "objects") and G a set (of "time granules") and $\Pi \subseteq P \times G$. Let $(\mathbf{T}, \mathbf{C})$ be a conceptual time system on Π and $R \subseteq \Pi \times \Pi$. Then the tuple $(P, G, \Pi, \mathbf{T}, \mathbf{C}, R)$ is called a *conceptual time system (on $\Pi \subseteq P \times G$) with actual objects and a time relation,* in short a CTSOT. For each object $p \in P$ the set $p^\Pi := \{g \in G \mid (p,g) \in \Pi\}$ is called the *time of p in Π* . Then the set $\mathbf{R}_p := \{(g,h) \mid ((p,g), (p,h)) \in R \}$ is called the set of *R-transitions of p* and the relational structure $(p^\Pi, \mathbf{R}_p)$ is called the *time structure of p*.

The subsystem of the "rows of a single person p" can be described as a CTST. The previously described definitions of situations, states, time states, transitions, and life tracks can be used to describe the corresponding notions for a CTSOT. The formal definitions are given in [Wol02a]. Here we mention the definition of the life track of an object.

Definition: "life track of an object"
Let $(P, G, \Pi, \mathbf{T}, \mathbf{C}, R)$ be a CTSOT, and $p \in P$. Then for any mapping f: $\{p\} \times p^\Pi \to X$ (into some set X) the set $f = \{((p,g),f(p,g)) \mid g \in p^\Pi\}$ is called the *f-life track of p.*

The two most useful examples for such mappings are the object concept mappings γ and γ_C of the derived contexts $\mathbb{K}_T | \mathbb{K}_C$ and $\mathbb{K}_C$ of the CTST $(\mathbf{T}, \mathbf{C},$

R) on Π, each of them restricted to the set $\{p\} \times p^{\Pi}$ of actual objects. They are called the *life track of p in the situation space* and the *life track of p in the state space* respectively.

Clearly, there are other possibilities for describing the subsystems of the "persons", for example by introducing a new many-valued attribute "PERSON" with the names of the persons as values, more precisely PERSON(p,g):= p. Then the scale for the many-valued attribute PERSON can be chosen to represent hierarchies for persons, for example the membership hierarchy of a family, where the family itself can be understood as a "general person" or "general object". That led the author recently to a conceptual understanding of particles, waves and wave packets in "Conceptual Semantic Systems" [Wol04a]. The connection between CTSOTs, Conceptual Semantic Systems, conceptual graphs and power context families as introduced by Wille [Wil97b] will be discussed elsewhere.

5   Conceptual Granularity Reasoning

Now we are ready to discuss some basic aspects of "conceptual granularity reasoning". First, we study an example. In colloquial speech we conclude from "John took a flight on Sunday to Frankfurt" that "John took a flight at the weekend to Germany". For that kind of reasoning we use our "background knowledge" that "Sunday belongs to the weekend" and "Frankfurt belongs to Germany". Clearly, we cannot conclude from any judgement by replacing some concepts by superconcepts that the new statement is also valid, for example the judgement that "the regions of two towns are disjoint" does not imply that "the regions of the counties of these towns are disjoint". Therefore, we take some first cautious steps to investigate granularity reasoning.

With respect to CTSOTs, we are interested in statements about life tracks and granularity. In the example of the life track of "John" in the situation space in Figure 5 we see that we get the life track of "John" in the time scale "by projection" from the life track of "John" in the situation space. This leads to the conjecture that "the life track of a person can be mapped by a suitable projection onto the life track of that person in some factor space". Indeed, there is such a general projection which is called the "closure function" in [Ern82].

Definition: "closure function"
Let $(V, \leq)$ be an ordered set and $T \subseteq V$ such that each subset $S \subseteq T$ has an infimum in T, i.e. $\forall_{S \subseteq T} \exists_{t \in T} t = \inf S$. Then the mapping
$$\pi: V \to T \text{ defined by } \pi(x) := \inf\{y \in T | x \leq y\}$$
is called the *closure function from* $(V, \leq)$ *onto* T.

Clearly, π is a projection from V onto T, i.e. $\pi^2 = \pi$, since $\pi(t) = t$ for all t $\in$ T. Furthermore, $\pi(x) \geq x$. In the special case that V is the power set $P(X)$ of a set X, and T is a closure system on X, then the corresponding closure function is just the closure operator of the closure system T.

Using the closure function we now prove the following Life Track Lemma:

Life Track Lemma:
Let $(P, G, \Pi, \mathbf{T}, \mathbf{C}, R)$ be a CTSOT and $p \in P$.
Let $\mathbb{K}_{TC} := \mathbb{K}_T | \mathbb{K}_C$ be the derived context of the given CTSOT. The object set of $\mathbb{K}_{TC}$ is Π, let M be its set of attributes, and I its incidence relation. Let $\mathbf{B}$ denote the set of all formal concepts of $\mathbb{K}_{TC}$ and for $N \subseteq M$ let $\mathbf{B}_N$ denote the set of all formal concepts of the subcontext $\mathbf{K}_N := (\Pi, N, I \cap (\Pi \times N))$ and γ, γ_N the object concept mappings of $\mathbb{K}_{TC}$ and $\mathbf{K}_N$ respectively. Let $\varphi: \mathbf{B}_N \to \mathbf{B}$ be the meet-preserving order embedding satisfying $\varphi(A,B) := (A, A^I)$. Then the closure function $\pi: \mathbf{B} \to \varphi \mathbf{B}_N$ satisfies
$$\pi\gamma = \varphi\gamma_N$$
and the *extended closure function* $\tau := \varphi^{-1}\pi$ satisfies
$$\tau\gamma = \gamma_N$$
and maps each object concept $\gamma(p,g)$ of the actual person (p,g) onto the object concept $\gamma_N(p,g)$ and therefore the γ-life track of p in the situation space onto the γ_N-life track of p in the factor space $\mathbf{B}_N$ obtained by restricting the attribute set M to the subset N.

Clearly, if we restrict the situation space to the state space by omitting all attributes of the time part, the corresponding extended closure function maps the life track of a person in the situation space onto the life track of the same person in the state space.

Proof of the Life Track Lemma:
First, we mention that $\varphi: \mathbf{B}_N \to \mathbf{B}$ is a meet-preserving order embedding ([GaWi99], p.98), hence the set $\varphi \mathbf{B}_N$ has the property that each of its subsets has an infimum in $\varphi \mathbf{B}_N$. Therefore, the closure function $\pi: \mathbf{B} \to \varphi \mathbf{B}_N$ exists and satisfies for any actual object (p,g) that the extent of $\pi(\gamma(p,g))$ can be described by the following formula (where we use $J := I \cap (\Pi \times N)$)
$$\bigcap \{C \mid (p,g)^{II} \subseteq C, (C,C^J) \in \mathbf{B}_N \} = \bigcap \{C \mid (p,g) \in C, (C,C^J) \in \mathbf{B}_N \} = (p,g)^{JJ}$$
since $(p,g) \in (p,g)^{II} \subseteq (p,g)^{JJ}$. Using that $\varphi(\gamma_N(p,g))$ has the same extent as $\gamma_N(p,g)$, namely $(p,g)^{JJ}$ we get $\pi(\gamma(p,g)) = \varphi(\gamma_N(p,g))$ and that proves the Life Track Lemma.

6   Applications and Computer Programs

Temporal Concept Analysis was developed by the author motivated by many applications of Formal Concept Analysis in practice [SpWo91, WoSt93, Wol95a]. To improve process control the formal representation of the temporal structure of processes was necessary. After having introduced the notions of conceptual time systems, states, and situations many previously studied examples could be represented much clearer. The introduction of transitions and life tracks led to valuable computer animations of processes. We demonstrate two examples, one from my long cooperation with the psychoanalyst Dr. Norbert Spangen-

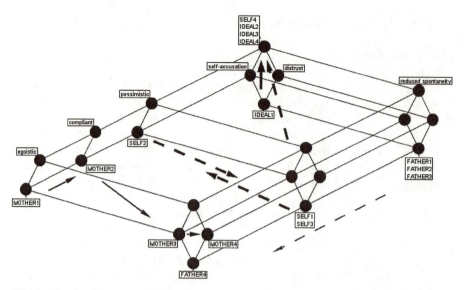

Fig. 6. The development of an anorectic young woman and her family over about two years

berg (then at the Sigmund Freud Institute Frankfurt, Germany) who is working in psychosomatic process research [Spa90]; the other example demonstrates an application in the multi-dimensional visualization of processes in a chemical distillation column. Finally, we briefly mention the main computer programs for TCA.

6.1 The Development of an Anorectic Young Woman

The following example in Figure 6 describes the development of an anorectic young woman (SELF), her father, mother, and her self ideal (IDEAL) during a period of about two years. The underlying formal context was constructed by the psychoanalyst Spangenberg on the basis of four repertory grids taken about each half year from the beginning (time granule 1) until the end (time granule 4) of the psychoanalytic treatment of his patient. SELF1, the self at the beginning of the treatment, has the attributes "distrust" and "reduced spontaneity", SELF2 "pessimistic" and "self-accusation", SELF3 is in the same state as SELF1, and SELF4 reaches the state of IDEAL2,3,4. Indeed, the patient was healthy again at this point in time. It is remarkable that the life tracks of FATHER and MOTHER start from quite different states and end in similar states, the FATHER having all negative attributes of that context. For further information the reader is referred to [Spa90, SpWo91, SpWo93].

6.2 A Chemical Process in a Distillation Column

The diagram in Figure 7 demonstrates a visualization of a chemical process in a distillation column over a period of 20 days.

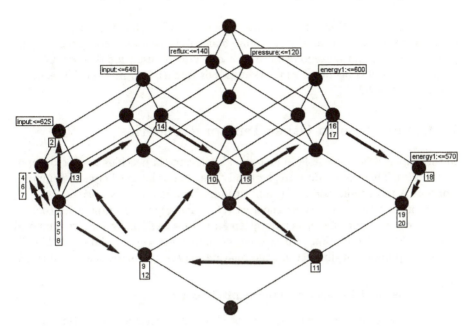

Fig. 7. A chemical process represented in a 4-dimensional state space

From such diagrams the process can be understood quite well taking the attributes used by the experts. A short and coarse description of that process might be:

> Starting on the first day from a state of low input, low reflux and low pressure, but high energy1, the system switched at low input between low and high pressure; from day 9 to 12 it visited in a circular movement states of middle and low energies; finally it came at middle resp. high input to states of middle resp. low energy1, low pressure and low reflux.

Typically, in such applications the experts suggest first a coarse granularity by a few "cuts" like "energy1$\leq$600". After having studied the concept lattice with a coarse granularity it is usually refined, depending on the data and on the interest of the experts. That leads in a few steps to valuable visualizations of multidimensional processes. For further information the reader is referred to [Wol95a, Wol00b].

6.3 Computer Programs

The state of the art in the graphical representation of concept lattices by computer programs is mainly represented by two tools. The first one is the NAVICON DECISION SUITE with the main programs ANACONDA, TOSCANA, and CERNATO from NAVICON AG (Frankfurt). Its extended Java-version TOSCANAJ contains the program ELBA for the construction of conceptual scales, which are used in the main program TOSCANAJ for the generation of nested line diagrams. For drawing transition diagrams as in Figure 6 and 7 the temporal component of the program SIENA can be employed; SIENA can also be used for the presentation of animations of conceptual time systems. For further information the reader is referred to [Bec95].

7   Connections to Other Temporal Theories

In this paper, it is impossible to mention all relevant connections to other temporal theories. Therefore, I describe here only the main relations between Temporal Concept Analysis and some of the most important temporal theories.

In contrast to the following theories, TCA has a general granularity tool allowing for a common conceptual notation for finite as well as for infinite temporal systems. The introduction of *actual objects* and their *time relation* in CTSOTs is a new approach to understand the relation between objects, space and time.

7.1   Classical Physics and Quantum Theory

In this subsection some basic aspects of classical physics and quantum theory are related to Temporal Concept Analysis. First, we discuss the roles of scales and objects.

The great success of classical physics is based on the Euclidean space together with its differentiable structure. The points in that space are used as "places for objects" showing that the Euclidean space is employed as a scale for the embedding of objects – but that is not made explicit by general theoretical notions for objects and scales. Clearly, the classical scale types on the real numbers [LKST90, Wol95b] are well-known also to many physicists; but a general investigation of not only infinite but also finite scales with the purpose of developing a physical granularity theory for combining the discrete measurements with the continuous theory in a theoretical way is not known to the author.

The "space occupied by an object" is described as a subset of the Euclidean space $\mathbb{R}^3$ and the "time of an event" as a subset of the time axis $\mathbb{R}$ – but such a granularity structure causes problems. Indeed, Einstein mentioned some of these problems in his "granularity remark" in the 1905 – paper introducing the theory of special relativity [Ein05], Footnote on page 893 (translated by the author):

> The inaccuracy which lies in the concept of simultaneity of two events at (about) the same place and which has to be bridged also by an abstraction, shall not be discussed here.

I believe that a theory (and not only a well-developed practice) of granularity in physics could lead to a better understanding of many problems related to the meaning of limits (like velocities and energies), and to the understanding of inaccuracy and Heisenberg's uncertainty relation. The *problem of time* as discussed in [But99] and [ButIsh99], page 147, could be embedded into a general granularity theory for objects in space and time. Recent investigations in

TCA might be a starting point for such a development: the introduction of a granularity not only for space and time but also for the objects, as for example, persons as members of a family, led the author to a mathematical definition of wave packets, yielding definitions of particles and waves as special examples [Wol04a]. These definitions cover the continuous as well as the discrete waves, as for example, electro-magnetic waves as well as waves of influenza represented in discrete data.

7.2   Mathematical System Theory, Turing Machines, and Automata Theory

The formalization of the ideas of Bertalanffy [Ber69] led to Mathematical System Theory (cf. Kalman [KFA69], Mesarovic [MeTa75], Lin [Lin99]). As pointed out by Zadeh [Zad64] Mathematical System Theory did not find a satisfactory notion of state, and Lin [Lin99] writes that there is no generally accepted notion of a system. The recent developments in TCA might be a first step towards a better understanding of *states* and *systems*.

The introduction of computers was accompanied by the development of a theory of computation initiated by Post [Pos36] and Turing [Tur36]. Their *computing machines* are now known as *Turing machines*. It was shown recently by Wolff and Yameogo [WoYa05] that any Turing machine can be represented by a suitable "Turing CTSOT" such that for each possible input of the Turing machine the uniquely determined sequence of computation steps is represented as the life track of the input word in the state space of that Turing CTSOT. The conceptual role of the instructions of the Turing machine is understood as a set of background implications of the derived context of the Turing CTSOT.

The investigation of *computing machines* led to the development of automata theory which is mainly concerned with finite automata as described for example by Arbib [Arb70] and Eilenberg [Eil74]. The *continuous* time of physics was replaced by a *discrete* time, the set of *states* was introduced axiomatically as a set of things without an explicit definition in terms of time, but these *states* can be connected by *labelled transitions*. Finite paths from an *initial state* to some *final* (or *terminal*) *states* are used to describe *runs* of the *machine*. Automata can be described by CTSOTs such that the states, the transitions, and the successful paths of an automaton are represented by the states, the transitions, and the life tracks of a suitable CTSOT. For further information the reader is referred to [Wol02b].

7.3   Temporal Logic and Conceptual Temporal Logic

Temporal Logic in the sense of Gabbay [GHR94] and van Benthem [vBe95] is developed as a general logic for temporal phenomena. After having introduced Temporal Concept Analysis as a theory for handling temporal phenomena on the basis of mathematically defined conceptual time systems, time granules, states, situations, and life tracks I could discuss at the 9th International Symposium on Temporal Representation and Reasoning (TIME'02) in July 2002 in Manchester with Dov Gabbay the relations between Temporal Logic and Temporal Concept Analysis. His central idea of a "branching time" is described in [GHR94], page 86:

> We should, therefore, pay special attention to discrete future branching past-linear flows of time.

This "tree structure" of time might be extended to a more general framework as for example to the temporal scales in TCA (which may be chosen as trees in their usual lattice representation). But the basic idea of a "branching time" is independent of the time scale since it is based on the idea of branching possible future life tracks; those life tracks can be easily represented in a CTSOT with an arbitrarily given time scale.

The main difference between Temporal Logic and TCA seems to be that Temporal Logic is designed as a "logic" for arbitrary temporal models while TCA yields a general description of temporal models. It seems to be desirable to combine the classical Temporal Logic with TCA towards a "Conceptual Temporal Logic" which, for instance, could include a tool for the representation of relational logic using for example power context families or relational conceptual time systems. Then the CTSOTs (or more general temporal structures) could be models in that Conceptual Temporal Logic having general logical tools for spatio-temporal granularity reasoning in those conceptual structures.

8   Conclusion and Future Research

Temporal Concept Analysis is based on mathematically defined notions of conceptual time systems, states, situations, transitions, and life tracks of objects such that continuous and discrete temporal phenomena can be described in the same conceptual framework.

Future research in TCA should develop not only the just mentioned Conceptual Temporal Logic but also the temporal aspects of relational logic. Furthermore, the connections to other temporal theories should be clarified. Especially, applications in physics might yield progress in understanding temporal phenomena as for example further discussions about particles and waves including interference of waves. The formal representation of granularity might be a powerful tool for understanding Heisenberg's uncertainty relation in a more general framework. The *problem of time* in quantum theory might become better understandable with the tools of TCA too.

References

[Arb70]   Arbib, M.A.: Theory of Abstract Automata. Prentice Hall, Englewood Cliffs, N.J., 1970.

[AriM95]   Aristoteles: Philosophische Schriften in sechs Bänden. Felix Meiner Verlag Hamburg 1995.

[AriBar95]   Aristotle: The complete works of Aristotle. Vol. I, II. Edited by J. Barnes. Bollingen Series; 71:2, Princeton University Press, 1995.

[Bec95]   Becker, P.: Multi-dimensional Representation of Conceptual Hierarchies. In: G. Stumme and G. Mineau (eds.): *Proceedings of the 9th International Conference on Conceptual Structures*, pp.33-46, Supplementary Proceedings ICCS, Department of Computer Science, University Laval, 2001.

[Ber69]   Bertalanffy, L.v.; General System Theory. George Braziller, New York, 1969.

[Bir67]   Birkhoff, G.: Lattice theory, 3rd ed., Amer.Math.Soc., Providence 1967.

[But99]   Butterfield, J. (ed): The Arguments of Time, Oxford University Press, 1999.

[ButIsh99] Butterfield, J., C.J. Isham: On the Emergence of Time in Quantum Gravity. In Butterfield, J. (ed.): The Arguments of Time, Oxford University Press, 1999.

[Cast98]   Castellani, E.(ed.): Interpreting Bodies: Classical and Quantum Objects in Modern Physics. Princeton University Press 1998.

[Eil74]   Eilenberg, S.: Automata, Languages, and Machines. Vol. A. Academic Press 1974.

[Ein05]   Einstein, A.: Zur Elektrodynamik bewegter Körper. Annalen der Physik 17 (1905): 891-921.

[Ein07]   Einstein, A.: Über das Relativitätsprinzip und die aus demselben gezogenen Folgerungen. In: *Jahrbuch der Radioaktivität und Elektronik 4* (1907) : 411-462.

[Ein89]   Einstein, A.: *The collected papers of Albert Einstein*. Vol. **2**: *The Swiss Years: Writings*, 1900-1909. Princeton University Press 1989.

[Ern82]   Erné, M.: *Einführung in die Ordnungstheorie*. B.I.Wissenschaftsverlag, Mannheim 1982.

[GHR94]   Gabbay, D.M., I. Hodkinson, M. Reynolds: *Temporal Logic – Mathematical Foundations and Computational Aspects*. Vol.1, Clarendon Press Oxford 1994.

[GSW86]   Ganter, B., J.Stahl, R.Wille: Conceptual measurement and many-valued contexts. In: W.Gaul, M.Schader (eds.): *Classification as a tool of research*. North-Holland, Amsterdam 1986, 169-176.

[GaWi89]   Ganter, B., R. Wille: Conceptual Scaling. In: F.Roberts (ed.) *Applications of combinatorics and graph theory to the biological and social sciences*, 139-167. Springer, New York, 1989.

[GaWi99]   Ganter, B., R. Wille: *Formal Concept Analysis: mathematical foundations*. (translated from the German by Cornelia Franzke) Springer, Berlin-Heidelberg 1999.

[Got90]   Gottwald, S. (ed.): *Lexikon bedeutender Mathematiker*. Bibliographisches Institut 1990.

[Haw88]   Hawking, S.: *A Brief History of Time: From the Big Bang to Black Holes*. Bantam Books, New York 1988.

[HaPe96]   Hawking, S., Penrose, R.: *The Nature of Space and Time*. Princeton University Press, 1996.

[Ish02]   Isham, C.J.: *Time and Modern Physics*. In: Ridderbos, K. (ed.): *Time*. Cambridge University Press 2002, 6-26.

[KFA69]   Kalman, R.E., Falb, P.L., Arbib, M.A.: *Topics in Mathematical System Theory*. McGraw-Hill Book Company, New York, 1969.

[Kan1781]   Kant, I.: *Kritik der reinen Vernunft.* In: Weischedel, W. (ed.): *Immanuel Kant – Werke in sechs Bänden.* Band II, Insel Verlag, Wiesbaden 1956 (first edition 1781).

[KSVW94]   Kollewe, W., M.Skorsky, F.Vogt, R.Wille: TOSCANA – ein Werkzeug zur begrifflichen Analyse und Erkundung von Daten.In: R.Wille und M.Zickwolff (Hrsg.), *Begriffliche Wissensverarbeitung – Grundfragen und Aufgaben.* B.I.-Wissenschaftsverlag, Mannheim 1994, 267-288.

[Kr98]   Krabs, W.: Dynamische Systeme: Steuerbarkeit und chaotisches Verhalten. B.G.Teubner Stuttgart, Leipzig, 1998.

[Lin99]   Lin, Y.: *General Systems Theory: A Mathematical Approach.* Kluwer Academic/ Plenum Publishers, New York, 1999.

[LKST90]   Luce, R.D., D.H.Krantz, P.Suppes, A.Tversky: *Foundations of Measurement,* Vol. 3, Akademic Press, San Diego, 1990.

[MeTa75]   Mesarovic, M.D., Y. Takahara: *General Systems Theory: Mathematical Foundations.* Academic Press, London, 1975.

[Paw91]   Pawlak, Z.: *Rough Sets: Theoretical Aspects of Reasoning About Data.* Kluwer Academic Publishers, 1991.

[Pos36]   Post, E.L.: Finite combinatory processes – Formulation. J.Symbolic Logic 1 (1936)103-105.

[Spa90]   Spangenberg, N.: *Familienkonflikte eßgestörter Patientinnen: Eine empirische Untersuchung mit Hilfe der Repertory Grid-Technik.* Habilitationsschrift am FB Humanmedizin der Justus-Liebig-Universität Gießen, 1990.

[SpWo91]   Spangenberg, N., K.E. Wolff: Comparison of Biplot Analysis and Formal Concept Analysis in the case of a Repertory Grid. In: *Classification, Data Analysis, and Knowledge Organization* (eds.: H.H. Bock, P. Ihm), Springer, Heidelberg 1991, 104-112.

[SpWo93]   Spangenberg, N., K.E. Wolff: Datenreduktion durch die Formale Begriffsanalyse von Repertory Grids. In: *Einführung in die Repertory Grid-Technik,* Band 2, Klinische Forschung und Praxis. (eds.: J.W. Scheer, A. Catina), Verlag Hans Huber, 1993, 38-54.

[Tur36]   Turing, A.M.: On computable numbers with an application to the Entscheidungsproblem. Proc. London Math. Soc.,2: 42, 230-265. A correction, ibid. 43, pp. 544-546, 1936.

[Tur36a]   Turing, A.M.: On computable numbers, with an application to the Entscheidungsproblem. http://www.abelard.org/turpap2/tp2-ie.asp#section-9

[vBe95]   van Benthem, J.: Temporal Logic. In: Gabbay, D.M., C.J. Hogger, J.A. Robinson: *Handbook of Logic in Artificial Intelligence and Logic Programming.* Vol. **4**, *Epistemic and Temporal Reasoning.* Clarendon Press, Oxford, 1995, 241-350.

[vHe72]   von Hentig, H.: *Magier oder Magister? Über die Einheit der Wissenschaft im Verständigungsprozess.* Klett-Verlag, Stuttgart 1972.

[Wil82]   Wille, R.: Restructuring lattice theory: an approach based on hierarchies of concepts. In: Rival, I. (ed.): *Ordered Sets.* Reidel, Dordrecht-Boston 1982, 445-470.

[Wil97a]   Wille, R.: Introduction to Formal Concept Analysis. In: G. Negrini (ed.): *Modelli e modellizzazione. Models and modelling.* Consiglio Nazionale delle Ricerche, Instituto di Studi sulli Ricerca e Documentatione Scientifica, Roma 1997, 39-51.

[Wil97b]   Wille, R.: Conceptual graphs and Formal Concept Analysis. In: D. Lucose, H. Delugach, M. Keeler, L. Searle, J.F. Sowa (eds.): *Conceptual structures: Fulfilling Peirce's dream.* LNAI **1257**. Springer, Heidelberg 1997, 290-303.

[Wil04]   Wille, R.: Dyadic Mathematics – Abstractions from Logical Thought. In: K. Denecke, M. Erné, S.L. Wismath (eds.): Galois Connections and Applications. Kluwer, Dordrecht 2004, 453-498.

[WoSt93]   Wolff, K.E., M. Stellwagen: Conceptual optimization in the production of chips. In: Janssen, J., Skiadas, C.H. (eds.) *Applied Stochastic Models and Data Analysis,* Vol. **2**, 1054-1064. World Scientific Publishing Co. Pte. Ltd. 1993.

[Wol94]   Wolff, K.E.: A first course in Formal Concept Analysis – How to understand line diagrams. In: Faulbaum, F. (ed.): SoftStat '93, *Advances in Statistical Software 4,* Gustav Fischer Verlag, Stuttgart 1994, 429-438.

[Wol95a]   Wolff, K.E.: Conceptual Quality Control in Chemical Distillation Columns. In: J. Janssen, S. McClean (eds.), *Applied Stochastic Models and Data Analysis.* University of Ulster 1995, 652-654.

[Wol95b]   Wolff, K.E.: Anwendungen der Formalen Begriffsanalyse in der Meßtheorie und der Meßpraxis. In: H. Hofmann, D. Richter, Ch. Zeidler (eds.) : *Informationsgewinnung aus Meßdaten.* 6. Arbeitsgespräch der Fachgruppe Physik/Informatik/Informationstechnik. 122. PTB-Seminar, Physikalisch-Technische Bundesanstalt, Berlin 1995.

[Wol00a]   Wolff, K.E.: Concepts, States, and Systems. In: Dubois, D.M. (ed.): *Computing Anticipatory Systems.* CASYS'99 – Third International Conference, Liège, Belgium, 1999, American Institute of Physics, Conference Proceedings **517**, 2000, pp. 83-97.

[Wol00b]   Wolff, K.E.: Towards a Conceptual System Theory. In: B. Sanchez, N. Nada, A. Rashid, T. Arndt, M. Sanchez (eds.): *Proceedings of the World Multiconference on Systemics, Cybernetics and Informatics, SCI 2000, Vol. II: Information Systems Development, International Institute of Informatics and Systemics,* 2000, 124-132.

[Wol00c]   Wolff, K.E.: A Conceptual View of Knowledge Bases in Rough Set Theory. In: Ziarko, W., Yao, Y. (eds.): *Rough Sets and Current Trends in Computing.* Second International Conference, RSCTC 2000, Banff, Canada, October 16-19, 2000, Revised Papers, 220-228.

[Wol01]   Wolff, K.E.: Temporal Concept Analysis. In: E. Mephu Nguifo & al. (eds.): *ICCS-2001 International Workshop on Concept Lattices-Based Theory, Methods and Tools for Knowledge Discovery in Databases,* Stanford University, Palo Alto (CA), 91-107.

[Wol02a]   Wolff, K.E.: Transitions in Conceptual Time Systems. In: D.M.Dubois (ed.): *International Journal of Computing Anticipatory Systems,* vol. **11**, CHAOS 2002, p.398-412.

[Wol02b]   Wolff, K.E.: Interpretation of Automata in Temporal Concept Analysis. In: U. Priss, D. Corbett, G. Angelova (eds.): *Integration and Interfaces.* Tenth International Conference on Conceptual Structures. LNAI **2393**, Springer 2002, 341-353.

[Wol02c]   Wolff, K.E.: Concepts in Fuzzy Scaling Theory: Order and Granularity. 7th European Congress on Intelligent Techniques and Soft Computing, Aachen 1999. *Fuzzy Sets and Systems* **132**, 2002, 63-75.

[WoYa03] Wolff, K.E., W. Yameogo: Time Dimension, Objects, and Life Tracks - A Conceptual Analysis. In: A. de Moor, W. Lex, B. Ganter (eds.): *Conceptual structures for knowledge creation and communication*. LNAI **2746**. Springer, Heidelberg 2003, 188-200.

[Wol04a] Wolff, K.E.: 'Particles' and 'Waves' as Understood by Temporal Concept Analysis. In: K.E. Wolff, H.D. Pfeiffer, H.S. Delugach (eds.): Conceptual Structures at Work. 12th International Conference on Conceptual Structures, ICCS 2004. Huntsville, AL, USA, July 2004. Proceedings. Springer Lecture Notes in Artificial Intelligence, LNAI 3127, Springer-Verlag, Berlin Heidelberg 2004, 126-141.

[WoYa05] Wolff, K.E., W. Yameogo: Turing Machine Representation in Temporal Concept Analysis. To appear in the Proceedings of the 3rd International Conference on Formal Concept Analysis 2005.

[Yam03] Yameogo, W.: Time Conceptual Foundations of Programming. Master Thesis. Department of Computer Science at Darmstadt University of Applied Sciences, 2003.

[Zad64] Zadeh, L.A.: The Concept of State in System Theory. In: M.D. Mesarovic: *Views on General Systems Theory*. John Wiley & Sons, New York 1964, 39-50.

[Zad65] Zadeh, L.A.: Fuzzy sets. *Information and Control 8*, 1965, 338 – 353.

[Zad75] Zadeh, L.A.: The concept of a linguistic variable and its application to approximate reasoning. Part I: *Inf. Science 8*,199-249; Part II: *Inf. Science 8*, 301-357; Part III: *Inf. Science 9*, 43-80, 1975.

Linguistic Applications of Formal Concept Analysis

Uta Priss

School of Computing, Napier University
u.priss@napier.ac.uk

1 Introduction

Formal concept analysis as a methodology of data analysis and knowledge representation has potential to be applied to a variety of linguistic problems. First, linguistic
applications often involve the identification and analysis of features, such as phonemes
or syntactical or grammatical markers. Formal concept analysis can be used to record
and analyze such features. The line diagrams of concept lattices can be used for communication among linguists about such features (see section 2).

Second, modeling and storage of lexical information is becoming increasingly important for natural language processing tasks. This causes a growing need for detailed
lexical databases, which should preferably be automatically constructed. Section 3 describes the role that formal concept analysis can play in the automated or semi-automated construction of lexical databases from corpora.

Third, lexical databases usually contain hierarchical components, such as hyponymy
or type hierarchies. Because formal concept lattices are a natural representation of hierarchies and classifications, lexical databases can often be represented or analyzed using
formal concept analysis. This is described in section 4.

It should be remarked that because this paper appears in a collection volume of papers on formal concept analysis, the underlying notions, such as formal concept, formal
object and attribute, and lattice, are not further explained in this paper. The reader is
referred to Ganter & Wille (1999) for detailed information on formal concept analysis.

2 Analyzing Linguistic Features with Formal Concept Analysis

Linguists often characterize datasets using distinct features, such as semantic components, phonemes or syntactical or grammatical markers, which can easily be interpreted
using formal concept analysis. An example is Kipke & Wille's (1987) paper which applies formal concept analysis to semantic fields. In that paper, data from an analysis of
the German and English words for the semantic field of "bodies of water" is modeled
using different types of concept lattices.

A different but also feature-based analysis is conducted by Großkopf (1996) who
analyzes verb paradigms of the German language and by Großkopf & Harras (1999)
who analyze speech act semantics of German verbs. Großkopf & Harras use formal
contexts which have verbs as objects and semantic features that characterize the speech
acts of the verbs as attributes. Figure 1 shows an English example which is equivalent
to of one of Großkopf and Harras's German examples. The lattice shows the attitudes

B. Ganter et al. (Eds.): Formal Concept Analysis, LNAI 3626, pp. 149–160, 2005.

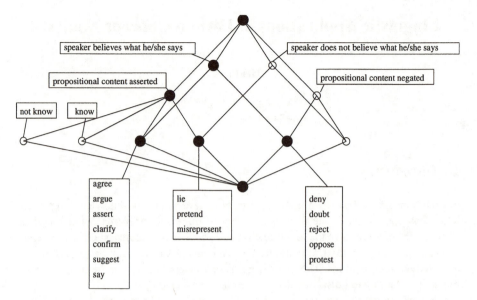

Fig. 1. Verb speech acts in analogy to Großkopf & Harras (1999)

of speakers for some verbs of assertion. For verbs, such as "agree" and "lie", a certain propositional content is asserted. But in one case (assert) the speaker believes the content, in the other case (lie) the speaker does not believe the content. Both cases are different from verbs, such as "deny", for which the speaker neither believes the content nor pretends to believe it.

The lattice in the example is a conceptual scale which can be applied to different sets of verbs. In the example, some of the attributes ("know" and "not know") are not relevant for the chosen set of verbs. But these attributes are relevant when comparing verbs of knowledge acquisition (e.g., "learn") with verbs of assertion. Großkopf and Harras built a Toscana system which facilitates browsing through a large selection of German verbs using different conceptual scales. A researcher can interactively analyze the semantic structures of sets of verbs using such a system. Großkopf and Harras's paper contains further examples and some linguistic results achieved with these methods.

It is somewhat surprising that formal concept analysis is not used more frequently for the analysis of linguistic features. It may be that linguists are still not widely aware of these methods. There has been a recent interest in formal concept analysis in areas of formal linguistics. In modern theories of grammar, such as head-driven phrase structure grammar, HPSG (Pollard & Sag, 1994), lexical knowledge is organized in hierarchies of feature structures of classes of lexical entries. According to Sporleder (2002) and Osswald & Petersen (2002), formal concept analysis may be a suitable method for deriving such hierarchies of lexical information automatically. But so far, their research is more theoretical than practical. No detailed examples of applications have yet been published in this area.

3   The Use of Formal Concept Analysis in Natural Language Processing Tasks

This section focuses on an approach developed by Basili et al. (1997). Because this approach has significance for many applications both in linguistics and in AI, we cover this paper in greater detail. We first explain the underlying problems of natural language processing before describing the actual implementation in section 3.3.

3.1   The Limits of Automated Word Sense Disambiguation

It is a central task for any natural language processing application to disambiguate the words encountered in a text and to determine to which lexical entries they belong. Normally this involves both the use of a lexicon as well as the use of a syntactic parser. But there is an inverse relationship between lexical and syntactic processing: a simple and shallow lexicon requires a detailed syntactic parser whereas a detailed lexicon requires only a shallow parser. The tendency in recent linguistic research has been to focus on improvements in the representation and detail of lexica. But there may be a limit to how much detail can be represented in a lexicon. For example, Basili et al. (1998) claim that "in any randomly chosen newspaper paragraph, each sentence will be likely to have an extended sense of at least one word, usually a verb, in the sense of a use that breaks conventional preferences and which might be considered extended or metaphorical use, and quite likely not in a standard lexicon."

The following examples from Pustejovsky (1991) illustrate the problem: in the sentence "the woman finished her beer", the verb "finish" is synonymous to "stopped drinking" whereas in "the man finished his cigarette" it is synonymous to "extinguished". From a common sense viewpoint, it is obvious that the shared meaning between both sentences is the act of finishing an activity. But a detailed lexical representation should include the logical inferences which are different. A natural language processing application might need to 'know' that an appropriate activity after "finishing" is "ordering a new one" in the case of beer, and "lighting another one" in the case of cigarettes. This information could either be stored with "cigarettes" and "beer" or with "finishing". It could also be stored with more general terms ("alcoholic beverage") or in rule format. This example shows that the issues are complicated.

A second example, is "she looked through the window" as opposed to "he painted the window". In the first case, "the window" stands for "the glass of the window". In the second case, it stands for "the frame of the window". A natural language processing application might need to 'know' that glass is transparent and more likely to break and that frames can be made of a variety of materials. Essentially, a lexical database might need to store detailed descriptions of windows. But because many different types of windows exist, this would be an enormous amount of information.

Priss (2002) asserts that the types of problems outlined in these examples may require more than linguistic knowledge because even humans do not store this information in a linguistic format. Humans parse such linguistic ambiguities by utilizing common sense non-linguistic knowledge about the world. Ultimately, automated natural language processing may need to be combined with other methods, such as perceptive, robotic interfaces or associative interfaces which facilitate better co-operation between humans and computers.

3.2   Lexical Databases and Ontological Knowledge

There is no clear dividing line between lexical databases and AI ontologies. Traditionally, lexical databases are often less formalized than AI ontologies. For example, WordNet (Fellbaum, 1998) is less formalized than the AI ontology, CYC (2003). But contemporary linguistic representations of lexical knowledge can be as formalized and logic-based as formal ontologies.

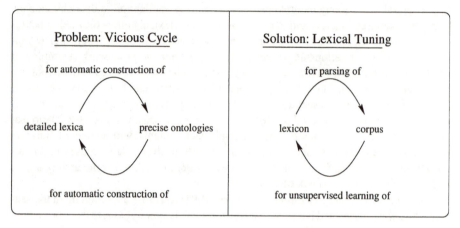

Fig. 2. A vicious cycle and a possible solution

As depicted in the left half of figure 2, lexical databases and AI ontologies form a vicious cycle because the construction of detailed lexica requires precise ontological information, and vice versa. Linguistic structures convey information about the knowledge domains which are expressed in a language. For example, Scottish people have more words for rain than people who live in dryer climates. Other often quoted examples are kinship terms, color terms (not just cultural differences but also the number of color terms used, for example, in the paint industry) and food vocabulary. Thus lexica can serve as a source of information for AI ontology engineers.

Ontologies are useful for natural language processing, because differences in meaning correspond to syntactic differences, such as whether a verb is used transitively or intransitively. Before processing text, it is helpful, to first construct a model of the underlying semantic structures and to describe an underlying AI ontology. Because of the large size of lexica and because of the ever changing nature of language, both the construction of lexica and of AI ontologies should be semi-automatic to achieve sufficient depth. Large corpora, such as the World Wide Web, provide an almost unlimited resource for lexical input. But the vicious cycle holds for them as well: sophisticated parsing presupposes already existing sophisticated lexical databases and vice versa. As Basili et al. (2000) state: words can be more easily disambiguated if one knows their selectional restrictions, but selectional restrictions can be more easily identified if one knows the disambiguated meanings of their words.

3.3   Bootstrapping of Linguistic Information from Large Corpora

Fortunately, vicious cycles can sometimes be dissolved by bootstrapping. In this case, Basili et al. (1997) claim that a lexicon of any quality can be used as a starting point. As depicted in the right half of figure 2, the lexicon is improved by corpus-driven unsupervised learning from a corpus. The improved lexicon is then re-applied to the parsing of the corpus which improves the learning from the corpus and so forth. This form of bootstrapping is used for "lexical tuning" (Basili et al., 1997), which is the process of refining lexical entries to adapt them to new domains or purposes. Basili's approach, which is described in the remainder of this section, is based on formal concept analysis. Formal concept analysis provides an excellent tool for lexical tuning because the duality between formal objects and attributes is used to represent linguistic dualities, such as verb frames and noun phrases or corpus-derived sentence parts and rule-based lexical structures, which drive the bootstrapping process.

Verbs are more difficult to disambiguate than nouns or adjectives. Verb subcategorization frames are the result of classifying verbs based on their argument structures. For example, the verb "multiply" usually has a mathematical meaning (e.g., "multiply 3 by 4"), if it occurs with a direct object, and a biological meaning (e.g., "the bacteria multiplied"), if it occurs without a direct object. The prepositional modifiers and syntactic constituents that co-occur with a verb indicate which subcategorization frame is used. The frames often correspond to different senses as recorded in a lexicon. A verb in a corpus can often be matched to its sense by comparing its subcategorization frame with the frames in the lexicon entry for that verb.

Basili et al. (1997) describe the following machine-learning method for extracting verb subcategorization frames from a corpus. After parsing the corpus, a formal context is derived for each verb. The objects of the formal context are all phrases that contain the verb. The attributes are the arguments of the verbs, such as direct or indirect objects and prepositions. Phrases that have the same argument structure are clustered in the extension of an object concept.

At this stage the concept lattices only represent possible subcategorization frames. Natural language processing as used for the parsing and for the automatic identification of the verb argument structures from the corpus does not yield error-free results. Furthermore because language is often ambiguous, there can never be absolutely perfect results. Thus after generating the lattices, it still needs to be identified which concepts in each lattice represent the senses of the verb which are most acceptable for the corpus.

Basili et al. describe methods to assign weights to the nodes of the concept lattices of each verb, which facilitate the selection of the most relevant subcategorization frames. These weights can be trained using a training subset of a corpus. Probabilistic models for prepositions can also be derived that correspond to the specific domain of the corpus. These can then be incorporated into the calculations. Using these methods, the most probable verb frames can be selected from the context.

For improved precision, Basili at al. point out that instead of automatic extraction, a human lexicographer could also use the lattice for each verb as a decision tool in manual lexicon construction. Since estimates state that constructing a lexical entry (in English) by hand takes at least one hour, using the automatically derived lattices for each verb as a guideline can be a significant time-saver.

3.4   Possible Applications for Ontology Engineering

Basili at al. (1997) experimentally evaluate their approach for the purpose of lexical tuning and conclude that it improves on other techniques. Their approach has subsequentially been successfully implemented in a variety of projects of lexical tuning (for example, Basili et al. (1998)). We believe that this approach is very promising and can have many more applications. For example, conceptual graphs (Sowa, 1984) are very similar to verb subcategorization frames. Concept lattices can be automatically derived, that have the core concepts from a set of conceptual graphs as objects and the relations as attributes. These can be used to analyze the text from which the conceptual graphs are derived. More importantly, they can be used to design concept hierarchies based on the structure of the arguments, which can be used by ontology engineers.

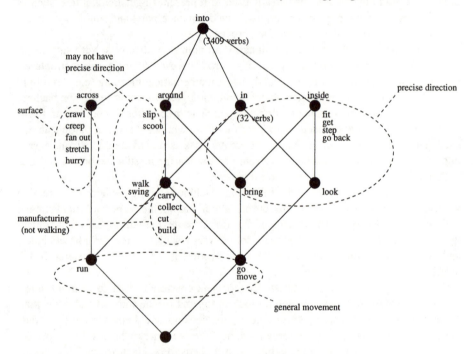

Fig. 3. Clustering of verbs from a corpus

Figure 3 shows an example, which is derived from the semantically tagged Brown corpus that is available for WordNet 1.6. The underlying algorithm is even simpler than Basili's approach: instead of determining the full argument structure of the phrases, only the first preposition that follows a verb is considered. While Basili's approach constructs a separate lattice for each verb, our algorithm is applied to all verbs from a corpus. The attributes of the lattice in this example are the prepositions "into", "across", "around", "in" and "inside". The objects of the lattice are all verbs in a subset of the corpus which are at least once followed by "into". Even this basic algorithm clusters the verbs in potentially meaningful groups, which are indicated by the circles. The circles are not meant to be automatically derivable. Instead they are open to human interpretation of the data.

It is conceivable to apply this algorithm to the complete Brown corpus (or other corpora) and to obtain an event hierarchy that can be useful for ontology engineering. Ontologies often focus on noun hierarchies and overlook hierarchies based on verbs, event structures or argument structures. A promising exception is Basili et al.'s (2002) paper which develops event hierarchies for information extraction and multilingual document analysis.

Another approach is presented by Petersen (2001). She uses formal concept analysis for the automatic acquisition of lexical knowledge from unstructured data. She reports on an application of computing a hierarchy of derivational information for English and German lemmas from the lexical database CELEX. Osswald & Petersen (2002) describe why formal concept analysis is suitable for automatic induction of classifications from linguistic data. A similar approach is also described by Sporleder (2002).

4   Representing Lexical Databases

Apart from bootstrapping, there are other possibilities to avoid the problem that common sense knowledge may not be completely representable in linguistic structures. Instead of attempting to represent as much information as possible in detailed logical formalisms, it may be sufficient to represent some information in an easily human-readable format, which can then be browsed through and explored by human users. Thus humans and machines co-operate in information processing tasks.

With respect to linguistic applications, an important but also challenging task is to construct interfaces for lexical databases. As mentioned before, the structures in lexical databases are often not very different from AI ontologies, such as CYC. For example, WordNet contains a noun hierarchy, which is similar in structure to taxonomies in ontologies, to classification systems or to object-oriented type hierarchies. Thus representations of hierarchies in lexical databases have similar applications as AI ontologies and classifications. They can be used to browse through collections of documents in information retrieval, to visualize relationships in textual information, to aid in the structuring and classification of scientific knowledge and they can serve as an interlingua.

The following sections describe how to formalize lexical databases in terms of formal concept analysis and how to use the formalizations for an analysis of semantic relations and for comparing and merging of lexical databases.

4.1   Formalizing Lexical Databases, Thesauri or Ontologies

Roget's Thesaurus and the lexical database, WordNet, have both been formalized with formal concept analysis methods. Roget's Thesaurus contains a six-level classification. At the lowest level, words are grouped that are either synonymous or closely related according to some semantic field, such as animals or food items. Because of polysemy, many words occur multiple times in the thesaurus. Thus one could construct a concept lattice using words as objects and their thesaurus classes as attributes to explore polysemy. But a lattice of the whole thesaurus would be far too large to be visualized. Therefore, Wille (1993) describes a method of constructing smaller, so-called "neighborhood" lattices. The semantic neighborhood of a word consists of all words that share some meanings with the word, that means all words which co-occur with the original word in at least one bottom-level class of the thesaurus. The set of objects of a neighborhood lattice consists of such a semantic neighborhood of a word. The attributes are either all bottom-level thesaurus classes of the original word, or all bottom-level thesaurus classes of all the words in the neighborhood. The choice depends on whether a larger or a smaller semantic environment is intended.

As an example, figure 4 shows the neighborhood lattice of "over" according to Old (1996). The lattice shows that there are two separate clusters of meanings of "over": the temporal/completion/direction senses ("it is over") are distinguished from the other senses. This is visible in the lattice because the six nodes in the right only share top and bottom with the rest of the lattice. The left side of the lattice shows that "over" can be used at different levels of intensity: "addition", "excess", "superiority" and "distance, past". Other similar examples can be found in Sedelow & Sedelow (1993).

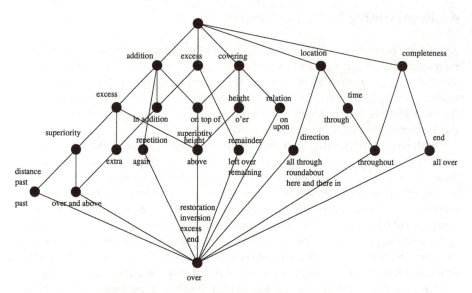

Fig. 4. Neighborhood lattice of "over" (Old, 1996)

In contrast to Roget's Thesaurus, the synonym sets in WordNet occur at all levels of the hierarchy not just at the bottom. Furthermore the hyponymy hierarchy is only one relational structure in WordNet. Not all of the other relations in WordNet are hierarchical. Although the part-whole relation is hierarchical, it is not necessarily meaningful to embed it into a lattice because intersections of parts may not be meaningful. For example, in a "substance-of" relation, ketchup and cake both contain sugar and salt, but is "sugar and salt" a meaningful grouping? For these reasons, WordNet requires a slightly different modeling than Roget's Thesaurus. Priss (1998) develops relational concept analysis as a means for modeling WordNet and similar lexical databases. One advantage of this approach is that semantic relations can be implemented using bases, that means a reduced set of relation instances from which the complete relation can be derived (Priss, 1999).

4.2 Analyzing Semantic Relations in Lexical Databases

The formalization of WordNet and Roget's Thesaurus as described in the previous section, can serve as a basis for a linguistic, cognitive or anthropological analysis of the structures and the knowledge that is encoded in such lexical databases. Priss (1996) uses relational concept analysis to analyze meronymy (part-whole) relations in Word-Net. Classifications of types of meronymy that are manually derived based on semantic analysis are often fuzzy and never agreed upon among different researchers. Priss shows that relational concept analysis facilitates the derivation of a classification of meronymy that is based on an entirely structural analysis. This classification is less fuzzy but still retains a significant amount of the information contained in manually derived classifications.

Another method of analyzing WordNet using formal concept analysis, identifies "facets" in the noun hierarchy. Facets are regular structures, such as "regular polysemy" (Apresjan, 1973). Figure 5 shows an example of family relationships in WordNet, which are arranged into facets (or scales) in figure 6.

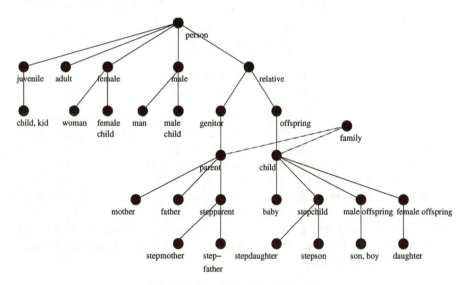

Fig. 5. Family relationships in WordNet

4.3 Comparing or Merging Lexical Databases

Once lexical databases are formalized as concept lattices, the lattices can serve as an interlingua, for creating multilingual databases or to identify lexical gaps among different languages. This idea was first mentioned in Kipke & Wille's (1987) concept lattices of the semantic fields of "bodies of water" in English and German. The lower half of the example in figure 7, which is taken from Old & Priss (2001), shows separate concept lattices for English and German words for "building". The main difference between English and German is that in English "house" only applies to small residential buildings

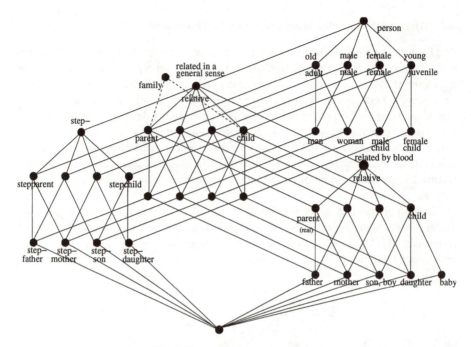

Fig. 6. Facets of family relationships

(denoted by letter "H"), whereas in German even small office buildings (denoted by let-ter "O") and larger residential buildings can be called "Haus". Only factories would not normally be called "Haus" in German. The lattice in the top of the figure constitutes an information channel in the sense of Barwise & Seligman (1997) between the German and the English concept lattice. In the process of manually creating such a channel a linguist must identify the attributes, which essentially describe the difference between the word use in the different languages. The concept lattices help to identify the rele-vant differences. A similar (but more automated) approach is implemented in Janssen's (2002) SIMuLLDA tool, which is a multilingual lexical database application that uses concept lattices as an interlingua.

5   Conclusion

This paper argues that there are many possibilities to use formal concept analysis for lin-guistic applications. So far in linguistics, formal concept analysis has been mainly used for the semi-automated construction of lexical databases and for analyses of seman-tic relations and lexical databases. Because of the close relationship between lexical databases and ontologies, any of these applications has relevance both for linguistics and AI. Hopefully this paper might stimulate some more research into these applica-tions. Specifically Basili's approach described in section 3.3 has a significant potential for further work.

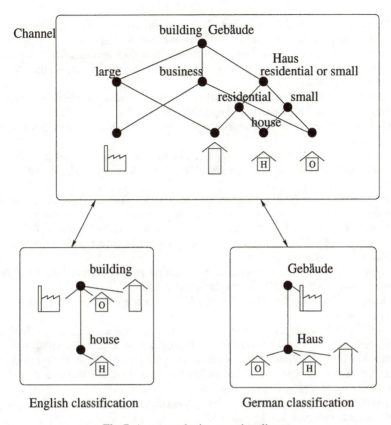

Fig. 7. A concept lattice as an interlingua

References

1. Apresjan, J. (1973). *Regular Polysemy.* Linguistics, 142.
2. Barwise, J.; Seligman, J. (1997). *Information Flow. The Logic of Distributed Systems.* Cambridge University Press.
3. Basili, R.; Pazienza, M.; Vindigni, M. (1997). *Corpus-driven unsupervised learning of verb subcategorization frames.* AI*IA-97.
4. Basili, R.; Catizone, R.; Padro, L.; Pazienza, M. T.; Rigau, G.; Setzer, A.; Webb, N.; Zanzotto, F. (2002). *Knowledge-Based Multilingual Document Analysis.* In: Proceedings of SemaNet02.
5. Basili, R.; Catizone, R.; Pazienza, M.; Stevenson, M.; Velardi, P.; Vindigni, M.; Wilks, Y. (1998). *An empirical approach to Lexical Tuning.* In: Proceedings of the Workshop "Adapting Lexical and Corpus Resources to Sublanguages and Applications", First International Conference on Language Resources and Evaluation, Granada, Spain.
6. CYC (2001). On-line available at http://www.cyc.com.
7. Fellbaum, C. (1998). *WordNet: An Electronic Lexical Database and Some of its Applications.* MIT press.
8. Ganter, B.; Wille, R. (1999). *Formal Concept Analysis.* Mathematical Foundations. Berlin-Heidelberg-New York: Springer, Berlin-Heidelberg.

9. Großkopf, A., (1996). *Formal concept analysis of verb paradigms in linguistics.* In: Diday; Lechevallier & Opitz (Eds.) Ordinal and Symbolic Data Analysis.

10. Großkopf, A.; Harras, G. (1999). *Begriffliche Erkundung semantischer Strukturen von Sprechaktverben.* In: Stumme & Wille (Eds.) Begriffliche Wissensverarbeitung: Methoden und Anwendungen.

11. Janssen, M. (2002). *SIMuLLDA. A Multilingual Lexical Database Application using a Structured Interlingua.* PhD Thesis, Universiteit Utrecht.

12. Kipke, U.; Wille, R. (1987). *Formale Begriffsanalyse erläutert an einem Wortfeld.* LDV–Forum, 5.

13. Old, L. J., (1996). *Synonymy and Word Equivalence.* In: Proceedings of the Midwest Artificial Intelligence and Cognitive Science Society Conference (MAICS96), Bloomington, IN.

14. Old, L. J.; Priss, U. (2001). *Metaphor and Information Flow.* In: Proceedings of the 12th Midwest Artificial Intelligence and Cognitive Science Conference, p. 99-104.

15. Osswald R.; Petersen, W. (2002). *Induction of Classifications from Linguistic Data.* In: Proceedings of the ECAI-Workshop on Advances in Formal Concept Analysis for Knowledge Discovery in Databases.

16. Petersen, W. (2001). *A Set-Theoretical Approach for the Induction of Inheritance Hierarchies.* Electronic Notes in Theoretical Computer Science 51.

17. Pollard, C.; Sag, I. (1994). *Head-Driven Phrase Structure Grammar.* CSLI Lecture Notes Series, Chicago.

18. Priss, U. (1996). *Classification of Meronymy by Methods of Relational Concept Analysis.* In: Proceedings of the 1996 Midwest Artificial Intelligence Conference, Bloomington, Indiana.

19. Priss, U. (1998). *The Formalization of WordNet by Methods of Relational Concept Analysis.* In: Fellbaum, Christiane (Ed.), WordNet: An Electronic Lexical Database and Some of its Applications, MIT press, 1998, p. 179-196.

20. Priss, U. (1999). *Efficient Implementation of Semantic Relations in Lexical Databases.* Computational Intelligence, Vol. 15, 1, p. 79-87.

21. Priss, U. (2002). *Associative and Formal Concepts.* In: Priss; Corbett; Angelova (eds.), Conceptual Structures: Integration and Interfaces. Proceedings of the 10th International Conference on Conceptual Structures, Springer Verlag, LNAI 2393, p. 354-368.

22. Pustejovsky, J. (1991). *The Generative Lexicon.* Computational Linguistics, 17, 4, p. 409-441.

23. Sedelow, S.; Sedelow, W. (1993). *The Concept concept.* Proceedings of the Fifth International Conference on Computing and Information, Sudbury, Ontario, Canada, p. 339-343.

24. Sowa, J. (1984). *Conceptual Structures: Information Processing in Mind and Machine.* Addison-Wesley, Reading, MA.

25. Sporleder, C. (2002). *A Galois Lattice based Approach to Lexical Inheritance Learning.* ECAI Workshop on ML and NLP for Ontology Engineering.

26. Wille, R. (1993). *The Formalization of Roget's International Thesaurus.* Unpublished manuscript.

Using Concept Lattices
for Text Retrieval and Mining

Claudio Carpineto and Giovanni Romano

Fondazione Ugo Bordoni
Via Baldassarre Castiglione 59, 00142, Rome, Italy
{carpineto,romano}@fub.it

Abstract. The potentials of formal concept analysis (FCA) for information retrieval (IR) have been highlighted by a number of research studies
since its inception. With the proliferation of small-size specialised text
databases available in electronic format and the advent of Web-based
graphical interfaces, FCA has then become even more appealing and
practical for searching text collections. The main advantage of FCA for
IR is the possibility of eliciting context, which may be used both to improve the retrieval of specific items from a text collection and to drive
the mining of its contents. In this paper, we will focus on the unique
features of FCA for building contextual IR applications as well as on its
most critical aspects. The development of a FCA-based application for
mining the web results returned by a major search engine is envisaged
as the next big challenge for the field.

1 Introduction

The (short) history of the applications of FCA to information retrieval can be
roughly split in three parts. In the 80's, some basic ideas were put forth – essentially that a concept can be seen as a query (the intent) with a set of retrieved
documents (the extent) and that neighbour concepts can be seen as minimal
query changes – and some preliminary study about the complexity of document
lattices (i.e., the concept lattices built from collections of documents) was performed, mostly by Robert Godin and his co-workers ([28], [24]).

In the 90's, FCA has been integrated with basic IR techniques to build more
comprehensive systems for information access. Concept lattices have been mainly
used as a support structure for interactive subject finding tasks, with some explorations of the possibilities of FCA for text mining. We have seen several running
prototypes, and some experimental evaluation comparing the performance of
FCA-based IR with that of conventional IR methods (e.g., [27], [11], [10]).

Over the last few years, the range of functionalities has been expanded to
include new tasks such as automatic text ranking and IR from semistructured
data (e.g., [13], [16]); at the same time, new IR domains have been investigated
including email messages, web documents, and file systems.

Although it might be argued that the impact made on mainstream IR has
not been dramatic, the interest in using concept lattices for IR has grown both
within and outside of the FCA community. On the other hand, there is nowadays

B. Ganter et al. (Eds.): Formal Concept Analysis, LNAI 3626, pp. 161–179, 2005.
© Springer-Verlag Berlin Heidelberg 2005

a much better awareness of the strengths and limitations of this technique for organizing and searching information. Furthermore, perhaps more importantly, we believe that the major changes faced by the information access world provide new unprecedented opportunities for FCA applications.

First of all, IR is no longer a boutique discipline for librarians, as the whole internet is being searched every day by billions of people around the world. Second, the availability of basic services for storing, networking, searching, and displaying information has led to a proliferation of specialized electronic databases for enterprises and individual users. Third, users may be interested in a variety of information searching tasks that go well beyond the capabilities of traditional IR systems dealing with the topic relevance task; examples include text data mining and several variants of the general topic relevance paradigm such as home page finding, question answering, IR from XML documents, news filtering, etc.

In the rest of the paper we will first discuss what makes FCA appealing to the development of more powerful search front ends. Next, we review how concept lattices have been used so far to improve specific traditional IR tasks or to handle new tasks that would be hardly dealt with by conventional systems. Then we will discuss the most difficult steps in the development of FCA-based IR applications, which may affect both the efficiency and the effectiveness of the overall system. Finally, we will argue for building a FCA-based system for visualizing web retrieval results on top of a major search engine.

2   Current Search Front Ends and FCA

The growth of the web has favoured the emergence of new search applications, usage patterns, data formats, and interaction paradigms. To cope with the new requirements, more advanced search interfaces are being developed that provide the user with a range of access methods.

For instance, the Digital Library of the Association for Computing Machinery (http://www.acm.org) has been recently equipped with a new front end that not only allows the user to browse the documents contained in each category (e.g., journals, magazines, transactons, proceedings) and to search the full text (or title, or abstract) of publications, but also to use a number of browsable views into the literature, e.g., by Authors, by the ACM Computing Classification System (CCS), by Subjects, by Technical Interest.

Using the currently available tools, a number of interesting search tasks can be easily accomplished. For instance, a view by Author allows the user to drill down to a page that might be called an "author's virtual bibliographic home page", listing all the works by that author known to the system, or, to take another example, a view by CCS may be used to look into a specific domain, with the user drilling down from more general categories to narrower subjects and then to specific topics, thus progressively reducing the set of complying results.

One disadvantage of this system, as well as of most currently available front ends, is that the user must use a specific search functionality for each task, with little or no possibility of combining the results.

An even greater limitation is that while the present range of functionalities seem to support pretty well the user for the case when she is interested in finding those documents which are described by certain terms or categories, the same tools may be of little help if the user wants to disclose the content of specific sections of the Digital Library or to mine the concepts contained in a set of articles which have been filtered out by using some criterion.

This is an instance of the dichotomy between information retrieval and data view on one side and text mining on the other side. The latter involves the discovery of previously unknown information [31], as opposed to finding the best element among many other valid pieces of information, and can be seen as a form of exploratory data analysis [32].

The use of FCA can effectively complement the existing search systems to address some of their main limitations. Basically, FCA exploits the interdocument similarity between documents to offer an automatically-built support structure (i.e., the document lattice) in which to place the information searching process. The document lattice can be used to improve basic individual search strategies, as well as to host multiple integrated search strategies.

Having at her disposal a range of methods (e.g., querying, navigation, combination of data views, thesaurus climbing, pruning by search constraints), the user can select those that best fit the goal of the search, her knowledge of domain, her skills and preferences, and the results of the past interaction with the system. The user may then form hybrid strategies to make the search more accurate or fast.

The features discussed are naturally supported by concept lattices. Other approaches either require some manual coding, or do not allow for multiple retrieval strategies, or integrate multiple strategies in a loose manner.

3   Search Functionalities Enhanced by FCA

In this section we will examine which search functionalities or which combinations of search functionalities may be improved through a concept lattice. Most of the examined functionalities can be used both for text retrieval and text mining.

3.1   Query Refinement

One of the most natural applications of concept lattices is query refinement, where the main objective is to recover from the null-output or the information overload problem.

This is not new, as some lattice representations were used in early IR [50] and even more recently [52] for refining queries containing Boolean operators. However, as these approaches typically rely on a Boolean lattice formalization of the query, the number of proposed refinements may grow too large even for a very limited number of terms and they may easily become semantically meaningless to the user.

These limitations can be overcome by using concept lattices. One example is the REFINER system [12].

In response to a Boolean query, REFINER builds and displays a portion of the concept lattice of the documents being searched which is centered around a query concept. Such a query concept is found by computing the set of documents that satisfy the query and then by determining the set of terms possessed by all previously found documents. At this point, the most general concept containing these terms is chosen as the query concept; if there are no concepts that contain all the terms specified in the query (i.e., there are no documents exactly matching the query), REFINER adds a virtual concept to the lattice, as if the query represented a new document.

The potentials of this approach have been confirmed in an experiment on the classical CISI test set – a bibliographical collection of 1460 information science documents described by a title and an abstract, which is available at $http : //www.dcs.gla.ac.uk/idom/ir_resources/test_collections/cisi/$ – showing that the effectiveness of information retrieval using REFINER was better than unrefined, conventional Boolean retrieval [12].

Concept lattices can be used also to refine queries expressed in natural language. The mapping of a query on to the lattice can be done by choosing the most general concept that contains all the query terms, similar to REFINER, or with some weaker criteria if such a concept coincides with the bottom of the lattice [54].

Once the query has been mapped on to a concept, the user may choose one of the neighbours of that concept, as in REFINER, or select one term from a list containing all the terms that are below that concept [36]. In the latter case, a substantial portion of the full concept lattice must be built.

3.2 Integrating Querying and Navigation

The effective integration of the query-based mode with the navigation paradigm has been the focus of much current research on information systems.

One typical choice is to maintain different retrieval methods in parallel (e.g., [39], [23]); in this case, the integrated system is, in practice, like a switch whereby the user may select either strategy. A tighter form of integration is achieved by cascading the two strategies, e.g., browsing prior to querying [42], or querying prior to browsing [38], or by having them coexist in the same search space ([1], [29]).

In these forms of integration the system may have to maintain several data structures possibly supporting different kinds of operations; when a single data structure is used consistency problems may arise.

Concept lattices take the hybrid searching paradigm one step further. As querying and navigation share the same data space and exchange their search results, they can be consistently integrated, without the need of mapping different representations on the part of the user. Furthermore, other search strategies such as thesaurus climbing, space pruning, and partial views, can be easily combined in the same framework.

To characterize this state of affairs, in [9] the metaphor of the GOMS user's cognitive model [5] and user activity [40] is used. At any given time, the system is in a certain state, characterised by a current retrieval space (usually a subset of the original document lattice) and by a focus concept within it. In each state, the user may select an operator (browsing, querying, bounding, thesaurus climbing) and apply it. As a result, a transition is made to a new state, possibly characterised by a new retrieval space and/or new focus. The new state is evaluated by the user for retrieval, and then the whole cycle may be iterated.

Therefore each interaction sequence may be composed of several operators, connected in various orders. For instance, the user may initially bound the search space exploiting her knowledge about the goal, then query the system to locate a region of interest within the bounded space, then browse through the region; also, at any time during this process, the user may take advantage of the feedback obtained during the interaction to make a jump to a different but related region (e.g., by thesaurus climbing), or to further bound the retrieval space.

The merits and performance of using concept lattices for supporting hybrid search strategies have been described in a number of papers (e.g., [27], [10], [19],[16]). They can be summarized as greater flexibility, good retrieval effectiveness, and mining capabilities.

Among the various pieces of information that can be easily mined in a collection D using a concept lattice-based method, are the following: (i) Find the most common or uncommon subjects in D, (ii) Find which subjects imply, or are implied by, other subjects in D, (iii) Find novel and unpredictable subject associations in D, (iv) Find which subjects allow gradual refinement of subsets of D. Several detailed examples of mining information that would be difficult to acquire using the traditional information retrieval methods are provided in the cited papers.

3.3   Context-Sensitive Use of Thesauri

In information retrieval applications, there often exist subsumption hierarchies on the set of terms describing the documents, in the form of a thesaurus.

A thesaurus can be integrated into a concept lattice either by explicitly expanding the original context with the implied terms or by taking into account the thesaurus ordering relation during the construction of the lattice ([8], [11]).

Using a thesaurus basically makes it possible to create new meaningful queries and guarantees that more general queries are indexed with more general terms, whereas in a standard concept lattice each query is strictly described by the terms present in the documents and possible semantic relationships between the terms themselves are ignored.

The user may thus locate the information of interest more effectively and quickly, partly because of enhanced navigation (the proximity of concepts in the lattice being related to semantic factors) and partly because of focused querying (as concept terms may be specialized/generalized using the thesaurus). An experimental evaluation of the retrieval effectiveness of a thesaurus-enhanced concept lattice is described in [11].

As stated above, the common approach is to (explicitly or implicitly) add the implied terms to each document according to the thesaurus ordering relation. Uta Priss [44] discusses other possible ways in which a context and a thesaurus can be merged into an expanded context. She also suggests that the user should be given the possibility of interactively combinining concepts from multiple thesauri, or thesaurus facets, using Boolean operators [45].

Improving the representation of the document collection at hand is not the only possible reason to use a thesaurus. One might integrate a thesaurus in a lattice with the goal of analyzing the appropriateness of the thesaurus classification for a specific collection of documents.

The latter approach draws an interesting analogy with the applications of concept lattices in object-oriented modelling (e.g., [25], [49]), where type or class hierarchies are merged into a lattice of software programs with the goal of restructuring the existing hierarchies.

3.4   Combining Multiple Views of Semi-structured Data

When the data can be classified along multiple axes (e.g., functional, geographical, descriptive), it may be convenient for the user to bring in new attributes in an incremental fashion, making decisions based on the information displayed by the system for the current choice of the attributes.

Think of a topic such as *Italian restaurants with a "dehors" near the Louvre Museum.* If there is no such restaurant, the user may find it useful to look for best matching restaurants by examining first the attributes that have higher priority to her and then moving on to the attributes with lower priority, e.g., geographical proximity first, then type of cuisine, and lastly possession of an open-air space.

In the FCA setting, this general approach has been implemented by a nesting & zooming technique, whereby the user may combine the lattices corresponding to each partial view and focus on the points of interest. To visualize the combination of partial views, a particular lattice visualization scheme is used, called nested line diagram, which will be discussed in Section 4.3.

Using partial views is most suitable for many-valued contexts, because it may be easier to identify valuable subcontexts. Indeed, in many cases, the lattices of certain subcontexts may be seen as conceptual scales of the given context, in the sense of [21]. In principle, partial views can be applied also to one-valued contexts by vertically slicing the context table (an example is described in [47]), but in the latter case it may be more difficult to select subcontexts that bear value, or just meaning. In fact, some of the most interesting applications have been developed in domains characterized by semistructured data, such as those for searching collections of emails ([17], [16]) or for analysing real-estate data extracted from the web [15].

3.5   Bounding the Search Space with User Constraints

Bounding is one of the functionalities implemented in the ULYSSES prototype ([9], [10]) to help the user focus the search on the relevant parts of a large concept lattice. Bounding allows users to prune the search space from which they are retrieving information during the search. The user may dynamically apply constraints with which the sought documents have to comply and the current search space is bounded accordingly. The constraints are expressed as inequality relations between the description of admissible concepts and a particular conjunction of terms, and the partitions induced over the search space by the application of such constraints present useful properties from the point of view of the information retrieval performance.

There are four possible constraints: $\uparrow c_1$, $\downarrow c_1$, $\neg \uparrow c_1$, $\neg \downarrow c_1$, where c_1 is the intent of some concept in the lattice. The constraint $\downarrow c_1$, for example, causes the system to prune away from the concept lattice all the concepts whose intent is either greater than or incomparable with c_1 (in other terms, all the concepts which are not below c_1).

To implement this framework, in [8] it is described an efficient algorithm based on two boundary sets – one containing the most specific elements of the admissible space (i.e., the lower boundary set) and the other containing the most general elements (i.e., the upper boundary set) – that can incrementally represent and update the constrained space. As more and more constraints are added, the admissible space shrinks, and the two boundary sets may eventually converge to the target class.

3.6 Overcoming the Vocabulary Problem in Text Ranking

Current best-matching information retrieval systems are limited by their inability of retrieving documents which contain the same concept as the query but are expressed with different words. A common solution to alleviate this vocabulary problem is to create a richer query context, mainly based on the first documents retrieved by the original query [6] or based on some form of terminological knowledge structure [18].

A more fundamental solution to word mismatch relies on the exploitation of inter-document similarity, following van Rijsbergen's cluster hypothesis that relevant documents tend to be more similar to each other than non-relevant documents. The best known approach is to rank a query not against individual documents but against a hierarchically grouped set of document clusters [58]. This approach, however, may involve the use of some heuristic decisions both to cluster the set of documents and to compute a similarity between individual document clusters and a query. As a result, hierarchical clustering-based ranking may easily fail to discriminate between documents that have manifestly different degrees of relevance for a certain query.

The limitations of hierarchical clustering-based ranking can be overcome by using the concept lattice of the document collection as the underlying clustering structure. The concept lattice may then be used to drive a transformation between the representation of a query and the representation of each document. This approach is described in [13].

Essentially, the query is merged into the document lattice and each document is ranked according to the length of the shortest path linking the query to the document concept. Of course, this is a quasi ordered retrieval output, because the documents that are equally distant from the query concept have the same score. We can think of the sets containing equally-ranked documents as concentric rings around the query node, the longer the radius, the lower the document score (of the associated documents).

An evaluation performed on two test document collections of small size, i.e., the abovementioned CISI (1460 documents) and CACM – a collection consisting of 3204 titles and abstracts from the journal CACM available at $http://www.dcs.gla.ac.uk/idom/ir_resources/test_collections/cacm/$ – showed that concept lattice-based ranking was comparable to best-matching ranking and better than hierarchical clustering-based ranking on the whole document set, whereas it clearly outperformed the other two methods when the specific ability to rank documents that did not match the query was measured.

4   Issues for FCA-Based IR Applications

Most FCA-based IR applications involve the following three steps: (a) extraction of a set of index terms that describe each document of the given collection, (b) construction of the concept lattice of the document-term relation generated at step (a), (c) visualization of the concept lattice built at step (b).

The solution to each step may crucially affect the efficiency and/or the effectiveness of the overall application. In the next subsections we will analyze each step in turn.

4.1   Automatic Generation of Index Units

This step is not necessary if each document is already equipped with a set of index terms. In most situations of interest, however, the index terms are not available and their manual generation is often impractical or even unfeasible (think of large text databases that change frequently over time).

Automatic indexing has long been studied in information retrieval. To automatically extract a set of index terms describing each document, the following method consisting of five steps can be used.

1. Text segmentation. The individual words occurring in a text collection are extracted, ignoring punctuation and case.

2. Word stemming. Each word is reduced to word-stem form. This may be done by using some large morphological lexicon that contains the standard inflections for nouns, verbs, and adjectives (e.g., [34]), or via some rule-based stemmer such as Porter's [43].

3. Stop wording. A stop list is used to delete from the texts the (root) words that are insufficiently specific to represent content. The stop list included in the CACM dataset, for instance, contains 428 common function words, such as "the", "of", "this", "on", etc. and some verbs, e.g., "have", "can", "indicate", etc.

4. Word weighting. This step is necessary to perform word selection, described in step 6; it may be also useful to discriminate between the documents that belong to a same concept, e.g., for automatic text ranking.

For each document and for each term, a measure of the usefulness of that term in that document is derived. The goal is to identify words that characterize the document to which they are assigned, while also discriminating it from the remainder of the collection. This has long been modeled by the well known $tf \cdot idf$ weighting scheme, which is now a bit outdated.

The two typical assumptions of the $tf\text{-}idf$ scheme – namely that multiple appearences of a term in a document are more important than the single appearence (tf) and that rare terms are more important than frequent terms (idf) – have been extended through a third length normalization assumption stating that for the same quantity of term matching, long documents are less important than short documents.

These three assumptions have been implemented using several approaches, most notably using Robertson's probabilistic model [46], statistical language modeling [60], and deviation from randomness [2]. These recent models have been shown to perform much better than the classical $tf\text{-}idf$ scheme on large, heterogeneous test collections, such as those used at TREC (Text REtrieval Conference, http://trec.nist.gov).

When the documents to be indexed are obtained in response to a query, it might be more effective to use term scoring functions that are based on the difference between the distribution of the terms in the set of retrieved documents and the distribution of the terms in the whole collection. In this way, the scores assigned to each term may more closely reflect the relevance of the term to the specific query at hand rather than the general importance of the term in the collection. Several term-scoring functions of this kind and possible ways of combining them to improve the quality of the generated terms are discussed in [14].

Also, for semi-structured or web documents, text-based indexing might be complemented with other techniques that take advantage of additional sources of knowledge, such as document fields, incoming or outgoing links, anchor texts, and url structure.

5. Word selection. This last step is not necessary for IR systems performing full-text indexing (in fact, it has not been included in the classic blueprint for automatic indexing suggested by Salton [48]), but it is very important for FCA-based systems to facilitate the subsequent process of lattice construction.

This problem is customarily addressed by using some heuristic threshold which restricts the index set. Among others, one can use as selection criterion the mean of weights in the document [13] or the value corresponding to one standard deviation above the mean [10]. A more elaborate approach is to choose the feature subset that maximizes the performance of a certain retrieval task or minimizes some involved error, but this might be too difficult or expensive in many cases.

Clearly, reducing the set of features may affect the retrieval effectiveness, although this does not necessarily result in performance degradation. The effects of feature selection on FCA-based text ranking are discussed in [13].

4.2   Efficient Lattice Construction

It is well known that the size of a concept lattice may grow exponentially with the number of objects. However, this situation occurs rarely in the information retrieval domain, as witnessed by a number of theoretical and experimental findings.

To gain some deeper insights into the actual order of magnitude of document lattices, one can hypothesize that the document description obeys some simple distribution of probability (estimated by term frequency). If each index term is assigned to each document with constant, independent probability $p = k/|M|$, the number of keywords per object follows a binomial distribution with a mean value of k and the mean number of concepts in the lattice, derived by Godin *etal.* [28], is given by:

$$|C| = \sum_{i=0}^{|G|} \sum_{j=0}^{|M|} \binom{|G|}{i} \binom{|M|}{j} p^{ij} (1 - p^i)^{|M|-j} (1 - p^j)^{|G|-i} \tag{1}$$

Here we plot Equation 1 for four values of k (5, 10, 20, and 50), choosing $|G| = 10000$. Figure 1 clearly shows that the number of concepts varies from linear to quadratic with respect to the number of documents (note that the y axis shows the ratio of the number of concepts to the number of documents), at least for the chosen parameter values. The size of the lattice grows as the number of terms per document increases; the upper bound is reached for $k = 50$, corresponding to a probability of assigning a term to a document equal to $50/1000 = 0.05$. This latter value accounts for a relatively dense context table, at least for information retrieval applications.

These findings agree with experimental observations. For instance, for the test collection CACM (3204 documents), it has been reported that the concept lattice contained some 40,000 concepts [13] , whereas for the test collection CISI (1460 documents), characterized by a larger number of terms per document (about 40), the size of the lattice grew to 250,000 ([10], [12], [13]).

Several algorithms have been developed for building the concept lattice of an input context (G, M, I) (e.g., [20], [4], [7], [26]). Usually, the efficiency of such algorithms critically depend on the number of concepts present in the lattice.

The best theoretical worst time complexity is $O(|C||M|(|G| + |M|))$, exhibited by the algorithm presented by Nourine and Raynaud [41]; in practice, the behaviour may significantly vary depending on a number of factors including the relative sizes of G and M, the size of I, and the density of the context, i.e., the size of I relative to the product $|G||M|$ (see [35] for an experimental comparison).

As the size of the document lattice may largely exceed the number of documents and because of the inherent complexity of the lattice-building algorithms,

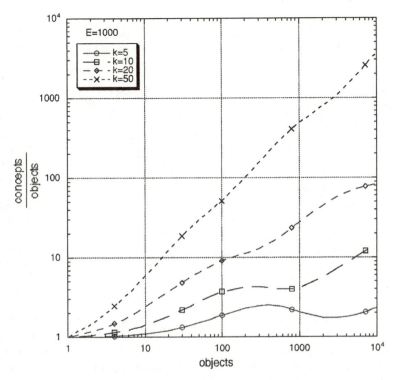

Fig. 1. Theoretical space complexity of document lattices. Both scales are logarithmic

the full document lattice may be constructed only for small to medium size collections, usually up to thousands of documents. For larger test collection, such as those containing millions of documents used at TREC, it is just unfeasible to build the complete associated concept lattice.

Fortunately, in many applications it is enough to compute a very small portion of the lattice, typically consisting of a focus concept and its neighbours. Such a focus concept, for instance, might be selected by the user through a point-and-click graphical user interface showing a partial lattice, or, as seen earlier, it might be computed by mapping a natural language or Boolean query on the document lattice. In this case, the system returns just the neighbours of a focus concept in the lattice.

The problem of generating all the nearest neighbours of a given concept has been addressed both to build a full lattice ([4], [37]) and to find just the portion of lattice centered around that concept [12]. As this is a very general and useful algorithm, we describe it here in a detailed manner.

Our version follows the same general strategy as the works cited above but differs in two main details, namely the generation of the candidate extent and the choice of the admissible candidates. To solve the latter subtasks, we borrow the more efficient procedures presented in [41].

Find Lower Neighbours
Input: Context (G, M, I), concept (X, Y) of context (G, M, I)
Output: The set of lower neighbours of (X, Y) in the concept lattice of (G, M, I)

1. $lowerNeighbours := \emptyset$
2. $testedCand := \emptyset$
3. **for** each $m \in M \backslash Y$
4. $X_1 := X \cap \{m\}'$
5. $Y_1 := X_1'$
6.   **if** $(X_1, Y_1) \notin testedCand$
          then
7.   Add (X_1, Y_1) to $testedCand$
8. $count(X_1, Y_1) := 1$
          else
9. $count(X_1, Y_1) := count(X_1, Y_1) + 1$
10.   **if** $(|Y_1| - |Y|) = count(X_1, Y_1)$ **then**
11.   Add (X_1, Y_1) to $lowerNeighbours$

Fig. 2. Find Lower Neighbours algorithm

Figure 2 describes the algorithm for determining the set of lower neighbours of a given concept; the determination of the upper neighbours is a dual problem and can be solved by easily adapting the given algorithm. The theoretical time complexity of the computation of the lower neighbours is $O(|G||M|^2)$; the time complexity of the algorithm for finding both the lower and upper neighbours is $O(|G||M|(|G| + |M|))$. The use of Nourine and Raynaud's procedures does not affect the theoretical complexity of the algorithm but they may produce a substantial efficiency gain in practical situations.

Although the possession of a fast algorithm for computing the underlying concept lattice or part of it may be an essential prerequisite for IR applications as well as for applications concerning rule mining or software analysis, the issue of an optimal selection of the available algorithms has not been adequately addressed. More research is needed on the evaluation of competing algorithms, both from a theoretical and an experimental point of view. We will return to this in the conclusion.

4.3   Effective Lattice Visualization

Except for automatic tasks such as document ranking, most of the IR applications based on concept lattices require some form of exploration of the graph diagram on the part of the user. However, forming useful visualizations of graph structures is notoriously difficult due to the conflicting issues of size, layout, and legibility on limited screen area. The problem is further compounded by the fact that the concept lattices of real applications are usually very large. The common approach is to show or hide parts of the lattice via interactive specification of a focus concept and/or subsets of terms.

One simple method consists of showing just the neighbours of a focus concept. Simple graphical interfaces of this kind have been suggested or implemented in several works, including [24], [27], [11], and the REFINER system discussed earlier [12].

To show a larger portion centered around a focus concept, we can resort to focus+context visualization tehniques. Focus+context viewers use as a general metaphor the effects observed when looking through fisheye lenses or magnifying glasses. A simple way to implement a fish eye view is to display the information contained in a lattice in varying levels of details depending on the distance from the focus; the size of the information at the focal point are increased whereas the information placed further away are reduced in scale.

In practice, a specific display format for each subset of concepts placed at the same distance from the focus concept can be used, the distance being the length of the shortest paths between the concepts. Such displays may involve different combinations of sizes, fonts, and types of information. A similar approach has been adopted in the ULYSSES prototype ([9], [10]).

In some cases, we are mainly interested in the portion of the lattice placed *below* a focus concept. A simple and useful approach is to use a tree, by making the focus concept the root and associating each sequence of concepts below the focus with a path. The tree representation has several advantages. As the metaphor of hierarchical folders is used for storing and retrieving files, bookmarks, menus items, etc., most users are familiar with it and hence no training on the part of the user is required. Furthemore, it takes little space on the screen and it may be drawn efficiently.

The main disadvantage is that there may be a considerable amount of duplication of information when the concepts have multiple parents. On the other hand, this is not very likely to happen if only some levels of the hierarchies are visualized. The tree-like representation surfaces in some more recent prototypes based on concept lattices such as HierMail [16] and its commercial follow-up Mail-Sleuth (*http://www.mail-sleuth.com*).

An alternative approach to lattice visualization is based on combining multiple partial views of the data represented in the context. A particular scheme termed nested line diagram has been developed within the FCA community and first implemented in the Toscana system ([57], [55],[56], [53]). In essence, (i) two or more subsets of attributes are chosen by the user, (ii) the concept lattices of the subcontexts identified by the attribute subsets of step 1 are found, and (iii) the full concept lattice is embedded in the direct product of the lattices of subcontexts as a join-semilattice. The overall effect is that of having several complete lattices of partial contexts nested into one another rather than a partial lattice of a complete context.

One advantage of nested line diagrams is that the size of each local diagram cannot exceed the number of possible combinations of the attribute values present in the corresponding subcontext, regardless of the number of objects in the database. Hence, it is possible to draw the full lattices of each subcontext even for large databases, provided that the subcontexts are sufficiently small.

Clearly, this approach is effective when the number of scales to combine is limited.

Before concluding this section we would like to emphasize that the fast advances in the field of graphical web interfaces may spur a renewed interest in the techniques for lattice visualization. In addition to exploring the use of alternative visual layouts proposed in the information visualization field [22], whether focused on more complex inherent graph substructures or on richer interactive or linking mechanisms, it would be useful to compare relative merits and drawbacks of each visualization scheme for specific performance tasks.

5   The Next Challenge

Current Web retrieval interfaces are limited by a lack of a concise representation of the content of all retrieved documents; conventional textual displays take much perusal time and screen space and do not enable inspection of more documents at a time. A related drawback is represented by the inability of providing good refinement terms for narrowing down the large set of results that are typically returned in response to a query.

Current research is addressing these shortcomings by attempting to provide visual or terminological cues for interpreting and manipulating retrieval results (e.g., [30], [3], [59], [33]), but most proposed approaches are still hampered by theoretical and practical limitations.

Concept lattices are a good candidate for extending user control over presentation and selection of web retrieval results. Among the anticipated advantages are the following. As the refinement terms in the lattice are based on all the words in the query concept (rather than on single words) and they are driven by the content of the documents being searched (rather than on predefined term-term associations), the suggested terms may work well not only for simple, popular topics – as with the "related keyword" feature provided by some commercial search engines – but also for specialized or ambiguous topics. Furthermore, concept lattices are more flexible than hierarchical clustering approaches, because it is easier for the user to recover from bad early decisions while traversing the structure. Finally, several search strategies can be integrated in the same framework.

Searching the web results using FCA is technically feasible. Now we sketch a possible architecture. The system takes as input a user query. The query is forwarded to a selected search engine, and the first pages retrieved by the search engine in response to the query are collected and parsed. At this point, a set of index units that describe each returned document is generated; such indices are next used to build the concept lattice corresponding to the retrieved results. The last steps consists of showing the lattice to the user and managing the subsequent interaction between the user and the system.

There are a number of design decisions involved here, which may affect both the efficiency and the effectiveness of the whole system. Following the data flow, the main decisions are: analyzing a small/large number of retrieved documents, using document snippets or full text documents, performing single- or multi-keyword indexing, constructing partial or full concept lattice, using simple or so-phisticated visualization schemes, allowing single or multiple interaction modes. In order to ensure fast response times and good overall retrieval effectiveness, each component should be carefully designed and engineered; also, the interactions with the other components should be studied.

In spite of such difficulties, building a system of this kind can help to address the issue of retrieving and mining web documents in a more principled and effective manner. We believe that the challenge is quite realistic, given the current state of the art of FCA and IR techniques, and that this might be a big success for the whole field of concept data analysis.

6   Conclusions

The advances in the methods for constructing and searching document lattices coupled with the unprecedented need for contextual text processing techniques make FCA a strong tool for building modern IR applications. In order to grasp this opportunity, some bigger effort and a few cautions are required on the part of FCA developers. Here we would like to make some recommendations that can help build successful applications.

Focus on appropriate IR tasks. The chosen tasks must be suitable for FCA and should not be easily solved by conventional IR techniques. For instance, natural language processing techniques could hardly demonstrate their usefulness as long as they were employed to improve the classical topic relevance task, whereas they have recently become an essential component of systems performing question answering on large text collections.

Integration with advanced IR techniques. To solve any nontrivial task, it may be necessary to integrate FCA methods with existing IR techniques. As the IR field is moving on fastly, it is important to pick up the most updated techniques. For instance, using the classical *tf-idf* weighting scheme rather than the much more effective methods that have been developed lately may seriously degrade the performance of the whole IR application.

Adoption of IR evaluation metrics. The effectiveness of the application should be measured using recognized evaluation metrics. This holds both for automatic and interactive tasks. Evaluation studies of the latter type of tasks, which is more relevant to FCA applications, are not frequent in the literature probably due to a combination of methodological, technological, organizational, and economical issues, although there are some significant exceptions (e.g., [51], [3]).

Engineering test collections. It would be very useful to have a set of test databases on which to run rigorous experimental comparisons. Test collections could be used to evaluate the efficiency (and perhaps the correctness) of the algorithms for constructing the document lattice and also to perform more controlled IR experiments. Engineering test collections may be an important step to take for the whole research community on FCA to encourage systems implementations and to measure advances.

Deployment of tools. Although a number of FCA papers have been published in major IR forums, the awareness of the utility of concept lattices for IR is still limited outside of the FCA community. The free availability of an on-line, concept lattice-based tool for mining Web retrieval results would probably greatly increase the scope of FCA for IR.

Acknowledgments

We would like to thank Gerd Stumme for his helpful comments on an earlier version of this paper.

References

1. M. Agosti, M. Melucci, and F. Crestani. Automatic authoring and construction of hypertexts for information retrieval. *ACM Multimedia Systems*, 3:15–24, 1995.
2. G. Amati, C. Carpineto, and G. Romano. FUB at TREC-10 Web Track: A Probabilistic Framework for Topic Relevance Term Weighting. In *Proceedings of the 10th Text REtrieval Conference (TREC-10), NIST Special Publication 500-250*, pages 182–191, Gaithersburg, MD, USA, 2001.
3. E. Berenci, C. Carpineto, V. Giannini, and S. Mizzaro. Effectiveness of keyword-based display and selection of retrieval results for interactive searches. *International Journal on Digital Libraries*, 3(3):249–260, 2000.
4. J.P. Bordat. Calcul pratique du treillis de Galois d'une correspondance. *Math. Sci. Hum.*, 96:31–47, 1986.
5. S. Card, T. Moran, and A. Newell. *The psychology of human-computer interaction.* Lawrence Erlbaum Associates, London, 1983.
6. C. Carpineto, R. De Mori, G. Romano, and B. Bigi. An information theoretic approach to automatic query expansion. *ACM Transactions on Information Systems*, 19(1):1–27, 2001.
7. C. Carpineto and G. Romano. An order-theoretic approach to conceptual clustering. In *Proceedings of the 10th International Conference on Machine Learning*, pages 33–40, Amherst, MA, USA, 1993.
8. C. Carpineto and G. Romano. Dynamically bounding browsable retrieval spaces: an application to Galois lattices. In *Proceedings of RIAO 94: Intelligent Multimedia Information Retrieval Systems and Management*, pages 520–533, New York, New York USA, 1994.
9. C. Carpineto and G. Romano. ULYSSES: A lattice-based multiple interaction strategy retrieval interface. In Unger Blumenthal, Gornostaev, editor, *Human-Computer Interaction, 5th International Conference, EWHCI, Selected Papers*, pages 91–104. Springer, Berlin, 1995.
10. C. Carpineto and G. Romano. Information retrieval through hybrid navigation of lattice representations. *International Journal of Human-Computer Studies*, 45(5):553–578, 1996.
11. C. Carpineto and G. Romano. A lattice conceptual clustering system and its application to browsing retrieval. *Machine Learning*, 24(2):1–28, 1996.
12. C. Carpineto and G. Romano. Effective reformulation of Boolean queries with concept lattices. In *Proceedings of the 3rd International Conference on Flexible Query-Answering Systems*, pages 83–94, Roskilde, Denmark, 1998.

13. C. Carpineto and G. Romano. Order-Theoretical Ranking. *Journal of the American Society for Information Science*, 51(7):587–601, 2000.

14. C. Carpineto, G. Romano, and V. Giannini. Improving retrieval feedback with multiple term-ranking function combination. *ACM Transactions on Information Systems*, 20(3):259–290, 2002.

15. R. Cole and P. Eklund. Browsing semi-structured web texts using formal concept analysis. In *Proceedings of the 9th International Conference on Conceptual Structures*, pages 319–332, Stanford, CA, USA, 2001.

16. R. Cole, P. Eklund, and G. Stumme. Document retrieval for e-mail search and discovery using formal concept analysis. *Applied Artificial Intelligence*, 17(3):257–280, 2003.

17. R. Cole and G. Stumme. CEM: A Conceptual Email Manager. In *Proceedings of the 8th International Conference on Conceptual Structures*, pages 438–452, Darmstadt, Germany, 2000.

18. E. Efthimiadis. Query expansion. In M. E. Williams, editor, *Annual Review of Information Systems and Technology, v31*, pages 121–187. American Society for Information Science, Silver Spring, Maryland, USA, 1996.

19. S. Ferré and O. Ridoux. A file system based on concept analysis. In *Proceedings of the 1st International Conference on Computational Logic*, pages 1033–1047, London, UK, 2000.

20. B. Ganter. Two basic algorithms in concept analysis. Technical Report FB4–Preprint No. 831, TU Darmstadt, Germany, 1984.

21. B. Ganter and R. Wille. *Formal Concept Analysis – Mathematical Foundations*. Springer, 1999.

22. N. Gershon, S. K. Card, and S. G. Eick. Information visualization tutorial. In *Proceedings of ACM CHI'98: Human Factors in Computing Systems*, pages 109–110, Los Angeles, CA, USA, 1998.

23. D. K. Gifford, P. Jouvelot, M. A. Sheldon, and J. W. Jr O'Toole. Semantic file systems. In *Proceedings of the 13th ACM Symposium on Operating Systems Principles*, pages 16–25, 1991.

24. R. Godin, J. Gecsei, and C. Pichet. Design of a browsing interfaces for information retrieval. In *Proceedings of the 12th Annual International ACM SIGIR Conference on Reasearch and Development in Information Retrieval*, pages 32–39, 1989.

25. R. Godin and H. Mili. Building and Maintaining Analysis Level Class Hierarchies Using Galois Lattices. In *Proceedings of the 8th Annual Conference on Object Oriented Programming Systems Languages and Applications*, pages 394–410, Washington, D.C., USA, 1993.

26. R. Godin, R. Missaoui, and H. Alaoui. Incremental concept formation algorithms based on Galois lattices. *Computational Intelligence*, 11(2):246–267, 1995.

27. R. Godin, R. Missaoui, and A. April. Experimental comparison of navigation in a Galois lattice with conventional information retrieval methods. *International Journal of Man-Machine Studies*, 38:747–767, 1993.

28. R. Godin, E. Saunders, and J. Jecsei. Lattice model of browsable data spaces. *Journal of Information Sciences*, 40:89–116, 1986.

29. B. Gopal and U. Manber. Integrating content-based access mechanisms with hierarchical file systems. In *Proceedings of 3rd Symposium on Operating Systems Design and Implementation*, pages 265–278, New Orleans, Louisiana, USA, 1999.

30. M. Hearst. User interfaces and visualization. In R. Baeza-Yates and B. Ribeiro-Neto, editors, *Modern Information Retrieval*, pages 257–322. ACM Press, New York, New York, USA, 1999.

31. M. A. Hearst. Untangling text data mining. In *Proceedings of the 37th Annual Meeting of the Association for Computational Linguistics (ACL'99)*, College Park, MD, USA, 1999.

32. D. C. Hoaglin, F. Mosteller, and J. W. Tukey. *Understanding robust and exploratory data analysis*. John Wiley & Sons, Inc., 1983.

33. H. Joho, M. Sanderson, and M. Beaulieu. Hierarchical approach to term suggestion device. In *Proceedings of the 25th Annual International ACM SIGIR Conference on Research and Development in Information Retrieval*, page 454, Tampere, Finland, 2002.

34. D. Karp, Y. Schabes, M. Zaidel, and D. Egedi. A freely available wide coverage morphological analyzer for English. In *Proceedings of the 14th International Conference on Computational Linguistics (COLING'92)*, pages 950–955, Nantes, France, 1992.

35. S.O. Kuznetsov and S.A. Obiedkov. Comparing performance of algorithms for generating concept lattices. *Journal of Experimental and Theoretical Artificial Intelligence*, 14(2–3):189–216, 2002.

36. C. Lindig. Concept-based component retrieval. In *Working notes of the IJCAI-95 workshop: Formal Approaches to the Reuse of Plans, Proofs, and Programs*, pages 21–25, Montreal, Canada, 1995.

37. C. Lindig. Fast concept analysis. In *Working with conceptual structures – Contribution to the 8th International Conference on Conceptual Structures*, pages 152–161, Darmstadt, Germany, 2000.

38. D. Lucarella, S. Parisotto, and A. Zanzi. MORE: Multimedia Object Retrieval Environment. In *Proceedings of ACM Hypertext'93*, pages 39–50, Seattle, WA, USA, 1993.

39. Y. Maarek, D. Berry, and G. Kaiser. An information retrieval approach for automatically constructing software libraries. *IEEE Transactions on software Engineering*, 17(8):800–813, 1991.

40. D. Norman. Cognitive engineering. In D. Norman and S. Draper, editors, *User centered system design*, pages 31–61. Lawrence Erlbaum Associates, Hillsdale, New Jersey, 1986.

41. L. Nourine and O. Raynaud. A fast algorithm for building lattices. *Information Processing Letters*, 71:199–204, 1999.

42. G. Pedersen. A browser for bibliographic information retrieval based on an application of lattice theory. In *Proceedings of the 16th Annual International ACM SIGIR Conference on Research and Development in Information Retrieval*, pages 270–279, Pittsburgh, PA, USA, 1993.

43. M. F. Porter. An algorithm for suffix stripping. *Program*, 14:130–137, 1980.

44. U. Priss. A graphical interface for document retrieval based on Formal Concept Analysis. In *Proceedings of the 8th Midwest Artificial Intelligence and Cognitive Science Conference*, pages 66–70, Dayton, Ohio, USA, 1997.

45. U. Priss. Lattice-based information retrieval. *Knowledge Organization*, 27(3):132–142, 2000.

46. S. E. Robertson, S. Walker, and M. M. Beaulieu. Okapi at TREC-7: Automatic Ad Hoc, Filtering, VLC, and Interactive track. In *Proceedings of the 7th Text REtrieval Conference (TREC-7), NIST Special Publication 500-242*, pages 253–264, Gaithersburg, MD, USA, 1998.

47. T. Rock and R. Wille. Ein Toscana-Erkundungssystem zur Literatursuche. In G. Stumme and R. Wille, editors, *Begriffliche Wissensverarbeitung. Methoden und Anwendungen*, pages 239–253. Springer, Berlin, Germany, 2000.

48. G. Salton. *Automatic Text Processing: The Transformation, Analysis, and Retrieval of Information by Computer.* Addison Wesley, 1989.

49. G. Snelting and F. Tip. Reengineering class hierarchies using concept analysis. In *Proceedings of ACM SIGSOFT 6th International Symposium on Foundations of Software Engineering*, pages 99–110, Lake Buena Vista, FL, USA, 1998.

50. D. Soergel. Mathematical analysis of documentation systems. *Information storage and retrieval*, 3:129–173, 1967.

51. A. Spink and T. Saracevic. Interaction in information retrieval: selection and effectiveness of search terms. *Journal of the American Society for Information Science*, 48(8):741–761, 1997.

52. A. Spoerri. InfoCrystal: Integrating exact and partial matching approaches through visualization. In *Proceedings of RIAO 94: Intelligent Multimedia Information Retrieval Systems and Management*, pages 687–696, New York, New York USA, 1994.

53. G. Stumme. Local scaling in conceptual data systems. In *Proceedings of the 6th International Conference on Conceptual Structures*, pages 308–320, Montpellier, France, 1998.

54. F. J. van der Merwe and D. G. Kourie. Compressed pseudo-lattices. *Journal of Experimental and Theoretical Artificial Intelligence*, 14(2–3):229–254, 2002.

55. F. Vogt, C. Wachter, and R. Wille. Data analysis based on a conceptual file. In H.-H. Bock, W. Lenski, and P. Ihm, editors, *Classification, Data Analysis and Knowledge Organization*, pages 131–140. Springer, Berlin, 1991.

56. F. Vogt and R. Wille. TOSCANA – A graphical tool for analyzing and exploring data. In R. Tammassia and I. G. Tollis, editors, *Graph Drawing'94*, pages 226–233. Springer, Berlin, 1995.

57. R. Wille. Line diagrams of hierarchical concept systems. *Int. Classif.*, 11(2):77–86, 1984.

58. P. Willet. Recent trends in hierarchic document clustering: a critical review. *Information Processing & Management*, 24(5):577–597, 1988.

59. O. Zamir and O. Etzioni. Grouper: A dynamic clustering interface to web search results. *WWW8/Computer Networks*, 31(11–16):1361–1374, 1999.

60. C. Zhai and J. Lafferty. A study of smoothing methods for language models applied to ad hoc information retrieval. In *Proceedings of the 24th Annual International ACM SIGIR Conference on Research and Development in Information Retrieval*, pages 334–342, New Orleans, LA, USA, 2001.

Efficient Mining of Association Rules Based on Formal Concept Analysis

Lotfi Lakhal[1] and Gerd Stumme[2]

[1] IUT d'Aix-en-Provence, Département d'Informatique
Avenue Gaston Berger, F-13625 Aix-en-Provence cedex, France
http://www.lif.univ-mrs.fr/EQUIPES/BD/
[2] Chair of Knowledge & Data Engineering,
Department of Mathematics and Computer Science, University of Kassel,
Wilhelmshöher Allee 73, D–34121 Kassel, Germany
http://www.kde.cs.uni-kassel.de

Abstract. Association rules are a popular knowledge discovery technique for warehouse basket analysis. They indicate which items of the warehouse are frequently bought together. The problem of association rule mining has first been stated in 1993. Five years later, several research groups discovered that this problem has a strong connection to Formal Concept Analysis (FCA). In this survey, we will first introduce some basic ideas of this connection along a specific algorithm, TITANIC, and show how FCA helps in reducing the number of resulting rules without loss of information, before giving a general overview over the history and state of the art of applying FCA for association rule mining.

1 Introduction

Knowledge discovery in databases (KDD) is defined as the non-trivial extraction of valid, implicit, potentially useful and ultimately understandable information in large databases [23]. For several years, a wide range of applications in various domains have benefited from KDD techniques and many work has been conducted on this topic. The problem of mining frequent itemsets arose first as a sub-problem of mining association rules [1], but it then turned out to be present in a variety of problems [24]: mining sequential patterns [3], episodes [32], association rules [2], correlations [12, 43], multi-dimensional patterns [26, 28], maximal itemsets [7, 29, 54], closed itemsets [37, 38, 41, 47]. Since the complexity of this problem is exponential in the size of the binary database input relation and since this relation has to be scanned several times during the process, efficient algorithms for mining frequent itemsets are required.

In Formal Concept Analysis (FCA), the task of mining frequent itemsets can be described as follows: Given a set G of objects, a set M of attributes (or items), a binary relation $I \subseteq G \times M$ (where $(g, m) \in I$ is read as "object g has attribute m"), and a threshold minsupp $\in [0, 1]$, determine all subsets X of M (also called *itemsets* or *patterns* here) where the *support* $\mathrm{supp}(X) := \frac{\mathrm{card}(X')}{\mathrm{card}(G)}$ (with $X' := \{g \in G \mid \forall m \in X : (g, m) \in I\}$) is above the threshold minsupp.

B. Ganter et al. (Eds.): Formal Concept Analysis, LNAI 3626, pp. 180–195, 2005.

The set of these *frequent itemsets* itself is usually not considered as final result of the mining process, but rather as intermediate step. Its most prominent use are certainly association rules. The task of mining association rules is to determine all pairs $X \to Y$ of subsets of M such that $\mathrm{supp}(X \to Y) := \mathrm{supp}(X \cup Y)$ is above the threshold minsupp, and the *confidence* $\mathrm{conf}(X \to Y) := \frac{\mathrm{supp}(X \cup Y)}{\mathrm{supp}(X)}$ is above a given threshold minconf $\in [0, 1]$. Association rules are for instance used in warehouse basket analysis, where the warehouse management is interested in learning about products frequently bought together.

Since determining the frequent itemsets is the computationally most expensive part, most research has focused on this aspect. Most algorithms follow the way of the well-known Apriori algorithm [2]. It is traversing iteratively the set of all itemsets in a levelwise manner. During each iteration one level is considered: a subset of candidate itemsets is created by joining the frequent itemsets discovered during the previous iteration, the supports of all candidate itemsets are counted, and the infrequent ones are discarded. A variety of modifications of this algorithm arose [13, 34, 42, 48] in order to improve different efficiency aspects. However, all of these algorithms have to determine the supports of *all* frequent itemsets and of some infrequent ones in the database.

Other algorithms are based on the extraction of maximal frequent itemsets (i. e., from which all supersets are infrequent and all subsets are frequent). They combine a levelwise bottom-up traversal with a top-down traversal in order to quickly find the maximal frequent itemsets. Then, all frequent itemsets are derived from these ones and one last database scan is carried on to count their support. The most prominent algorithm using this approach is Max-Miner [7]. Experimental results have shown that this approach is particularly efficient for extracting maximal frequent itemsets, but when applied to extracting all frequent itemsets, performances drastically decrease because of the cost of the last scan which requires roughly an inclusion test between each frequent itemset and each object of the database. As for the first approach, algorithms based on this approach have to extract the supports of *all* frequent itemsets from the database.

While all techniques mentioned so far count the support of all frequent itemsets, this is by no means necessary. Using basic results from Formal Concept Analysis (FCA), it is possible to derive from some known supports the supports of all other itemsets: it is sufficient to know the support of all frequent concept intents. This observation was independently made by three research groups around 1997/98: the first author and his database group in Clermont–Ferrand [37], M. Zaki in Troy, NY [52], and the second author in Darmstadt [44].

The use of FCA allows not only an efficient computation, but also to drastically reduce the number of rules that have to be presented to the user, without any information loss. We present therefore some 'bases' for association rules which are non-redundant and from which all the others can be derived. Interestingly, up to now most researchers focus on the first aspect of efficiently computing the set of all frequent concept intents (also called *frequent closed sets*) and related condensed representations, but rarely consider condensed representations of the association rules.

This survey consists of two parts: In the next four sections, we provide an introduction to mining association rules using FCA, which mainly follows our own work[1], before we provide a more general overview over the field in Section 6. First, we will first briefly relate the notions of Formal Concept Analysis to the association rule mining problem. Then we discuss how so-called iceberg concept lattices represent frequent itemsets, and how they can be used for visualizing the result. In Section 4, we will sketch one specific algorithm, called TITANIC, as an illustrative example for exploiting FCA theory for efficient computation. The reduction of the set of asociation rules to the relevant ones is the topic of Section 5. Section 6 gives an overview over the current state of the art.

2   Mining Frequent Itemsets with Formal Concept Analysis

Consider two itemsets X and Y such that both describe exactly the same set of objects, i. e., $X' = Y'$. So if we know the support of one of them, we do not need to count the support of the other one in the database. In fact, we can introduce an equivalence relation θ on the powerset $\mathfrak{P}(M)$ of M by $X \theta Y \iff X' = Y'$. If we knew the relation from the beginning, it would be sufficient to count the support of one itemset of each class only – all other supports can then be derived. Of course one does not know θ in advance, but one can determine it along the computation. It turns out that one usually has to count the support of more than one itemset of each class, but normally not of all of them. This observation leads to a speed-up in computing all frequent itemsets, since counting the support of an itemset is an expensive operation.

We will see below that it is not really necessary to compute all frequent itemsets for solving the association rule problem: it will be sufficient to focus on the frequent concept intents (here also called *closed itemsets* or *closed patterns*). As well known in FCA, each concept intent is exactly the largest itemset of the equivalence class of θ it belongs to. For any itemset $X \subseteq M$, the concept intent of its equivalence class is the set X''. The concept intents can hence be considered as 'normal forms' of the (frequent) itemsets. In particular, the concept lattice contains all information to derive the support of all (frequent) itemsets.

3   Iceberg Concept Lattices

While it is not really informative to study the set of all frequent itemsets, the situation changes when we consider the closed itemsets among them only. We call the concepts they belong to *frequent concepts*, and the set of all frequent concepts *iceberg concept lattice* of the context $\mathbb{K}$ for the threshold minsupp. We illustrate this by a small example. Figure 1 shows the iceberg concept lattice of the MUSHROOM database from the *UCI KDD Archive* [6] for a minimum support of 85 %.

---

[1] This part summarizes joint work with Yves Bastide, Nicolas Pasquier, and Rafik Taouil as presented in [5, 40, 45, 46]

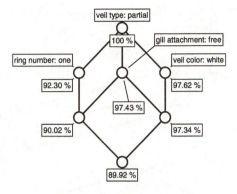

Fig. 1. Iceberg concept lattice of the mushroom database with minsupp = 85 %

The MUSHROOM database consists of 8,416 objects (mushrooms) and 22 (nominally valued) attributes. We obtain a formal context by creating one (Boolean) attribute for each of the 80 possible values of the 22 database attributes. The resulting formal context has thus 8,416 objects and 80 attributes. For a minimum support of 85 %, this dataset has 16 frequent itemsets, namely all 2^4 possible combinations of the attributes 'veil type: partial', 'veil color: white', 'gill attachment: free', and 'ring number: one'. Only seven of them are closed. The seven frequent concepts are shown in Figure 1.

In the diagram, each node stands for formal concept. The intent of each concept (i. e., each frequent closed itemset) consists of the attributes labeled at or above the concept. The number shows its support. One can clearly see that all mushrooms in the database have the attribute 'veil type: partial'. Furthermore the diagram tells us that the three next-frequent attributes are: 'veil color: white' (with 97.62 % support), 'gill attachment: free' (97.43 %), and 'ring number: one' (92.30 %). There is no other attribute having a support higher than 85 %. But even the combination of all these four concepts is frequent (with respect to our threshold of 85 %): 89.92 % of all mushrooms in our database have one ring, a white partial veil, and free gills. This concept with a quite complex description contains more objects than the concept described by the fifth-most attribute, which has a support below our threshold of 85 %, since it is not displayed in the diagram.

In the diagram, we can detect the implication

{ring number: one, veil color: white}⇒ {gill attachment: free} .

It is indicated by the fact that there is no concept having 'ring number: one' and 'veil color: white' (and 'veil type: partial') in its intent, but not 'gill attachment: free'. This implication has a support of 89.92 % and is globally valid in the database (i. e., it has a confidence of 100 %).

If we want to see more details, we have to decrease the minimum support. Figure 2 shows the MUSHROOM iceberg concept lattice for a minimum support of 70 %. Its 12 concepts represent all information about the 32 frequent itemsets for this threshold. One observes that, of course, its top-most part is just the

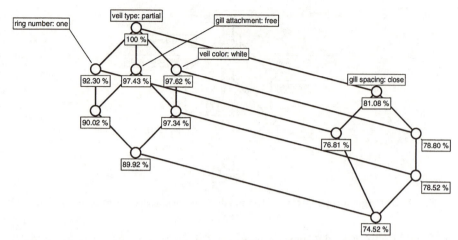

Fig. 2. Iceberg concept lattice of the mushroom database with minsupp = 70 %

iceberg lattice for minsupp = 85 %. Additionally, we obtain five new concepts, having the possible combinations of the next-frequent attribute 'gill spacing: close' (having support 81.08 %) with the previous four attributes. The fact that the combination {veil type: partial, gill attachment: free, gill spacing: close} is not realized as a concept intent indicates another implication:

$$\{\text{gill attachment: free, gill spacing: close}\} \Rightarrow \{\text{veil color: white}\} \qquad (*)$$

This implication has 78.52 % support (the support of the most general concept having all three attributes in its intent) and – being an implication – 100 % confidence.

By further decreasing the minimum support, we discover more and more details. Figure 3 shows the MUSHROOMS iceberg concept lattice for a minimum support of 55 %. It shows four more partial copies of the 85 % iceberg lattice, and three new, single concepts.

The Mushrooms example shows that iceberg concept lattices are suitable especially for strongly correlated data. In Table 1, the size of the iceberg concept lattice (i. e., the number of all frequent closed itemsets) is compared with the number of all frequent itemsets. It shows for instance, that, for the minimum support of 55 %, only 32 frequent closed itemsets are needed to provide all information about the support of all 116 frequent itemsets one obtains for the same threshold.

4   Computing the Iceberg Concept Lattice with Titanic

For illustrating the principles underlying the algorithms for mining frequent (closed) itemsets using FCA, we sketch one representative called TITANIC. For a more detailed discussion of the algorithm, we refer to [45].

TITANIC is counting the support of so-called key itemsets (and of some candidates for key itemsets) only: A *key itemset* (or *minimal generator*) is every

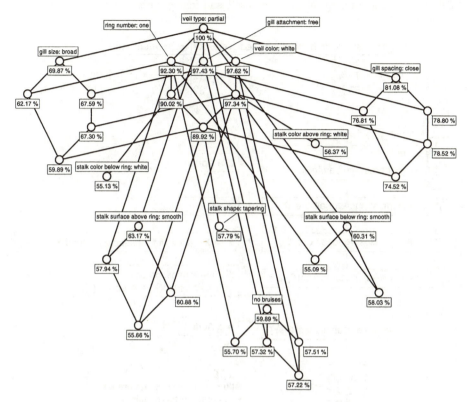

Fig. 3. Iceberg concept lattice of the mushroom database with minsupp = 55 %

minimal itemset in an equivalence class of θ. TITANIC makes use of the fact that the set of all key itemsets has the same property as the set of all frequent itemsets: it is an order ideal in the powerset of M. This means that each subset of a key itemset is a key itemset, and no superset of a non-key itemset is a key itemset. Thus we can reuse the pruning approach of Apriori for computing the supports of all frequent key itemsets. Once we have computed them, we have computed the support of at least one itemset in each equivalence class of θ, and we know the relation θ completely. Hence we can deduce the support of all frequent itemsets without accessing the database any more.

Figure 4 shows the principle of TITANIC. Its basic idea is as the original Apriori algorithm: At the ith iteration, we consider only itemsets with cardinality i (called i–itemsets for short), starting with $i = 1$ (step 1). In step 2, the support of all candidates is counted. For $i = 1$, the candidates are all 1–itemsets, later they are all i–itemsets which are potential key itemsets.

Once we know the support of all i–candidates, we have enough information to compute for all $(i-1)$–key itemsets their closure, i. e., the concept intent of their equivalence class. This is done in step 3, using the equation $X'' = X \cup \{x \in M \setminus X \mid \text{supp}(X) = \text{supp}(X \cup \{x\})\}$.

Table 1. Number of frequent closed itemsets and frequent itemsets for the Mushrooms example

minsupp	# frequent closed itemsets	# frequent itemsets
85 %	7	16
70 %	12	32
55 %	32	116

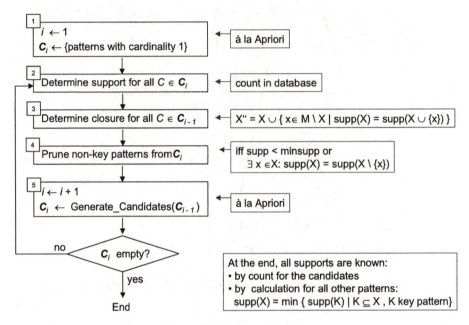

Fig. 4. The TITANIC algorithm

In step 4, all itemsets which are either not frequent or non-key are pruned. For the latter we use a characterization of key itemsets saying that a itemset is a key itemset iff its support is different from the support of all its immediate subsets. In strongly correlated data, this additional condition helps pruning a significant number of itemsets.

At the end of each iteration, the candidates for the next iteration are generated in step 5. The generation procedure is basically the same as for Apriori: An $(i+1)$–itemset is a candidate iff all its i–subitemsets are key itemsets. As long as new candidates are generated, the next iteration starts. Otherwise the algorithm terminates.

It is important to note that – especially in strongly correlated data – the number of frequent key itemsets is small compared to the number of all frequent itemsets. Even more important, the cardinality of the largest frequent key itemset is normally smaller than the one of the largest frequent itemset. This means that the algorithm has to perform fewer iterations, and thus fewer scans of the database. This is especially important when the database is too large for main memory, as each disk access significantly increases computation time. A theoretical and experimental analysis of this behavior is given in [45], further experimental results are provided in [40].

5   Bases of Association Rules

One problem in mining association rules is the large number of rules which are usually returned. But in fact not all rules are necessary to present the information. Similar to the representation of all frequent itemsets by the frequent closed itemsets, one can represent all valid association rules by certain subsets, so-called *bases*. The computation of the bases does not require all frequent itemsets, but only the closed ones.

Here we will show by an example (taken from [46]), how these bases look like. A general overview is given in the next section.

We have already discussed how implications (i. e., association rules with 100 % confidence) can be read from the line diagram. The Luxenburger basis for approximate association rules (i. e., association rules with less than 100 % confidence) can also be visualized directly in the line diagram of an iceberg concept lattice. It makes use of results of [30] and contains only those rules $B_1 \rightarrow B_2$ where B_1 and B_2 are frequent concept intents and where the concept (B_1', B_1) is an immediate subconcept of (B_2', B_2). Hence there corresponds to each approximate rule in the Luxenburger base exactly one edge in the line diagram. Figure 5 visualizes all rules in the Luxenburger basis for minsupp $= 70$ % and minconf $= 95$ %. For instance, the rightmost arrow stands for the association rule {veil color: white, gill spacing: close} $\rightarrow$ {gill attachment: free}, which holds with a confidence of

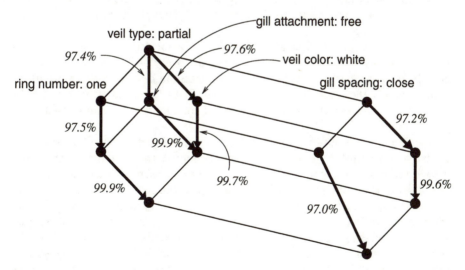

Fig. 5. Visualization of the Luxenburger basis for minsupp $= 70$ % and minconf$= 95$ %

99.6 %. Its support is the support of the concept the arrow is pointing to: 78.52 %, as shown in Figure 2. Edges without label indicate that the confidence of the rule is below the minimum confidence threshold. The visualization technique is described in more detail in [46]. In comparison with other visualization techniques for association rules (as for instance implemented in the IBM Intelligent Miner), the visualization of the Luxenburger basis within the iceberg concept lattice benefits of the smaller number of rules to be represented (without loss of information!), and of the presence of a 'reading direction' provided by the concept hierarchy. It might be worth to combine this advantage with the large effort spent in professional data mining tools to adher to human-computer interaction principles.

6   History and State of the Art in FCA-Based Association Rule Mining

Within the research area of Formal Concept Analysis, important results have been achieved for the association mining problem even before it has been stated. Because of their history, these results obviously did not consider all frequent itemsets, but rather the closed ones. At that point in time, the frequency was not used for pruning, i.e., implicitly a minimum frequency of 0 % was assumed. Early algorithms for computing the concept lattice / all closed itemsets were developed by Fay [19], Norris [33], Ganter [20], and Bordat [9].

Guigues and Duquenne [18] described a minimal set of *implications* (exact rules) from which all rules can be derived, the *Duquenne Guigues base* or *stem base*. In 1984, Ganter developed the Next Closure algorithm ([20], see also [21]) for its computation. Luxenburger was working on bases for *partial implications* (approximative rules) [30, 31]. From today's data mining perspective, all what these results are missing is the (rather straightforward) incorporation of the frequency threshold.

Five years after the first statement of the association mining problem in 1993, apparently the time was ripe for the discovery of its connection to Formal Concept Analysis – it came up independently in several research groups at approximately the same moment [37, 44, 52]. Since then, the attention of FCA has largely increased within the data mining community, and many researchers joined the field.

Inspired by the observation that the frequent closed itemsets are sufficient to derive all frequent itemsets and their supports, the search for other *condensed / concise representations* began. The *key itemsets* as introduced in Section 4 serve the same purpose: from them (plus the minimal non-frequent key itemsets) one can derive the support of any frequent itemset. The key itemsets are also referred to as *minimal generators*, and nowadays more often as *free sets*.

A trend which goes beyond the notions discussed above aims at further condensing the set of free itemsets. The set of *disjunction-free sets* [15] / *disjunction-free generators* [27] is extending the set of free itemsets. A disjunction-free set is an itemset I where there do not exist i_1 and i_2 with $\text{supp}(I) = \text{supp}(I \setminus \{i_1\}) +$

$\text{supp}(I \setminus \{i_2\}) - \text{supp}(I \setminus \{i_1, i_2\})$.[2] The frequent disjunction-free sets together with their support and the minimal non-frequent ones allow to compute the support of all frequent itemsets. This approach is further extended in [16] to *non-derivable itemsets*, where the previous equation is extended from the two elements i_1 and i_2 to an arbitrary number of elements.

A second trend is the analysis of *approximate representations*. One of them are δ–*free sets* [11], where an itemset is called δ–free if there are 'almost no dependencies' between its attributes[3]. They lead to *almost closures* [10], where an attribute m is in the almost–closure of an itemset X iff the supports of X and $X \cup \{m\}$ differ not more than a given threshold. (The usual closure is obtained with this threshold set to zero). The support of any frequent itemset can then be computed up to a certain error.

Amazingly, most of the recent work is focussing on condensed representations of the set of frequent itemsets rather than on condensed representations of the set of association rules. One reason may be that the former is the computationally expensive part, and thus considered as the more interesting problem. At least in the classical association rule setting, however, the analyst is confronted with a list of rules (and not a list of frequent itemsets), hence ultimately *this* list has to be kept small.

In the remainder of this section, we will present some results that have been obtained in FCA-based association rule mining since the late 1990ies. In the next subsection, we discuss algorithms for computing frequent closed and frequent key itemsets. While of course the development of algorithms for computing complete concept lattices goes on, they are out of the focus of the present survey. In the second subsection, we address condensed representations of association rules.

6.1   Algorithms for Computing Frequent Closed / Key Itemsets

The first set of algorithms that were explicitly designed to compute *frequent closed* itemsets were Close [39], Apriori–Close [37] and A-Close [38]. Inspired by Apriori [2], all these algorithms traverse the database in a level-wise approach. A-Close follows a two-step approach. In the first step, it performs a level-wise search for identifying the frequent key sets; in the second step, their closures are determined. Apriori-Close computes simultaneously the frequent closed and all frequent itemsets. Close computes first all frequent closed itemsets by computing the closures of all minimal generators. In a second step, it derives all frequent (closed or non-closed) itemsets. These algorithms put the emphasis on datasets which cannot be kept completely in the main memory. Their major concern is thus to decrease the number of necessary accesses to the hard disk.

CHARM [52] is also following a bottom-up approach. Contrary to A-Close it performs a depth-first search in the powerset of itemsets. It stores the frequent

---

[2] Free sets are a special case with $i_1 = i_2$

[3] A δ-free set is an itemset X such that all non-trivial rules $Y \rightarrow Z$ with $Y, Z \subseteq X$ have at least δ exceptions. Free sets are δ–free sets with $\delta = 0$, as there are no dependencies between their attributes. Because of this property which is known from database keys, free sets were originally called key sets

itemsets in a prefix tree in main memory. The algorithm traverses both the itemset and transaction search spaces. As soon as a frequent itemset is generated, the set of corresponding transactions is compared with those of the other itemsets having the same parent in the prefix tree. If they are equal, then their nodes are merged in the prefix tree, as both generate the same closure. The follow-up CHARM-L [56] of CHARM also computes the covering relation of the iceberg concept lattice.

Closet [41] and Closet+ [50] also compute the frequent closed sets and their supports in a depth-first manner, storing the transactions in a tree structure, called FP-tree, inherited from the FP–Growth algorithm [24]. Each branch of the FP–tree represents the transactions having the same prefix. They use the same merging strategy as CHARM.

Mafia [14] is an algorithm which mainly is intended for computing all maximal frequent itemsets (which, by basic results from FCA, are all closee). It also has an option to compute all frequent closed itemsets. Another very early algorithm for computing the maximal frequent itemsets which makes use of FCA is MaxClosure [17], but it does not provide a significant speed-up compared to Apriori.

TITANIC [45] – as introduced above in detail – continues the tradition of the Close family. It computes in a level-wise manner all frequent key sets, and in the same step their closures. PASCAL [5] differs from TITANIC in that it additionally produces all frequent itemsets. [4] discusses efficient data structures for the algorithms PASCAL and TITANIC.

There are also some approaches which combine the closeness constraint with other constraints. In [25], an algorithm is presented which determines, for given natural numbers k and min_l, the k most frequent closed itemsets which have no less than min_l elements. In [8], only itemsets within a pre-defined order interval of the powerset of the set of items are considered; the closure operator is modified to this scenario (but is no longer a closure operator in the usual sense). Bamboo [51] is an algorithm for mining closed itemsets following a 'length-decreasing support constraint', which states that large itemsets need a lower support to be considered relevant than small ones.

[49] provides an overview over some FCA based algorithms for mining association rules and discusses the data structures underlying the implementations. In particular, they address updates and compositions of the database / the formal context.

6.2   Bases of Association Rules

All existing proposals for bases of association rules distinguish between exact and approximate rules, and present separate bases for each of them. From the combination of both bases one can derive all frequent association rules – by some calculus which is not always stated explicitly by the authors.

As mentioned above, Duquenne/Guigues [18] and Luxenburger [30] presented bases of exact and approximate (frequent or not) association rules rather early. Their results were rediscovered from the data mining community [36, 53] and adapted to the frequency constraint [37, 44, 47, 55]. This approach is described in detail in [46], and sketched above in Section 5. Other approaches comprise the following:

The *min-max basis* [35, 40] is an alternative to the couple of Duquenne/Guigues and Luxenburger base. It is based both on free and on closed itemsets. The min-max base is also divided into two parts. For the exact rules, the set of all rules of the form $A \rightarrow B$ with B being a closed set and being A a generator of B with $A \neq B$ is a basis. For the approximate rules, the set of all rules of the form $A \rightarrow B$ with B being a closed set and being A a generator of some closed set strictly contained in B is a basis. As the min-max basis needs both the free and the closed sets, TITANIC is an appropriate algorithm for its computation.

Depending on the derivation calculus one allows, both the Luxenburger base (as discussed here) and the min-max base can still be reduced further. In fact, Luxenburger showed in [31] that it is sufficient to consider a set of rules which is a spanning tree of the concept lattice. E. g., the rule labeled by 99.7 % in Fig. 5 can be deduced from the rest. Since the deduction rules for this computation turn out to be rather complex for a knowledge discovery application, we decided in [46] to work with a simpler calculus.

In [22], the non-derivable itemsets [16] as described above are used to define a set of *non-derivable association rules*. This set is a lossless subset of the min-max basis, but the reduction comes with the cost of more complex formulae for computing support and confidence of the derivable rules by determining upper and lower bounds, based on set inclusion-exclusion principles.

Based on the disjunction-free generators mentioned above, [27] extends the notion of bases of association rules to rules where disjunction is allowed in the premise.

7   Conclusion

In this survey, we have shown that Formal Concept Analysis provides a strong theory for improving both performance and results of association rule mining algorithms. As the current discussion of 'condensed representations' and – more general – of other applications of Formal Concept Analysis within Knowledge Discovery (e. g., conceptual clustering or ontology learning) show, there remains still a huge potential for further exploitation of Formal Concept Analysis for data mining and knowledge discovery. On the other hand we expect that this interaction will also stimulate further results within Formal Concept Analysis.

References

1. R. Agrawal, T. Imielinski, and A. Swami. Mining association rules between sets of items in large databases. In *Proceedings of the 1993 ACM SIGMOD international conference on Management of Data (SIGMOD'93)*, pages 207–216. ACM Press, May 1993.
2. R. Agrawal and R. Srikant. Fast algorithms for mining association rules in large databases. In *Proceedings of the 20th international conference on Very Large Data Bases (VLDB'94)*, pages 478–499. Morgan Kaufmann, September 1994.

3. R. Agrawal and R. Srikant. Mining sequential patterns. In *Proceedings of the 11th International Conference on Data Engineering (ICDE'95)*, pages 3–14. IEEE Computer Society Press, March 1995.

4. Y. Bastide. *Data Mining: algorithmes par niveau, techniques d'implementation et applications*. PhD thesis, Université de Clermont-Ferrand II, 2000.

5. Y. Bastide, R. Taouil, N. Pasquier, G. Stumme, and L. Lakhal. Mining frequent patterns with counting inference. *SIGKDD Explorations, Special Issue on Scalable Algorithms*, 2(2):71–80, 2000.

6. S.D. Bay. The UCI KDD Archive. Technical report, University of California, Department of Information and Computer Science, Irvine, 99. http://kdd.ics.uci.edu.

7. R. J. Bayardo. Efficiently mining long patterns from databases. In *Proceedings of the 1998 ACM SIGMOD international conference on Management of Data (SIGMOD'98)*, pages 85–93. ACM Press, June 1998.

8. Francesco Bonchi and Claudio Lucchese. On closed constrained frequent pattern mining. In *Proceedings of the 4th IEEE International Conference on Data Mining (ICDM 2004)*, pages 35–42. IEEE Computer Society, 2004.

9. J. P. Bordat. Calcul pratique du treillis de galois d'une correspondance Galois. *Math. Sci. Hum.*, 96:31–47, 1986.

10. Jean-Francois Boulicaut and Artur Bykowski. Frequent closures as a concise representation for binary data mining. In *PADKK '00: Proceedings of the 4th Pacific-Asia Conference on Knowledge Discovery and Data Mining, Current Issues and New Applications*, pages 62–73, London, UK, 2000. Springer-Verlag.

11. Jean-Francois Boulicaut, Artur Bykowski, and Christophe Rigotti. Approximation of frequency queries by means of free-sets. In *PKDD '00: Proceedings of the 4th European Conference on Principles of Data Mining and Knowledge Discovery*, pages 75–85, London, UK, 2000. Springer-Verlag.

12. S. Brin, R. Motwani, and C. Silverstein. Beyond market baskets : Generalizing association rules to correlation. In *Proceedings of the 1997 ACM SIGMOD international conference on Management of Data (SIGMOD'97)*, pages 265–276. ACM Press, May 1997.

13. S. Brin, R. Motwani, J. D. Ullman, and S. Tsur. Dynamic itemset counting and implication rules for market basket data. In *Proceedings of the 1997 ACM SIGMOD international conference on Management of Data (SIGMOD'97)*, pages 255–264. ACM Press, May 1997.

14. D. Burdick, M. Calimlim, and J. Gehrke. Mafia: A maximal frequent itemset algorithm for transactional databases. In *Proc. of the 17th Int. Conf. on Data Engineering*. IEEE Computer Society, 2001.

15. Artur Bykowski and Christophe Rigotti. A condensed representation to find frequent patterns. In *PODS '01: Proceedings of the twentieth ACM SIGMOD-SIGACT-SIGART symposium on Principles of database systems*, pages 267–273, New York, NY, USA, 2001. ACM Press.

16. Toon Calders and Bart Goethals. Mining all non-derivable frequent itemsets. In *PKDD '02: Proceedings of the 6th European Conference on Principles of Data Mining and Knowledge Discovery*, pages 74–85, London, UK, 2002. Springer-Verlag.

17. D. Cristofor, L. Cristofor, and D.A. Simovici. Galois Connections and Data Mining. *Journal of Universal Computer Science*, 6(1):60–73, January 2000.

18. V. Duquenne and J.-L. Guigues. Famille minimale d'implications informatives résultant d'un tableau de données binaires. *Mathématiques et Sciences Humaines*, 24(95):5–18, 1986.

19. G. Fay. An algorithm for finite Galois connections. Technical report, Institute for Industrial Economy, Budapest, 1973.
20. Bernhard Ganter. Two basic algorithms in concept analysis. FB4–Preprint 831, TH Darmstadt, 1984.
21. Bernhard Ganter and Klaus Reuter. Finding all closed sets: a general approach. *Order*, 8:283–290, 1991.
22. Bart Goethals, Juhu Muhonen, and Hannu Toivonen. Mining non-derivable association rules. In *Proc. SIAM International Conference on Data Mining*, Newport Beach, CA, April 2005.
23. J. Han and M. Kamber. *Data Mining: Concepts and Techniques*. Morgan Kaufmann, Sept. 2000.
24. J. Han, J. Pei, and Y. Yin. Mining frequent patterns without candidate generation. In *Proc. ACM SIGMOD Int'l Conf. on Management of Data*, pages 1–12, May 2000.
25. Jiawei Han, Jianyong Wang, Ying Lu, and Petre Tzvetkov. Mining top-k frequent closed patterns without minimum support. In *Proceedings of the 2002 IEEE International Conference on Data Mining (ICDM 2002)*, pages 211–218. IEEE Computer Society, 2002.
26. M. Kamber, J. Han, and Y. Chiang. Metarule-guided mining of multi-dimensional association rules using data cubes. In *Proc. of the 3rd KDD Int'l Conf.*, August 1997.
27. Marzena Kryszkiewicz. Concise representation of frequent patterns based on disjunction-free generators. In *ICDM '01: Proceedings of the 2001 IEEE International Conference on Data Mining*, pages 305–312, Washington, DC, USA, 2001. IEEE Computer Society.
28. B. Lent, R. Agrawal, and R. Srikant. Discovering trends in text databases. In *Proceedings of the 3rd international conference on Knowledge Discovery and Data mining (KDD'97)*, pages 227–230. AAAI Press, August 1997.
29. D. Lin and M. Kedem. A new algorithm for discovering the maximum frequent set. In *Proceedings of the 6th Int'l Conf.on Extending Database Technology (EDBT)*, pages 105–119, March 1998.
30. M. Luxenburger. Implications partielles dans un contexte. *Mathématiques, Informatique et Sciences Humaines*, 29(113):35–55, 1991.
31. Michael Luxenburger. *Implikationen, Abhängigkeiten und Galois–Abbildungen.* PhD thesis, TH Darmstadt, 1993. Shaker Verlag, Aachen, 1993. In english language, beside the introduction).
32. H. Mannila. Methods and problems in data mining. In *Proceedings of the 6th biennial International Conference on Database Theory (ICDT'97)*, Lecture Notes in Computer Science, Vol. 1186, pages 41–55. Springer-Verlag, January 1997.
33. Eugene M. Norris. An algorithm for computing the maximal rectangles in a binary relation. *Rev. Roum. Math. Pures et Appl.*, 23(2):243–250, 1978.
34. J. S. Park, M.-S. Chen, and P. S. Yu. An efficient hash based algorithm for mining association rules. In *Proceedings of the 1995 ACM SIGMOD international conference on Management of Data (SIGMOD'95)*, pages 175–186. ACM Press, May 1995.
35. N. Pasquier. Extraction de bases pour les règles d'association à partir des itemsets fermés fréquents. In *Actes du 18ème congrès sur l'Informatique des Organisations et Systèmes d'Information et de Décision (INFORSID'2000)*, May 2000.
36. N. Pasquier, Y. Bastide, R. Taouil, and L. Lakhal. Pruning closed itemset lattices for association rules. In *Actes des 14èmes journées Bases de Données Avancées (BDA'98)*, pages 177–196, Octobre 1998.

37. N. Pasquier, Y. Bastide, R. Taouil, and L. Lakhal. Closed set based discovery of small covers for association rules. In *Actes des 15èmes journées Bases de Données Avancées (BDA'99)*, pages 361–381, Octobre 1999.
38. N. Pasquier, Y. Bastide, R. Taouil, and L. Lakhal. Discovering frequent closed itemsets for association rules. In *Proceedings of the 7th biennial International Conference on Database Theory (ICDT'99)*, Lecture Notes in Computer Science, Vol. 1540, pages 398–416. Springer-Verlag, January 1999.
39. N. Pasquier, Y. Bastide, R. Taouil, and L. Lakhal. Efficient mining of association rules using closed itemset lattices. *Information Systems*, 24(1):25–46, 1999.
40. Nicolas Pasquier, Rafik Taouil, Yves Bastide, Gerd Stumme, and Lotfi Lakhal. Generating a condensed representation for association rules. *Journal of Intelligent Information Systems*, 24(1):29–60, 2005.
41. J. Pei, J. Han, and R. Mao. Closet: An efficient algorithm for mining frequent closed itemsets. In *ACM SIGMOD Workshop on Research Issues in Data Mining and Knowledge Discovery*, pages 21–30, 2000.
42. A. Savasere, E. Omiecinski, and S. Navathe. An efficient algorithm for mining association rules in larges databases. In *Proceedings of the 21st international conference on Very Large Data Bases (VLDB'95)*, pages 432–444. Morgan Kaufmann, September 1995.
43. C. Silverstein, S. Brin, and R. Motwani. Beyond market baskets : Generalizing association rules to dependence rules. *Data Mining and Knowledge Discovery*, 2(1):39–68, January 1998.
44. G. Stumme. Conceptual knowledge discovery with frequent concept lattices. FB4-Preprint 2043, TU Darmstadt, 1999.
45. G. Stumme, R. Taouil, Y. Bastide, N. Pasqier, and L. Lakhal. Computing iceberg concept lattices with titanic. *J. on Knowledge and Data Engineering*, 42(2):189–222, 2002.
46. G. Stumme, R. Taouil, Y. Bastide, N. Pasquier, and L. Lakhal. Intelligent structuring and reducing of association rules with formal concept analysis. In F. Baader, G. Brewker, and T. Eiter, editors, *KI 2001: Advances in Artificial Intelligence Proc. KI 2001*, volume 2174 of *LNAI*, pages 335–350. Springer, Heidelberg, 2001.
47. R. Taouil. *Algorithmique du treillis des fermés : application à l'analyse formelle de concepts et aux bases de données*. PhD thesis, Université de Clermont-Ferrand II, 2000.
48. H. Toivonen. *Discovery of frequent patterns in large data collection*. PhD thesis, University of Helsinki, 1996.
49. P. Valtchev, R. Missaoui, and R. Godin. Formal concept analysis for knowledge discovery and data mining: The new challenges. In Peter W. Eklund, editor, *Concept Lattices*, volume 2961 of *Lecture Notes in Computer Science*, pages 352–371. Springer, 2004.
50. Jianyong Wang, Jiawei Han, and Jian Pei. Closet+: searching for the best strategies for mining frequent closed itemsets. In *KDD '03: Proceedings of the ninth ACM SIGKDD international conference on Knowledge discovery and data mining*, pages 236–245, New York, NY, USA, 2003. ACM Press.
51. Jianyong Wang and George Karypis. Bamboo: Accelerating closed itemset mining by deeply pushing the length-decreasing support constraint. In Michael W. Berry, Umeshwar Dayal, Chandrika Kamath, and David B. Skillicorn, editors, *Proceedings of the Fourth SIAM International Conference on Data Mining*. SIAM, 2004.
52. M. J. Zaki and C.-J. Hsiao. Chaarm: An efficient algorithm for closed association rule mining. technical report 99–10. Technical report, Computer Science Dept., Rensselaer Polytechnic, October 1999.

53. M. J. Zaki and M. Ogihara. Theoretical foundations of association rules. In *DMKD'98 workshop on research issues in Data Mining and Knowledge Discovery*, pages 1–8. ACM Press, June 1998.
54. M. J. Zaki, S. Parthasarathy, M. Ogihara, and W. Li. New algorithms for fast discovery of association rules. In *Proceedings of the 3rd international conference on Knowledge Discovery and Data mining (KDD'97)*, pages 283–286. AAAI Press, August 1997.
55. Mohammed J. Zaki. Generating non-redundant association rules. In *Proc. KDD 2000*, pages 34–43, 2000.
56. Mohammed J. Zaki and Ching-Jui Hsaio. Efficient algorithms for mining closed itemsets and their lattice structure. *IEEE Transactions on Knowledge and Data Engineering*, 17(4):462–478, April 2005.

Galois Connections in Data Analysis: Contributions from the Soviet Era and Modern Russian Research

Sergei O. Kuznetsov

All-Russia Institute for Scientific and Technical Information (VINITI)
Usievicha 20, 125190 Moscow, Russia
`serge@viniti.ru`

Abstract. A retrospective survey of several research directions at the All-Soviet (now All-Russia) Institute for Scientific and Technical Information (VINITI), as well as research represented in several VINITI editions, is proposed. In a number of papers of the 1970-1980s, taxonomies (classifications) were naturally considered as lattices. Several problems of classification required consideration of tolerance relations as a model of similarity of objects. Such relations define symmetric formal contexts. A JSM-method of inductive plausible reasoning, which has been developed at VINITI since the early 1980s, is considered in terms of Galois connections and concept lattices. Mathematical research around the JSM-method and its applications is discussed.

1 Introduction

The research on Galois connections in the classification school at the All-Soviet (now All-Russia) Institute for Scientific and Technical Information (VINITI), Moscow, was first motivated by problems of classification and storage of documents, which needed formal models of similarity of objects. Later motivations came from problems of data analysis in various applied domains and their solution by means of the JSM-method of hypothesis generation.

In this article we give a review of the research activity in VINITI and/or in its journals, mainly *Nauchno-Tekhnicheskaya Informatsiya* (NTI)[1], series 2, translated to English by Allerton Press under the name *Automated Documentation and Mathematical Linguistics*, and also in *Semiotika i Informatika* (Semiotics and Computer Science), *Itogi Nauki i Tekhniki* (Reviews in Science and Technology).

Around the mid 1960s, Yulii A. Shreider (1927-1998), one of the leading researchers of VINITI, considered the problem of automatic classification of documents and their retrieval by means of a model consisting of a triple (M, L, f), where M is a set of objects (documents), L is a set of attributes and $f: M \to \mathcal{P}(\mathcal{L})$ is a mapping taking each object to a set attributes from L [77]. Similarity of

---

[1] With Prof. Ruggero S. Gilyarevsky being the editor-in-chief for four decades, first *de facto*, and later also *de jure*

B. Ganter et al. (Eds.): Formal Concept Analysis, LNAI 3626, pp. 196–225, 2005.

documents x and y was defined by nonemptyness of the set of their common attributes: $f(x) \cap f(y) \neq \emptyset$. Defined in this way similarity is reflexive and symmetric, i.e., similarity is a tolerance relation on the set of objects.

Shreider mentioned the relevance of lattices to problems of classification and mathematical retrieval in his early paper [77], where he also cited the work of Soergel [82] on this issue. In [80] Shreider wrote about classifications of objects described by attributes, where each classification is given by an idempotent commutative semigroup (which is actually a semilattice) uniquely specified by bases (actually, by sets of irreducible elements). Implication between single attributes, analogous to that in Formal Concept Analysis (FCA), was defined. Together with Sergei V. Meien, a biological methodologist from St. Petersburg, he wrote on the duality of taxonomies and meronomies (the latter term, denoting a hierarchy of parts, was coined by Meien from the Greek word $\mu\epsilon\rho o\zeta$, part) [59][2]. This was almost like the starting point of FCA [89], however, no systematic theory appeared. Two directions of thought, the one of them related to the (semi)lattice nature of classification and the other one, which considered tolerance relations on the set of objects and their classes given by Galois correspondences, developed independently. An analogue of concept lattice theory appeared later, in mid 1980s, in works by O.M. Polyakov and V.V. Dunaev [14, 72–74].

To provide a general framework for the overview of research in different groups, we will use the standard definitions of Formal Concept Analysis [27, 89], which we will briefly recall below.

Let G and M be sets and $I \subseteq G \times M$ be a relation. The elements of G and M are called the objects and attributes, respectively, and gIm (i.e., $(g, m) \in I$) is read: the object g has the attribute m. The triple $\mathbb{K} = (G, M, I)$ is called a *formal context*. The *derivation operators*, defined for any $A \subseteq G$ and $B \subseteq M$ by

$$A^I := \{m \in M \mid gIm \text{ for all } g \in A\}, \quad B^I := \{g \in G \mid gIm \text{ for all } m \in B\}$$

induce a *Galois connection* between the ordered powersets $(\mathcal{P}(G), \subseteq)$ and $(\mathcal{P}(M), \subseteq)$. In the case of a fixed relation I one usually writes $(\cdot)'$ instead of $(\cdot)^I$. Any pair of sets (A, B) such that $A \subseteq G$, $B \subseteq M$, $A' = B$ and $B' = A$ is called a *formal concept* of the context $\mathbb{K}$ with *(formal) extent* A and *(formal) intent* B. The set of all formal concepts of a formal context $\mathbb{K}$ forms a complete lattice called *(formal) concept lattice* $\mathfrak{B}(\mathbb{K})$. Moreover, for an arbitrary complete lattice L, there is a concept lattice isomorphic to it. For $A, C \subseteq M$ the *implication* $A \to C$ holds if $A' \subseteq C'$ (or $C \subseteq A''$), i.e., if all objects from G that have the set of attributes A also have the set of attributes C.

The issues around Galois connections and functional dependencies motivated many other researchers. For example, in [11] the equivalence of the category of functional dependencies to the category of closure systems (Moore families) [6] was considered. In [84] the author studies implications between binary attributes. The definition of implication was not extended to sets of attributes as it had been done previously in FCA. The author considers the equivalence relation on objects (described by same sets of attributes) and studies the cases where the

---

[2] See also www.ento.vt.edu/~sharov/biosem/shreidr/shreidr.html

order defined by implications induces a Boolean algebra, as well as the possibility of embedding elements of a binary relation with implication in Boolean algebras by mappings called strict homomorphisms by the author (more known under the name *order embeddings*): for two ordered sets A and B a mapping $f: A \to B$ is called a *strict homomorphism* if $f(a_1) \le f(a_2)$ iff $a_1 \le a_2$ for any $a_1, a_2 \in A$.

The rest of the paper is organized as follows. In the second section we consider models of taxonomies and their relation to certain type of dependencies in databases. In the third section we consider models of similarity based on tolerance relations, classes of tolerance, and their relation to FCA. In the fourth section we give a review of the research around the JSM-method of hypothesis generation, a machine learning method naturally formalized in terms of Galois connections and FCA.

2 Taxonomies and Dependencies

After [80] the next step in the development of Shreider's classification model was made in [76], where objects are described by attributes, taxonomies are defined as refining sequences of coverings of object sets, meronomies are defined as refining sequences of coverings of attribute sets (see exact definitions below). An archetype was defined as a description common to all objects from a taxon (i.e., a member of classification). The authors relate this construction to the notion of a concept, its intent and extent, noticing that the size of the former decreases with the growth of size of the latter. However, no further mathematical theory was proposed. A theory that would pass completely to these methodological considerations is exactly that of FCA. Together with [59], where the duality of taxonomies and meronomies was underlined, the paper [76] is actually a prolegomena to FCA. So it is not surprising that a counterpart of FCA notions appeared later (and few years later than similar work in French and German schools) in the classification theory by Polyakov and Dunaev [14, 72–74].

They started from object-attribute representation, defined Galois correspondence as it is done in FCA and obtained two antiisomorphic complete lattices on sets of objects (called taxonomy) and sets of attributes (called meronomy). They stated that both lattices can be generated by the sets of corresponding irreducibles. Polyakov and Dunaev also considered relations between sets of objects, i.e., of the form $I \subseteq G \times G$, since to their minds, relations between objects induce taxonomies (see below) and often come before attributes. Moreover, attributes often result from the observation of relations between objects, for example the Mendeleev periodic table of chemical elements, discussed in [14], resulted from ordering of the objects (chemical elements) with respect to their atomic weights. This ordering motivated further study of properties of classes of chemical elements.

Several results from FCA were repeated by Dunaev and Polyakov: Representing concept lattices by products of concept lattices of subcontexts ([73, Theorem 2]) which allowed them to draw lattices in the way as it is done with nested line diagrams in FCA; describing morphisms of concept lattices specified by subsets of the set of all attributes ([73, Theorem 3]).

An aspect of their research that has not been previously covered by research in FCA is related to the study of multivalued and mutual dependencies, which are generalizations of functional dependencies. They showed how dependencies of this kind allow for decomposition of taxonomies into products. Below we present some definitions and results from [74].

Let $D(U)$ be a finite set of objects (U is the *name* of this set). A *taxonomy* $\mathcal{T}$ consisting of a system of *taxons*, which are subsets $D(U)$, is given as follows: $D(U)$ is a taxon; and if $T_1 \in \mathcal{T}$, $T_2 \in \mathcal{T}$ are taxons, then $T_1 \cap T_2$ is a taxon.

Obviously, taxonomy defined in this way is a *closure system* [27] (or, equivalently, a *Moore family* [6]) and the set of all taxons induces a lattice. In terms of FCA this is the lattice of extents. When objects are described in terms of attributes, then the dual lattice, called *meronomy*, on closed sets of attributes arises. In terms of FCA this is the lattice of intents.

A *product of taxonomies* [74] $\mathcal{T}_1$ and $\mathcal{T}_2$ on the same set of objects, denoted by $\mathcal{T}_1 \cdot \mathcal{T}_2$, is a taxonomy such that $T \in \mathcal{T}_1 \cdot \mathcal{T}_2$ iff $T = T_1 \cap T_2$ for some $T_1 \in \mathcal{T}_1$, $T_2 \in \mathcal{T}_2$. The order on taxonomies $\leq$ defined as $\mathcal{T}_1 \leq \mathcal{T}_2$ iff $\mathcal{T}_1 \cdot \mathcal{T}_2 = \mathcal{T}_1$ induces a lattice on the set K of all taxonomies of the set $D(U)$. This order on taxonomies is obviously related to refinement order on closure systems [27] and apposition of contexts.

An *attribute* X is given in [74] as a pair $\langle D(X), sX \rangle$, where $D(X)$ is the set of attribute values, sX is the "object-attribute value" relation ($sX \subseteq D(U) \times D(X)$). In terms of FCA, X is a *many-valued attribute* with the set of values $D(X)$. The relation sX defines its *scaling* already at the many-valued level: in contrast to FCA, there is no implicit dependencies of values that are specified by the choice of a scaling, i.e., a method of reduction to one-valued attributes. All possible (object, attribute value) pairs are given explicitly.

Criteria of decomposition of a taxonomy lattice resulting from a set of many-valued attributes into products of taxonomy lattices arising from single attributes are given in [74]. These criteria were given in terms of multi-valued and mutual dependencies.

Recall from [57] that a *functional dependency* $U \to X$ holds if for any $u \in D(U)$, $x, \tilde{x} \in D(X)$ the relations $(u, x) \in sX$, $(u, \tilde{x}) \in sX$ imply $x = \tilde{x}$. *Multivalued dependency* $U \to X$ holds if for any $u \in U$ the facts $(u, x, y) \in sV$ and $(u, x', y') \in sV$ imply $(u, x, y') \in sV$. The functional dependency $U \to X$ obviously implies $U \to X$. *Mutual dependency* [66] $U \simeq X$ holds if for every $u \in D(U)$ the facts $(u, x, y) \in sV$, $(u, x', y') \in sV$ and $(x, y') \in sV[XY]$ imply $(u, x, y') \in sV$. Multivalued dependency is obviously a particular type of mutual dependency.

For two attributes X and Y their (*natural*) *join* $sX \bowtie sY$ is defined as follows: $(u, x, y) \in sX \bowtie sY$ iff $(u, x) \in sX$ and $(u, y) \in sY$. By a theorem from [57], the decomposition $sV = \bowtie_{i=1}^{n} sX_i$ is possible iff there exists a set of multivalued dependencies $U \to X_i$, $i = \{1, \ldots, n\}$.

The decomposition $sV = sX \bowtie sY \bowtie sV[XY]$ is possible iff there exists mutual dependency $U \simeq X$ [66]. The following propositions from [74] give criteria of decomposition of a taxonomy lattice arising from the whole set of attributes into products of taxonomy lattices arising from single attributes.

Proposition 1 *Let* $sX_i = sV[UX_i]$, *where* $V = X_1 \ldots X_n$, $X_i \cap X_j = \emptyset$ *for any* $i \neq j$; $i, j = 1, \ldots, n$. *Mutual dependencies* $U \simeq X_i$ *hold for all* $i = \{1, \ldots, n\}$ *iff for any* $(x_1, \ldots, x_n) \in sV[V]$ *the relation*

$$\{u \in D(U) \mid (u, x_1, \ldots, x_n) \in sV\} = \bigcap_{i=1}^{n} \{u \in D(U) \mid (u, x_i) \in sX_i\}$$

holds.

In terms of taxonomies this result can be recast in the following form.

Proposition 2 *Let* $sX_i = sV[UX_i]$, *where* $V = X_1 \ldots X_n$, $X_i \cap X_j = \emptyset$ *for all* $i \neq j$, $i, j = 1, \ldots, n$. *If mutual dependencies* $U \simeq X_i$ *hold for all* $i = \{1, \ldots, n\}$, *then* $\mathcal{T}(V) \subseteq \mathcal{T}(X_1) \cdot \ldots \cdot \mathcal{T}(X_n)$.

As a corollary one has the following

Proposition 3 *If* $sX_i = sV[UX_i]$, *where* $V = X_1 \ldots X_n$, $X_i \cap X_j = \emptyset$ *for* $i \neq j$ *and multivalued dependencies* $U \rightarrow X_i$ *hold for all* $i = \{1, \ldots, n\}$, *then* $\mathcal{T}(V) = \mathcal{T}(X_1) \cdot \ldots \cdot \mathcal{T}(X_n)$.

In fact, if multivalued dependencies $U \rightarrow X_i$ hold for all $i \in \{1, \ldots, n\}$, then $\mathcal{T}(V) \supseteq \mathcal{T}(X_1) \cdot \ldots \cdot \mathcal{T}(X_n)$.

3   Tolerance Relations: Symmetric Contexts

In the works of V.Ya. Gusakov and S.M. Gusakova (Yakubovich) classes of a tolerance relation were studied. This study of tolerance was motivated first by modeling similarity of documents in document retrieval systems [34, 35, 91, 92].

It was Zeeman [98] who proposed first to formalize similarity as a tolerance (reflexive and symmetric) relation. The relation of similarity, being naturally reflexive and symmetric, should not be transitive: e.g., children are often similar to both their parents, the latter being very different. Although some authors, like Tversky [85] doubt that similarity is naturally symmetric and reflexive, this seems to be adequate to model similarity between documents.

Definition 3.1 For a set G a binary relation $T \subseteq G \times G$ is called *tolerance* if
   (1) $\forall x \in G \ xTx$ (reflexivity)
   (2) $\forall x, y \in G \ xTy \rightarrow yTx$ (symmetry)
A set G with tolerance T is called the *space of tolerance* and denoted by G_T.

Definition 3.2 A subset $K \subseteq G$ is called a *class of tolerance* if
   (1) $\forall x, y \in K, \ xTy$,
   (2) $\forall z \notin K \ \exists u \in K \neg (zTu)$
An arbitrary subset of a class of tolerance is called a *preclass*.

Definition 3.3 A set $\mathcal{A} = \{A_j\}_{j \in J}$ of preclasses is called a *system of preclasses preserving* T if

$$T = \bigcup_{j \in J} A_j \times A_j.$$

The most important preserving system of preclasses for the tolerance T is the system of all classes, which is denoted by $\mathcal{K}(G_T)$.

Tolerance classes defined by a tolerance relation are cliques (inclusion-maximal complete subgraphs) of the graph (G, T). On the other hand, a tolerance relation can be considered as an origin of formal context representation. First, some objects are observed to be pairwise similar. Then all pairs of the tolerance relation, and further on, the set all of maximal classes of similarity (classes of tolerance) is constructed. Eventually, the classes are given names, which are further used as attributes that describe objects.

By symmetry of the tolerance relation T, the Galois connection associated with the context (G, G, T) is given by a single mapping $(\cdot)^T$, where x^T is a set of all elements from G tolerant to x and X^T is the set of all elements from G tolerant to each $x \in X$.

Let $\mathcal{L}$ be a system of preclasses preserving tolerance T on the set G, then the context $(G, \mathcal{L}, I)$ is defined as usual: for an object $g \in G$ and a preclass $L \in \mathcal{L}$ one has gIL iff $g \in L$. The Galois connection given by the derivation operator $(\cdot)^I$ is called the *Galois connection that agrees with the tolerance T by the preserving system $\mathcal{L}$*.

The following relation from [37] recast in FCA terms gives so called *canonical representation of similarity*.

Proposition 4 *Let G be a set and $T \subseteq G \times G$ a tolerance relation and let $\mathcal{A}$ be a system of preclasses preserving T. Then $(G, \mathcal{A}, \in)$ is a formal context satisfying*

$$(g, h) \in T \Longleftrightarrow g' \cap h' \neq \emptyset \text{ for all } g, h \in G.$$

Conversely, if (G, M, I) is a context with $g' \neq \emptyset$ for all $g \in G$, then $T := \{(g, h) \mid g' \cap h' \neq \emptyset\}$ is a tolerance relation and $\mathcal{A} := \{m' \mid m \in M\}$ is a system of preclasses preserving T.

Thus, each tolerance can be obtained from some formal context and in turn, an arbitrary tolerance gives rise to a formal context: Starting from a tolerance relation, one finds classes of tolerance, which, after being named, can be further used as attributes.

The results obtained for tolerances were partially extended to the case of n-ary relations in [36], where the notions of n-ary tolerance relation and the corresponding definitions of a class, preclass, preserving system of preclasses, and basis are introduced.

In the 1980s a motivation for the further study of tolerance [36–39] came from the theory of plausible reasoning [17] based on similarity operation (see next section about the JSM-method). In [36, 37] the relationship between so-called global and local similarities was studied.

In [36] the following two definitions of similarity arising from formal contexts (called *karta*, map) were considered.

Definition 3.5. Objects $g_1, \ldots, g_n$ are *n-locally similar* in the context $\mathbb{K} = (G, M, I)$ if $g_i \in G$ $(i = 1, \ldots, n)$ and $\{g_1, \ldots, g_n\}^I \neq \emptyset$.

In terms of FCA, locally similar objects are exactly those that occur together in a nontrivial formal extent of the formal context $\mathbb{K} = (G, M, I)$. The sets of n-locally similar objects induce a tolerance on $G \times G$, these sets being preclasses of the tolerance.

Definition 3.6. Objects $g_1, \ldots, g_k$ are *globally similar* in the context $\mathbb{K} = (G, M, I)$ if $m' = \{g_1, \ldots, g_k\}$ for some $m \in M$.

So, a set of globally similar objects is an attribute extent of the formal context $\mathbb{K} = (G, M, I)$. Global similarity is not a relation, because it involves tuples of varying length. A global similarity on the set G can be represented by a covering $\pi = \{\pi_j\}_{j \in J}$ of G, where each π_j is a set of globally similar elements. A global similarity $\langle G, \pi \rangle$ *is represented by n-local similarity* if $\pi = \mathcal{K}$, where $\mathcal{K}$ is the set of all classes of the tolerance T induced by the n-local similarity [36].

To satisfy this condition, the global similarity should not give rise to new classes of the tolerance induced by the n-local similarity, and the maximal sets of n-locally similar objects should be globally similar.

In [92] the notion of a conjugate tolerance to a tolerance $T \subseteq G \times G$ was defined as a tolerance relation on the set $\mathcal{K}(G_T)$ of classes of T: a pair of classes belong to the conjugate tolerance if they are not disjoint. The relation between conjugated spaces to the initial tolerances, as well as sequences of conjugations, were studied in [92].

A further generalization of the similarity models was nonsymmetric similarity relation considered in [38], where criteria for canonical representation of nonsymmetric relation was given in terms of preclasses preserving relation.

4   JSM-Method

The initial motivation for the first version of the JSM-method proposed by Viktor K. Finn in late 1970s was the intention to describe induction in purely deductive form and give at least partial justification of induction. The method was named in honor of the English philosopher John Stuart Mill, who proposed schemes of inductive reasoning in the 19th century. Most well-known are the first and second canons of inductive logic [63].

The first canon, also called *Method of Agreement*, was formulated as follows: "If two or more instances of the phenomenon under investigation have only one circumstance in common, ... [it] is the cause (or effect) of the given phenomenon."

The second canon or *Method of Difference* sounds like: "If an instance in which the phenomenon under investigation occurs, and an instance in which it does not occur, have every circumstance in common save one, that one occurring only in the former; the circumstance in which alone the two instances differ, is the effect, or the cause, or an indispensable part of the cause, of the phenomenon."

To formalize the Mill's methods, Finn and colleagues used the principle of two-layered logics of Dmitrii A. Bochvar[3] [10]: Several truth types, including "empirical contradiction" between generalizations of data, were allowed at the internal logical level and classical logical values are used at the external level. More precisely, the JSM-method was described by means of a many-valued many-sorted extension of the First-Order Predicate Logic

with quantifiers over tuples of variable length (this logic is a proper part of the second order logic, often called *weak second order logic*).

The motivation for the use of quantifiers over tuples of variable length is as follows: Induction is based on the observation of similarity of objects. Since the number of objects with a particular similarity is not known in advance, quantification over tuples of variable length is necessary in the case of infinite number of objects, to express their similarity.

For example, the Mill's Method of Agreement is the formalized by the following predicate $\mathcal{M}_{a,n}^{+}(V,W)$ (some other Mill's canons, e.g., the method of differences, as well as new methods of inductive reasoning were described in similar way):

$$\mathcal{M}_{a,n}^{+}(V,W) := \exists k \, \widetilde{\mathcal{M}}_{a,n}^{+}(V,W,k),$$

$$\widetilde{\mathcal{M}}_{a,n}^{+}(V,W,k) := \exists Z_1 \ldots \exists Z_k \, \exists U_1 \ldots U_k \, (\overset{k}{\underset{i=1}{\&}} \, J_{\langle 1,n \rangle}(Z_i \Rightarrow_1 U_i) \, \& $$

$$\& \, \forall U (J_{\langle 1,n \rangle}(Z_i \Rightarrow_1 U) {\to} \mathcal{U} \subset U_i)) \, \& (Z_1 \cap \ldots \cap Z_k) = V \, \& \, V \neq \emptyset \, \& \, W \neq \emptyset \, \& $$

$$\& \, \forall i \, \forall j ((i \neq j) \, \& \, 1 \leq i,j \leq k){\to} Z_i \neq Z_j) \, \& \, \forall X \, \forall Y ((J_{\langle 1,n \rangle}(X \Rightarrow_1 Y) \, \& $$

$$\& \, \forall \mathcal{U}(J_{\langle 1,n \rangle}(X \Rightarrow_1 U) {\to} \mathcal{U} \subseteq Y) \, \& \, \& \, V \subset X) {\to} (W \subseteq Y \, \& (\overset{k}{\underset{i=1}{\vee}}(X = Z_i)))) \, \& \, k \geq 2).$$

Here, $J_{\langle \varepsilon,n \rangle}$ is a Rosser-Turquette operator taking formulas of many-valued "internal" logic to two classical logic values: $\varepsilon \in \{-1,0,1,\tau\}$, -1 denotes "empirically false," 1 denotes "empirically true," 0 denotes "empirical contradiction", and τ denotes "empirically undeterminate". n denotes the number of iteration step, which is an important feature of the JSM-method.

From the agreement (similarity) predicate one can construct some other predicates imposing additional conditions, like the following one:

$$\forall X \, \forall Y ((V \subseteq X \, \& \, W \subseteq Y) \to (J_{\langle 1,n \rangle}(X \Rightarrow_1 Y) \vee (J_{\langle \tau,n \rangle}(X \Rightarrow_1 U))),$$

which is called "no counterexample" or "counterexample forbidding". Additional conditions (conjunctively added to the main agreement predicate) make the "lattice of methods".

Upon construction of all pairs (V,W) by a certain method, one uses them for classification of new examples. When the latter are classified, they are added

---

[3] In his 1938 paper Bochvar proposed one of the first many-valued logics for the treatment of the liar paradox, where there were two types of logical values: the inner values were "true", "false" and "contradiction", whereas the external values were classical "true" and "false"

(now, as new positive or negative examples) to the initial sets of positive and negative examples, and the whole procedure is iterated.

Algebraic redefinitions of inductive methods started from the observation that the agreement predicate defines a Moore family w.r.t. $\bigcap$ with the set of generators given by sets of attributes each of which describes a positive example (the operation $\bigcap$ is a means of expressing "similarity" of objects described by attribute sets). This observation allowed redefinition of hypotheses [45] as pairs of the form

$$\langle V, \{Z_1, \ldots, Z_k\}\rangle : V = Z_1 \cap \ldots \cap Z_k, \quad \forall Z \in D \setminus \{Z_1, \ldots, Z_k\} \quad V \not\subseteq X, \qquad (1)$$

where $V, Z_1, \ldots, Z_k \subseteq U$ for some set of attributes U and $D = \{Z_1, \ldots, Z_n\}$ is the set of all positive examples (given as sets of attributes that describe them) of the phenomenon W.

In [45] the equivalence of pairs $\langle V, \{Z_1, \ldots, Z_k\}\rangle$ to bicliques (inclusion-maximal complete bipartite graphs) of a bipartite graph was shown. Some years later the equivalence between pairs of this form (with components interchanged) and formal concepts was recognized [47].

The following definition of a hypothesis ("no counterexample-hypothesis") in FCA terms was given in [24]:

Let a context $\mathbb{K} = (G, M, I)$ be given. In addition to attributes of M, a *target attribute* $\omega \notin M$ is considered. This partitions the set G of all objects into three subsets: The set G_+ of those objects that are known to have the property ω (these are the *positive examples*), the set G_- of those objects of which it is known that they do not have ω (the *negative examples*) and the set G_τ of *undetermined examples*, i.e., of those objects, of which it is unknown if they have property ω or not. This gives three subcontexts of $\mathbb{K} = (G, M, I)$, the first two staying for the training sample:

$$\mathbb{K}_+ := (G_+, M, I_+), \quad \mathbb{K}_- := (G_-, M, I_-), \quad \text{and } \mathbb{K}_\tau := (G_\tau, M, I_\tau),$$

where for $\varepsilon \in \{+, -, \tau\}$ we have $I_\varepsilon := I \cap (G_\varepsilon \times M)$ and the corresponding derivation operators are denoted by $(\cdot)^+$, $(\cdot)^-$, $(\cdot)^\tau$, respectively.

A subset $h \subseteq M$ is a *simple positive hypothesis* for ω if it satisfies the positive agreement predicate (see above) and does not satisfy the (symmetrically formulated) negative predicate). In terms of FCA,

$$h^{++} = h \quad \text{and} \quad h^{--} \neq h.$$

Another type of hypothesis which are mostly used in practice, namely "no counterexample hypothesis" [17, 18] (in what follows, we call it just a *positive hypothesis*), is an intent of $\mathbb{K}_+$ such that $h^+ \neq \emptyset$ and $h \not\subseteq g^- := \{m \in M \mid (g, m) \in I_-\}$ for any negative example $g \in G_-$. Equivalently,

$$h^{++} = h \quad \text{and} \quad h' \cap G_- \neq \emptyset,$$

where $(\cdot)'$ is taken in the whole context $\mathbb{K} = (G, M, I)$. An intent of $\mathbb{K}_+$ that is contained in the intent of a negative example is a *falsified (+)-generalization*.

Negative hypotheses and falsified generalizations are defined similarly. Hypotheses can be used to classify undetermined examples: If the intent

$$g^\tau := \{m \in M \mid (g,m) \in I_\tau\}$$

of an object $g \in G_\tau$ contains a positive, but no negative hypothesis, then g^τ is *classified positively*. Negative classifications are defined similarly. If g^τ contains hypotheses of both kinds, or if g^τ contains no hypothesis at all, then the classification is contradictory or undetermined, respectively. In this case one can apply standard probabilistic techniques known in machine learning and data mining (majority vote, Bayesian approach, etc.). Obviously, for classification purposes it suffices to have only *minimal* (w.r.t. inclusion $\subseteq$) hypotheses, positive as well as negative.

Example 1. Consider the following data table

G \ M	color	firm	smooth	form	target
1 apple	yellow	no	yes	round	+
2 grapefruit	yellow	no	no	round	+
3 kiwi	green	no	no	oval	+
4 plum	blue	no	yes	oval	+
5 toy cube	green	yes	yes	cubic	−
6 egg	white	yes	yes	oval	−
7 tennis ball	white	no	no	round	−

This dataset or *manyvalued context* can be reduced to a context of the form presented above by *scaling* [27], e.g., as follows (scaling 1):

G \ M	w	y	g	b	f	$\bar{f}$	s	$\bar{s}$	r	$\bar{r}$	target
1 apple	×					×	×		×		+
2 grapefruit	×					×		×	×		+
3 kiwi			×			×		×		×	+
4 plum				×		×	×			×	+
5 toy cube			×		×		×			×	−
6 egg	×				×		×			×	−
7 tennis ball	×					×		×	×		−

Here we use the following abbreviations: "w" for white, "y" for yellow, "g" for green, "b" for blue, "s" for smooth, "f" for firm, "r" for round, "o" for oval, and "$\bar{m}$" for $m \in \{w, y, g, b, s, f, r, o,\}$.

This context gives rise to the positive concept lattice in Fig. 1, where we marked minimal (+)-hypotheses and falsified (+)-generalizations. If we have an undetermined example `mango` with $\texttt{mango}^\tau = \{y, \bar{f}, s, \bar{r}\}$ then it is classified positively, since $\texttt{mango}^\tau$ contains the minimal hypothesis $\{\bar{f}, \bar{r}\}$ and does not contain any negative hypothesis. For this scaling we have two minimal negative hypotheses: $\{w\}$ (supported by examples `egg` and `tennis ball` and $\{f, s, \bar{r}\}$ (supported by examples `toy cube` and `egg`.

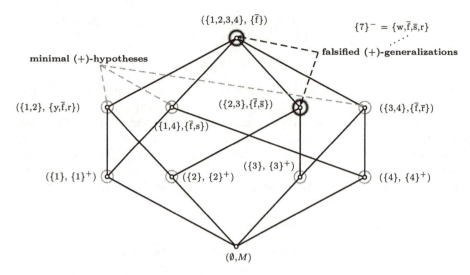

Fig. 1. Positive concept lattice for scaling 1

The context can be scaled differently, e.g. in this way (scaling 2):

G \ M	w	y	g	b	$\overline{w}$	$\overline{y}$	$\overline{g}$	$\overline{b}$	f	$\overline{f}$	s	$\overline{s}$	r	o	$\overline{r}$	$\overline{o}$	target
1 apple	×		×		×	×		×	×		×	×				+
2 grapefruit	×		×		×	×		×		×	×	×			+
3 kiwi		×		×	×		×		×		×		×	×			+
4 plum			×	×	×	×		×	×			×	×			+
5 toy cube		×		×	×		×	×		×		×	×		−
6 egg	×			×	×	×	×		×		×	×			−
7 tennis ball	×		×	×	×		×		×	×	×			−

This scaling gives rise to another positive concept lattice, all intents of which are (+)-hypotheses. The unique minimal hypothesis (corresponding to the top element of the concept lattice) is $\{\overline{w}, \overline{f}, o\}$. Two minimal negative hypotheses are $\{\overline{y}, \overline{b}, \overline{r}, f, s\}$ (supported by examples 5 and 6) and $\{\overline{y}, \overline{g}, \overline{b}, w, o\}$ (supported by examples 6 and 7).

The definitions of JSM-hypotheses can be varied, e.g., as follows:

- by imposing other logical conditions (e.g. of the "Difference method" of J.S. Mill), which gives rise to the "lattice of methods" [18],
- by allowance for $\alpha\%$ of counterexamples (for hypotheses and/or classifications) [29, 30],
- by using nonsymmetric classification (e.g., (−)-hypotheses are selected by stronger conditions than (+)-hypotheses) [18, 20],
- by varying "similarity operation" (see Section 4.2).

4.1 Various Hypotheses of the JSM-Method in Terms of Galois Connections

Various types of hypotheses expressed via respective plausible reasoning predicates of the JSM-method were supposed to capture different aspects of the relationship between structural and functional (target) attributes of objects. In the previous sections we considered representation of JSM-hypotheses by means of Galois connections for the case with a single target attribute. Here we give a description of various types of JSM-hypotheses from [3, 17, 18, 20, 21, 40], assuming that there are several target attributes.

For the sake of simplicity we also assume that each example (object) in the training dataset is either a positive or a negative example with respect to each attribute. Then the situation can be represented by two formal contexts: a *structural* context $\mathbb{K}_M = (G, M, I)$ and a *target* context $\mathbb{K}_P = (G, P, J)$, where P is the set of target attributes (properties) and M is the set of structural attributes. The resulting context (called the *apposition* of the two contexts) can be given by the following matrix:

	M	P
G	I	J

Derivation operators $(\cdot)^I$ and $(\cdot)^J$ are defined in the usual way. Complements of relations I and J are defined naturally as $\bar{I} : g\bar{I}m \iff \neg gIm$, $\bar{J} : g\bar{J}m \iff \neg gJm$. These relations also define derivation operators $(\cdot)^{\bar{I}}$ and $(\cdot)^{\bar{J}}$. Now the definitions of various hypothesis types of the JSM-method can be represented by the following table (here $V \subseteq M$ and $W \subseteq P$):

hypothesis name	expression for (+)	expression for (-)
agreement	$V = (V^I \cap W^+)^I,\ W = V^{I+}$	$V = W^{-I},\ W = V^{I-}$
no counterexample	$V = V^{II},\ V^I \subseteq W^+$	$V = V^{II},\ V^I \subseteq W^-$
inverse	$W = W^{++},\ V = W^{+I}$	$W = W^{--},\ V = W^{-I}$
situational	$V = ((V \cup S)^I \cap W^+)^I$	$W^{-I} = V \cup S$
Mill's difference	$(\bigcup_{j=0}^{m} V_j)^I \subseteq W^{\{-,\tau\}}$	symmetric
generalized	$\mathcal{X} = \min\{B \mid V \subset B = (B^I \cap W^-)^I\}$	symmetric

Here "symmetric" means that the expression for (-) is obtained from the expression for (+) by replacing "+" with "-". Each hypothesis type defines the set of all pairs (V, W) such that the set of structural attributes V is a hypothetical cause of the set of target attributes W. Above we have considered the methods of agreement (also with additional "no counterexample condition") and difference (as formulated by J.S. Mill). Its JSM-formalization requires that the effect W does not occur in the absence of causes from $\{V_j\}_j$ (determined by other methods, e.g., by agreement). The intuitive meaning of other methods in this table is as follows. The *inverse method* is applied for "effect-cause reasoning" [20], usually when the number of attributes in P is larger than that in M. In the *situational method* [21] the importance of situation S for establishing relation between cause V and effect W is underlined. In the *generalized method* [20] it is assumed that each hypothetical cause V of effect W can have specific hindrances from the set $\mathcal{X}$, so V plausibly causes W only in the absence of elements of $\mathcal{X}$. Note that for "no counterexample" hypotheses there can be no hindrances as defined in the table.

Another specific feature of the JSM-method is the so-called *condition of causal completeness* [20], which states that for chosen methods and a dataset the generated positive and negative hypotheses should classify the initial data correctly:

$$\bigcup_{M_x^+(V,W)} V^I = W^+, \qquad \bigcup_{M_y^-(V,W)} V^I = W^-,$$

where $M_x^+(V,W)$ and $M_y^-(V,W)$ denote some positive and negative methods, respectively. The condition is supposed to be tested each time hypotheses are generated.

The invariant feature of hypotheses w.r.t. different types of predicates is that they are sought among closed subsets of attributes.

4.2   Similarity Operation

Initially, similarity of object descriptions was defined in the JSM-method by means of set-intersection $\cap$. However, this definition suggested an obvious generalization: defining similarity as an idempotent, commutative and associative operation, i.e., as a meet operator in a semilattice. So, for each application domain with its specific data structure, a "similarity" operation was to be defined. This approach is equivalent to *scaling* in FCA where each many-valued attribute is turned to a set of related binary atributes.

An example of a similarity operation different from $\cap$ is the following interval algebra on real numbers. For two intervals $[a,b]$ and $[c,d]$ with $a,b,c,d \in \mathbb{R}$ and $a \leq c$ their meet can be defined as

$$[a,b] \wedge [c,d] = [max(a,c), min(b,d)] \text{ if } b \geq c, \text{ otherwise empty.}$$

This operation on intervals is often used in life-science applications, where, e.g., a number stays for a dose of a substance introduced [69] or a characteristic activation energy of a substance [55]. From the very beginning, the most important application of the JSM-method was the study of "chemical structure - biological activity" relationship. For this problem, adequate representation of chemical structure is essential. A special encoding scheme, called Fragmentary Code of Substructure Superposition (FCSS) (see, e.g., [9]), which turns molecular graphs to sets of binary attributes, was used. This encoding scheme allows efficient search for molecular similarities, however it leads also to the loss of information on connection between molecular parts. This problem motivated the search for mathematical means that would help dealing directly with graph representation of molecules. A solution was proposed in the form of a semilattice of graph sets [43, 44, 46, 52].

This semilattice is based on the following ordered set P of graphs with labels from the set $\mathcal{L}$ with partial order $\preceq$. Each labeled graph Γ from P is a triple of the form $((V, l), E)$, where V is a set of vertices, E is a set of edges and $l\colon V \to \mathcal{L}$ is a label assignment function, taking a vertex to its label.
For two graphs $\Gamma_1 := ((V_1, l_1), E_1)$ and $\Gamma_2 := ((V_2, l_2), E_2)$ from P Γ_1 *dominates* Γ_2 or $\Gamma_2 \leq \Gamma_1$ if there exists a one-to-one mapping $\varphi\colon V_2 \to V_1$ such that it

- respects edges: $(v, w) \in E_2 \Rightarrow (\varphi(v), \varphi(w)) \in E_1$,
- fits under labels: $l_2(v) \preceq l_1(\varphi(v))$.

Example 2. Let $\mathcal{L} = \{C, NH_2, CH_3, OH, x\}$, then we have the following domination relations:

vertex labels are unordered $x \preceq A$ for any vertex label $A \in \mathcal{L}$

A meet operation $\sqcap$ on graph sets can then be defined as follows: For two graphs X and Y from P

$$\{X\} \sqcap \{Y\} := \{Z \mid Z \leq X, Y, \quad \forall Z_* \leq X, Y \quad Z_* \not\geq Z\},$$

i.e., $\{X\} \sqcap \{Y\}$ is the set of all maximal common subgraphs of X and Y up to substitution of a vertex label by a vertex label smaller w.r.t. $\preceq$. The meet of nonsingleton sets of graphs is defined as

$$\{X_1, \ldots, X_k\} \sqcap \{Y_1, \ldots, Y_m\} := \mathrm{MAX}_{\leq}\left(\bigcup_{i,j}(\{X_i\} \sqcap \{Y_j\})\right)$$

for details see [25, 46, 48]. Here is an example of applying $\sqcap$ defined above:

Let positive examples be described by graphs $\Gamma_1, \Gamma_2, \Gamma_3, \Gamma_4$ and negative examples be described by graphs $\Gamma_5, \Gamma_6, \Gamma_7$:

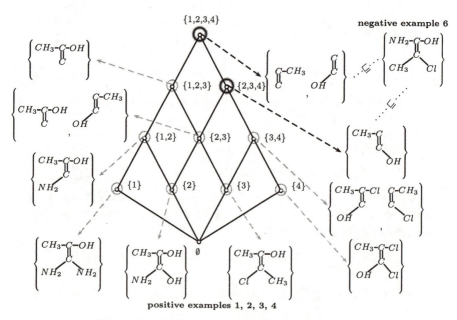

then the lattice of graph sets generated by positive examples and their graph descriptions is given in Fig. 2, where (+)-hypotheses and falsified (+)-generalizations are highlighted:

Fig. 2. The lattice of the positive pattern structure

The same approach is realizable for arbitrary data descriptions with generality (subsumption) order $\leq$. The general idea is to consider the (distributive) lattice of order ideals of $\leq$, distinguish the elements of it that correspond to descriptions of examples (objects) and consider these elements as generators of a meet-subsemilattice of the lattice of order ideals. Being supplied with a dummy top element (which is feasible, e.g., in the case when the number of objects is finite) this subsemilattice turns into a sublattice of the lattice of order ideals of $\leq$ (which is not necessarily distributive). These ideas were proposed and developed in [25], where semilattices of patterns were considered. Let G be a set (elements of which are called objects), let $(D, \sqcap)$ be a meet-semilattice and let $\delta : G \to D$ be a mapping. Then $(G, \underline{D}, \delta)$ with $\underline{D} = (D, \sqcap)$ is called a *pattern structure*, provided that the set

$$\delta(G) := \{\delta(g) \mid g \in G\}$$

generates a complete subsemilattice $(D_\delta, \sqcap)$ of $(D, \sqcap)$, i.e., every subset X of $\delta(G)$ has an infimum $\sqcap X$ in $(D, \sqcap)$

and D_δ is the set of these infima. Each such complete semilattice has lower and upper bounds, which we denote by $\mathbf{0}$ and $\mathbf{1}$, respectively. There are two natural situations where the condition on the complete subsemilattice is automatically satisfied: when $(D, \sqcap)$ is complete, and when G is finite. If $(G, \underline{D}, \delta)$ is a pattern structure, we define the derivation operators as

$$A^\diamond := \underset{g \in A}{\sqcap}\, \delta(g) \qquad \text{for } A \subseteq G$$

and

$$d^\diamond := \{g \in G \mid d \sqsubseteq \delta(g)\} \qquad \text{for } d \in D.$$

The elements of D are called *patterns*. The natural order on them is given, as usual, by

$$c \sqsubseteq d :\iff c \sqcap d = c,$$

and is called the *subsumption* order.

An algebraic model of approximation was proposed in [25] in the form of projection (or kernel, i.e., idempotent, monotone and contracting) operator and the reduction to standard concept lattices was discussed. Since projections (kernels) preserve meet operator, hypotheses in projected data have preimages in the original data that are hypotheses too. The research of pattern structures and their approximations led to further practical applications in chemistry with the approximation level being controlled by a parameter [23].

4.3   Mathematical Activity Around Concept-Based Hypotheses

Here we give a partial list of references to some research around JSM-method, concept-based hypotheses and implication bases.

Logics. Construction of quasi-axiomatic theories of plausible JSM-reasoning, completeness problem of the theory of plausible reasoning based on rules that generate hypotheses were considered in [3, 5]. Argumentation logic (where a proof of a statement takes into account arguments for and against the statement) were considered in [19]. In [86, 88] the author studies (partial) expressibility of plausible reasoning rules in Prolog and the expressibility in first-order predicate logic was studied in [87]. Logics of causal reasoning in the JSM-method were studied in [2]. A modal logic of incomplete contexts was studied in [67].

Algebraic Issues. Similarity operation on sets of labeled graphs, which is an infimum (meet) operation in a corresponding semilattice, was defined and studied in [1, 25, 43, 44, 46, 48]. A distributive lattice (of order ideals) of data for JSM-method was studied in [1]. Further generalization of the graph set semilattice and its translation in FCA terms was realized in [25], where general pattern semilattices were studied (see the previous subsection).

Algorithmic Issues. First algorithms for computing JSM-hypotheses were proposed in [61, 94, 95], a recent review which includes theoretical and experimental comparison of various algorithms for computing closed sets and concept lattices is found in [54]. Polynomial tractability and intractability of certain decision problems related to generation of hypotheses was considered in [45, 46, 96]. In [45] it was proved that the problem of computing the number of hypotheses is #P-complete, In [46, 51] same result was proved for the number of minimal hypotheses. In [49, 51] similar results were demonstrated for concepts. In particular, it was shown that the problems of computing the number all concepts is #P-complete. A very efficient incremental algorithm for computing concept lattices was proposed in [60]. A fast incremental algorithm for computing Duquenne-Guigues implication bases (with the best known experimental performance) was proposed in [68].

4.4   Applications of the JSM-Method

Starting from the early 1980s JSM-hypotheses were used in several applied domains, including bioscience analysis of biological activity of chemicals (see reviews [7, 97]) predicting metabolic pathways [15, 58]), medical diagnostics, technical diagnostics, sociology, document dating, spam filtering, and so on. JSM-hypotheses were used successfully for making predictions at two international competitions: that for predictive toxicology [9] (where JSM-hypotheses resulted in optimal classifications in all test groups) and that for spam filtering [13]. A freeware system QuDA [32, 33], which incorporates several data mining techniques also presents a possibility of generating JSM-hypotheses.

Life Sciences. Most numerous experiments were carried out in applied pharmacology or Structure-Activity Relationship domain, which deals with predicting biological activity of chemical compounds with known molecular structure. JSM-hypotheses were generated for antitumor [71], antibacterial, antileprous, hepatoprotective [12], plant growth-stimulating, cholesterase-inhibitine, toxic and carcinogenic activities, see reviews [7, 97]. JSM-method was many times applied to problems of medical diagnostics, e.g., the results of the study of human papilloma are found in [69]. Recent results in the study of toxicity of different substances, including alcohols and halogen-substituted hydrocarbons by means of learning in pattern structures on graph sets are found in [55].

Sociology and Humanities. In [21] strike readiness at joint-stock factories in St. Petersburg and Elets was analysed. The advantages of the JSM-based approach as compared to statistical methods resided in the fact that the former allowed for creating taxonomies of socio-psychological types and enabled creating "social portraits". In paleography [41] the JSM-method was applied to dating birch-bark documents of 10–16 centuries of the Novgorod republic. Here there were five types of attributes describing individual letter features, features common to several letters, handwriting, language features: morphology, syntax, and typical errors, style: letter format, addressing formulas and their key words. Time was considered as many-valued target attribute, with 20 nointersecting time intervals as attribute values. A model for analyzing human conflicts that uses for similarity of labeled graphs was studied in [22].

Spam Filtering. A first successful application of the JSM-like (concept-based) hypotheses for filtering spam was reported in [16]. In April-May 2003 Technical University Chemnitz, European Knowledge Discovery Network, and PrudSys AG organized the Data Mining Cup (DMC) competition for students specializing in Machine Learning [13]. Among 514 participants from 199 universities of 38 countries the sixth place was taken by a model that combined "Naive Bayes" approach with JSM-hypotheses.

5   Machine Learning in Terms of Galois Connection and FCA

In recent years some progress was done in describing various learning models like version spaces, decision trees in terms of Galois connection and concept lattices [26, 50].

5.1   Decision Trees Embedded in Concept Lattices

As input, a system constructing a decision tree (see, e.g., [75]) receives descriptions of positive and negative examples (or positive and negative contexts, in terms of the previous section). The root of the tree corresponds to the beginning of the process and is not labeled. Other vertices of the decision tree are labeled by attributes and edges are labeled by values of the attributes (e.g., 0 or 1 in case of binary contexts), each leaf is additionally labeled by a class + or −, meaning that all examples with attribute values from the path leading from the root to the leaf belong to a certain class, either + or −.

Systems like ID3 [75] (see also [65]) compute the value of the *information gain* (or negentropy) for each vertex and each attribute not chosen in the branch above. The attribute with the greatest value of the information gain (with the smallest entropy, respectively) "most strongly separates" objects from classes + and −. The algorithm sequentially extends branches of the tree by choosing attributes with the highest information gain. The extension of a branch stops when a next attribute value together with attributes above in the branch uniquely classify examples with this value combination in one of classes $+$ or $-$. In some algorithms, the process of extending a branch stops before this in order to avoid *overfitting*, i.e., the situation where all or almost all examples from the training sample are classified correctly by the resulting decision tree, but objects from test datasets are classified with many errors.

Now we consider decision trees more formally. Let the training data be described by the context $\mathbb{K}_{+-} = (G_+ \cup G_-, M, I_+ \cup I_-)$ with the derivation operator denoted by $(\cdot)'$. In FCA terms this context is called the *subposition* of $\mathbb{K}_+$ and $\mathbb{K}_-$. Assume for simplicity sake that for each attribute $m \in M$ there is an attribute $\overline{m} \in M$, a "negation" of m: $\overline{m} \in g'$ iff $m \notin g'$. A set of attributes M with this property is called *dichotomized* in FCA. We call a subset of attributes $A \subseteq M$ *noncontradictory* if either $m \notin A$ or $\overline{m} \notin A$. We call a subset of attributes $A \subseteq M$ *complete* if for every $m \in M$ one has $m \in A$ or $\overline{m} \in A$.

First no optimization functional (like information gain) for selecting attributes is involved and construction of all possible decision trees is considered. The construction of an arbitrary decision tree proceeds by sequentially choosing attributes. If different attributes $m_1, \ldots, m_k$ were chosen one after another, then the sequence $\langle m_1, \ldots, m_k \rangle$ is called a *decision path* if $\{m_1, \ldots, m_k\}$ is noncontradictory and there exists an object $g \in G_+ \cup G_-$ such that $\{m_1, \ldots, m_k\}' \subseteq g'$ (i.e., there is an example with this set of attributes). A decision path $\langle m_1, \ldots, m_i \rangle$ is a (proper) subpath of a decision path $\langle m_1, \ldots, m_k \rangle$ if $i \leq k$ ($i < k$, respectively). A decision path $\langle m_1, \ldots, m_k \rangle$ is called *full* if all objects having attributes $\{m_1, \ldots, m_k\}$ are either positive or negative examples (i.e., have either $+$ or $-$ value of the target attribute).

We call a full decision path *irredundant* if none of its subpaths is a full decision path. The set of all chosen attributes in a full decision path can be considered as a sufficient condition for an object to belong to a class $\varepsilon \in \{+, -\}$. A decision tree is then a set of full decision paths.

In what follows, we use the one-to-one correspondence between vertices of a decision tree and the related decision paths, representing the latter, when this does not lead to ambiguity, by their last chosen attributes. By *closure of a decision path* $\langle m_1, \ldots, m_k \rangle$ we mean the closure of the corresponding set of attributes, i.e., $\{m_1, \ldots, m_k\}''$. Now we relate decision trees with the covering relation graph of the concept lattice of the context $\mathbb{K} = (G, M, I)$, where the set of objects G is of size $2^{|M|/2}$ and the relation I is such that the set of object intents is exactly the set of complete noncontradictory subsets of attributes. In terms of FCA [27] the context $\mathbb{K}$ is the *semiproduct* of $|M|/2$ *dichotomic scales* or $\mathbb{K} = D_1 \amalg \ldots \amalg D_{|M|/2}$ (denoted by $\amalg_M D$ for short), where each dichotomic scale D_i stays for the pair of attributes $(m, \overline{m})$.

In a concept lattice a sequence of concepts with decreasing extents we call a *descending chain*. If the chain starts at the top element of the lattice, we call it *rooted*.

Proposition 5 *Every decision path is a rooted descending chain in $\underline{\mathfrak{B}}(\amalg_M D)$ and every rooted descending chain consisting of concepts with nonempty extents in $\underline{\mathfrak{B}}(\amalg_M D)$ is a decision path.*

To relate decision trees to hypotheses introduced above we consider again the contexts $\mathbb{K}_+ = (G_+, M, I_+)$, $\mathbb{K}_- = (G_-, M, I_-)$, and $\mathbb{K}_{+-} = (G_+ \cup G_-, M, I_+ \cup I_-)$. The context $\mathbb{K}_{+-}$ can be much smaller than $\mathbb{X}_M D$ because the latter always has $2^{|M|/2}$ objects while the number of objects in the former is the number of examples. Also the lattice $\mathfrak{B}(\mathbb{K}_{+-})$ can be much smaller than $\mathfrak{B}(\mathbb{X}_M D)$.

The relation between decision trees and (minimal "no counterexample") hypotheses from the previous section is given by the following

Proposition 6 *A full decision path $\langle m_1, \ldots, m_k \rangle$ corresponds to a rooted descending chain $\langle (m_1'', m_1'), \ldots, (\{m_1, \ldots, m_k\}'', \{m_1, \ldots, m_k\}') \rangle$ of the line diagram of $\mathfrak{B}(\mathbb{K}_{+-})$ and the closure of each full decision path $\langle m_1, \ldots, m_k \rangle$ is a hypothesis, either positive or negative. Moreover, for each minimal hypothesis h, there is a full irredundant path $\langle m_1, \ldots, m_k \rangle$ such that $\{m_1, \ldots, m_k\}'' = h$.*

By the proposition, hypotheses correspond to the "most cautious" (most specific) learning strategy in the sense that they are least general generalizations of descriptions of positive examples (or object intents, in terms of FCA). The shortest decision paths (for which in no decision tree there exist full paths with proper subsets of attribute values) correspond to the "most courageous" ("most discriminating") learning strategy: being the shortest possible rules, they are most general generalizations of positive example descriptions. However, it is not guaranteed that for a given training set resulting in a certain set of minimal hypothesis there is a decision tree such that minimal hypotheses are among closures of its paths (see Example 3 below). In general, to obtain all minimal hypotheses as closures of decision paths one needs to consider several decision trees, not all of them being optimal w.r.t. a procedure based on the information gain functional (like ID3 or C4.5). The issues of generality of generalizations and, in particular, the relation between most specific and most general generalizations, are naturally captured in terms of version spaces, which we consider in the next section.

In real systems for the construction of decision trees like ID3 or C4.5 the process of constructing a decision path is driven by the information gain functional: a next chosen attribute should have maximal information gain. For dichotomized attributes the information gain is defined for a pair of attributes $m, \overline{m} \in M$.

Given a decision path $\langle m_1, \ldots, m_k \rangle$

$$\mathrm{IG}(m) := -\frac{|A_m'|}{|G|} \mathrm{Ent}(A_m) - \frac{|A_{\overline{m}}'|}{|G|} \mathrm{Ent}(A_{\overline{m}}),$$

where $A_m := \{m_1, \ldots, m_k, m\}$, $A_{\overline{m}} := \{m_1, \ldots, m_k, \overline{m}\}$, and for $A \subseteq M$

$$\mathrm{Ent}(A) := - \sum_{\varepsilon \in \{+, -\}} p(\varepsilon \mid A) \cdot \log_2 p(\varepsilon \mid A),$$

$\{+, -\}$ are values of the target attribute and $p(\varepsilon \mid A)$ is the conditional sample probability (for the training set) that an object having a set of attributes A belongs to a class $\varepsilon \in \{+, -\}$.

If the derivation operator $(\cdot)'$ is associated with the context $(G_+ \cup G_-, M, I_+ \cup I_-)$, then, by definition of the conditional probability, we have

$$p(\varepsilon \mid A) = \frac{|A' \cap G_\varepsilon|}{|A'|} = \frac{|(A'')' \cap G_\varepsilon|}{|(A'')'|} = p(\varepsilon \mid A'')$$

by the property of the derivation operator $(\cdot)'$: $(A'')' = A'$. This observation implies that instead of considering decision paths, one can consider their closures without affecting the values of the information gain. In terms of lattices this means that instead of the concept lattice $\mathfrak{B}(\mathbb{X}_M D)$ one can consider the concept lattice of the context $\mathbb{K}_{+-} = (G_+ \cup G_-, M, I_+ \cup I_-)$. Another consequence of the invariance of IG w.r.t. closure is the following fact: If implication $m \to n$ holds in the context $\mathbb{K}_{+-} = (G_+ \cup G_-, M, I_+ \cup I_-)$, then an IG-based algorithm will not choose attribute n in the branch below chosen m and will not choose m in the branch below chosen $\bar{n}$.

Example 3. Consider the training set from Example 1. The decision tree obtained by the IG-based algorithm is given in Fig. 3. Note that attributes f and w has the same IG value (a similar tree with f at the root is also optimal), the IG-based algorithms usually take the first attribute with the same value of IG.

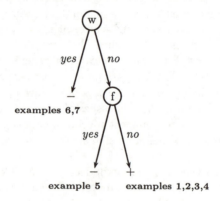

Fig. 3. A decision tree for Example 1

The decision tree in Fig. 3 corresponds to three implications $\{w\} \to -$, $\{\bar{w}, f\} \to -$, $\{\bar{w}, \bar{f}\} \to +$, such that closures of their premises make the corresponding negative and positive hypotheses for the second scaling from Example 1. Note that the hypothesis $\{\bar{w}, f\}''$ is not minimal, since there is a minimal hypothesis $\{f\}''$ contained in it. The minimal hypothesis $\{f\}''$ corresponds to a decision path of the mentioned IG-based tree with the attribute f at the root.

5.2   Version Spaces vs. Concept-Based Hypotheses

The term *version space* was proposed by T. Mitchell [64, 65] to denote a variety of models compatible with the training sample of positive and negative examples. Version spaces can be defined in different ways. Here they are described in terms somewhat different to those in [64], in order to avoid collision with FCA terminology.

- An *example language* L_e (elsewhere also called *instance language*) by means of which the examples (instances) are described. This language describes a *set E of examples*.
- A *classifier language* L_c describing the possible classifiers (elsewhere called *concepts*). This language describes a set C of *classifiers*.
- A *matching predicate* $M(c,e)$ that defines if a classifier c does or does not *match* an example e: We have $M(c,e)$ iff e is an example of classifier c. The set of classifiers is (partially) ordered by a *subsumption order*: for $c_1, c_2 \in L_c$ the classifier c_1 subsumes c_2 or $c_1 \sqsupseteq c_2$ if c_1 corresponds to a more specific description and thus, covers less objects than c_2:

$$c_1 \sqsupseteq c_2 : \iff \forall_{e \in E} \; M(c_1, e) \to M(c_2, e).$$

The corresponding strict order $\sqsupset$ is called *proper subsumption*.
- Sets E_+ and E_- of *positive* and *negative examples* of a *target attribute* with $E_+ \cap E_- = \emptyset$. The target attribute is not explicitly given.
- *consistency predicate* cons(c):
  cons(c) holds if for every $e \in E_+$ the matching predicate $M(c,e)$ holds and for every $e \in E_-$ the negation $\neg M(c,e)$ holds. The set of all consistent classifiers is called the *version space*

$$\mathrm{VS}(L_c, L_e, M(c,e), E_+, E_-).$$

The learning problem is then defined as follows:

Given $L_c, L_e, M(c,e), E_+, E_-$.
Find the version space $\mathrm{VS}(L_c, L_e, M(c,e), E_+, E_-)$.

In the sequel, we shall usually fix L_c, L_e, and $M(c,e)$ and write $\mathrm{VS}(E_+, E_-)$ or even just VS for short. Version spaces are often considered in terms of *boundary sets* proposed in [64]. They can be defined if the language L_c is *admissible*, i.e., if every chain in the subsumption order has a minimal and a maximal element. In this case,

$$\mathrm{GVS}(L_c, L_e, M(c,e), E_+, E_-) := \mathrm{MIN}_{\sqsubseteq}(\mathrm{VS}) := \{c \in \mathrm{VS} \mid \neg \exists c_1 \in \mathrm{VS} \, c_1 \sqsubset c\},$$

$$\mathrm{SVS}(L_c, L_e, M(c,e), E_+, E_-) := \mathrm{MAX}_{\sqsubseteq}(\mathrm{VS}) := \{c \in \mathrm{VS} \mid \neg \exists c_1 \in \mathrm{VS} \, c \sqsubset c_1\}.$$

If a version space VS is fixed, we also use notation G(VS) and S(VS) for short.

The elements of the version space can be used as potential classifiers for the target attribute: A classifier $c \in \mathrm{VS}$ *classifies* an example positively if c matches e and negatively else. Then, all positive examples are classified positively, all negative examples are classified negatively, and undetermined examples may be classified either way. If it is assumed that E_+ and E_- carry sufficient information about the target attribute, we may expect that an undetermined example is likely to have the target attribute if it is classified positively by a large percentage of the version space (cf. [65]). We say that an example e is p%-classified (for $0 \leq$

$p \leq 100$) if no less than $p\%$ classifiers of the version space classify it positively. This means, e.g., that 100%-classification of e takes place if e is matched by all elements of SVS and negative classification of e (0%-classification) takes place if e is not matched by any element of GVS.

As showed in [26] the basic properties of general version spaces can easily be expressed with Galois connections. Consider the formal context (E, C, I), where E is the set of examples containing the disjoint sets of observed positive and negative examples: $E \supseteq E_+ \cup E_-$, $E_+ \cap E_- = \emptyset$, C is the set of classifiers and the relation I corresponds to the matching predicate $M(c, e)$: for $c \in C$, $e \in E$ the relation eIc holds iff $M(c, e) = 1$. The complementary relation, $\bar{I}$, corresponds to the negation: $e\bar{I}c$ holds iff $M(c, e) = 0$. As shown in [26]

$$\mathrm{VS}(E_+, E_-) = E_+{}^I \cap E_-{}^{\bar{I}}.$$

This characterization of version spaces implies immediately the property of *merging version spaces*, proved in [42]: For fixed L_c, L_e, $M(c, e)$ and two sets E_{+1}, E_{-1} and E_{+2}, E_{-2} of positive and negative examples one has

$$\mathrm{VS}(E_{+1} \cup E_{+2}, E_{-1} \cup E_{-2}) = \mathrm{VS}(E_{+1}, E_{-1}) \cap \mathrm{VS}(E_{+2}, E_{-2}).$$

This follows from the relation $(A \cup B)' = A' \cap B'$, which holds for a derivation operator $(\cdot)'$ of an arbitrary context.

The classifications produced by classifiers from the version space are characterized as follows. The set of all 100%-classified examples w.r.t the version space $\mathrm{VS}(E_+, E_-)$ is given by

$$(E_+{}^I \cap E_-{}^{\bar{I}})^I.$$

In particular, if one of the following conditions is satisfied, then there cannot be any 100%-classified undetermined example:

1. $E_- = \emptyset$ and $E_+{}^{II} = E_+$,
2. $(E_+{}^I \cap E_-{}^{\bar{I}})^I = E_+$.

The set of examples that are classified positively by at least one element of the version space $\mathrm{VS}(E_+, E_-)$ is given by

$$E \setminus (E_+{}^I \cap E_-{}^{\bar{I}})^{\bar{I}}.$$

Consider a very important special case where the ordered set $(C, \leq)$ of classifiers given in terms of some language L_c makes a meet-semilattice w.r.t. $\wedge$ meet operation, like in Section 4.2. This also covers the case of attributes with values.

In [26] it was shown that in the case where the classifiers, ordered by subsumption, form a complete semilattice, the version space is a complete subsemilattice for any sets of examples E_+ and E_-. If the set of classifiers C makes a complete semilattice $(C, \sqcap)$, we can consider a *pattern structure* $(E, (C, \sqcap), \delta)$, where E is a set (of "examples"), δ is a mapping $\delta : E \to C$, $\delta(E) := \{\delta(e) \mid e \in E\}$. The subsumption order can be reconstructed from the semilattice operation: $c \sqsubseteq d \iff c \sqcap d = c$.

The version space may be empty, in which case there are no classifiers separating positive examples from negative ones. This happens, e.g., if there is a *hopeless* positive example (an outlier), by which we mean an element $e_+ \in E_+$ having a negative counterpart $e_- \in E_-$ such that every classifier which matches e_+ also matches e_-. An equivalent formulation of the hopelessness of e_+ is that $(e_+)^{\diamond\diamond} \cap E_- \neq \emptyset$. The following relation between the version space with lattice-ordered classifiers and minimal hypotheses was shown in [26]:

Suppose that the classifiers, ordered by subsumption, form a complete meet-semilattice $(C, \sqcap)$, and let $(E, (C, \sqcap), \delta)$ denote the corresponding pattern structure.

Proposition 7 *The following statements are equivalent:*

1. *The version space* $\mathrm{VS}(E_+, E_-)$ *is not empty.*
2. $(E_+)^{\diamond\diamond} \cap E_- = \emptyset$.
3. *There are no hopeless positive examples and there is a unique minimal positive hypothesis* h_{min}.

In this case, $h_{min} = (E_+)^{\diamond}$, and the version space is a convex set in the lattice of all pattern intents, ordered by subsumption, with maximal element h_{min}.

In case where conditions 1-3 are satisfied, the set of training examples is often referred to as *separable* in machine learning. The theorem gives access to generation of the version space, e.g., with the use of a standard Next Closure [27] algorithm.

According to [27] a subset $A \subseteq M$ can be defined as a *proper premise of an attribute* $m \in M$ if $m \notin A$, $m \in A''$ and for any $A_1 \subset A$ one has $m \notin A_1''$. In particular we can define a *positive proper premise* as a proper premise of the target attribute ω. In [26] we generalized this notion to include the possibility of the unknown value of a target attribute (for an undetermined example): $d \in L_c$ is a *positive proper predictor with respect to examples* E_+, E_-, and E_τ if the following conditions 1-3 are satisfied:

1. $d^{\diamond} \subseteq E_+ \cup E_\tau$,
2. $\exists g \in E_+ : g \in d^{\diamond}$ (or $d^{\diamond} \cap E_+ \neq \emptyset$),
3. $\forall d_1$ such that $d \sqsubseteq d_1$ and $d \neq d_1$, the relation $d_1^{\diamond} \not\subseteq E_+ \cup E_\tau$ holds.

In the case where $E_\tau = \emptyset$, condition 2 of the definition follows from condition 1 and a proper predictor is just a *proper premise* [27] of the target attribute.

The proper predictors and hypotheses are related to the boundaries of the version space as follows [26]:

Proposition 8 *Let* $PP_+(\Pi, E_+, E_-)$ *denote the set of all positive proper predictors for the pattern structure* $\Pi = (E, (C, \sqcap), \delta)$ *and sets of positive and negative examples* E_+ *and* E_-. *Let* $H_+(\Pi, E_+, E_-)$ *denote the set of positive hypotheses and* $VS(\Pi, E_+, E_-)$ *denote the version space for the pattern structure* $\Pi = (E, (C, \sqcap), \delta)$ *and sets of examples* E_+ *and* E_-. *Then the following holds:*

1. $PP_+(\Pi, E_+, E_-) = MAX_{\sqsubseteq}(\bigcup_{F_+ \subseteq E_+} GVS(\Pi, F_+, E_-))$,
2. $H_+(\Pi, E_+, E_-) = \bigcup_{F_+ \subseteq E_+} SVS(\Pi, F_+, E_-)$.

In contrast to version spaces with purely conjunctive classifiers, hypotheses propose a sort of "context-restricted" disjunction (which, hence is not so "loose" as purely syntactical disjunction over conjunction of attribute values): not all disjunctions are possible, but only those of minimal hypotheses (that are equivalent to certain conjunctions of attributes), which express similarities of examples.

6 Conclusion

We considered activity in classification, data analysis, and machine learning around the VINITI Institute in Moscow and its NTI journal that used models naturally described in terms of Galois connections and FCA. Early research was related to the models of taxonomies and meronomies, which are naturally recast in terms of concept lattices. Galois connection are very helpful in modeling similarity given by tolerance relation and its classes.

Recasting the JSM-method in FCA terms motivated further activity in describing well-known models of machine learning and knowledge discovery, such as version spaces and decision trees, in terms of Galois connections and concept lattices. Translations of this kind often provide with a unified view, simpler definitions, and simpler proofs of the results. Further work in this direction will be related to other widely used models of learning, such as Naive Bayes, induction of ripple-down rules, support vector machines, and so on. The language of Galois connections provide with standard algorithmic machinery from FCA and new developments from Data Mining related to finding (closed) frequent itemsets.

Concept lattices that seem from the first glance to be a tool for processing binary tables, actually provide with means for dealing with complex structure such as logical formulas, labeled graphs (e.g., concept graphs, molecular graphs), texts, 3D-structures. This aspect indicates yet another direction of further study: fast algorithms for models with complex and/or large data. Successful applications in chemistry and conflict modelling give hope for future results.

Acknowledgments

This work was supported by the Russian Foundation for Humanities, project no. 05-03-03019a, by the project *Problem solver for the analysis of causal dependencies* of the Russian Academy of Science, and by the Alexander-von-Humboldt Foundation. The author thanks Bernhard Ganter for helpful discussions.

References

1. O.M. Anshakov, On a data lattice for the JSM-method of automated hypothesis generation, *Nauchno-Tekhnicheskaya Informatsiya*, Ser. 2, 1996, No. 5-6, pp. 33-36 [in Russian].
2. O.M. Anshakov, Causal models of subject domains, *Nauchno-Tekhnicheskaya Informatsiya*, Ser. 2, 2000, No. 3, pp. 3-16 [in Russian], 2000.
3. O.M. Anshakov, D.P. Skvortsov, V.K. Finn, Logical Means of Expert Systems of JSM-type, *Semiotika i Informatika*, **28**, 1986, pp. 65-101 [in Russian].

4. O.M. Anshakov, D.P. Skvortsov, V.G. Ivashko, and V.K. Finn, Logical Means of the JSM-method of Automated Hypothesis Generation: Main Notions and System of Inference Rules, *Nauchno-Tekhnicheskaya Informatsiya*, Ser. 2, 1987, No. 9, pp. 10-18 [in Russian].

5. O.M. Anshakov, V.K. Finn, and D.P. Skvortsov, On axiomatization of many-valued logics associated with the formalization of plausible reasonings, *Stud. Log.*, **25**, No. 4, 23-47 (1989).

6. G.D. Birkhoff, *Lattice Theory*, Amer. Math. Soc. (1979).

7. V.G. Blinova, Results of Application of the JSM-method of Hypothesis Generation to Problems of Analyzing the Relation "Structure of a Chemical Compound - Biological Activity," *Autom. Docum. Math. Ling.*, vol. 29, no. 3, pp. 26-33, 1995.

8. V.G. Blinova, D.A. Dobrynin, Languages for Representing Chemical Structures in Intelligent Systems of Drug Design *Nauchno-Tekhnicheskaya Informatsiya*, Ser. 2, 2000, No. 6, pp. 14-21 [in Russian].

9. V.G. Blinova, D.A. Dobrynin, V.K. Finn, S.O. Kuznetsov, and E.S. Pankratova, Toxicology analysis by means of the JSM-method, *Bioinformatics* 2003, vol. 19, pp. 1201-1207.

10. D.A. Bochvar, On a three-valued calculus and its application to the analysis of paradoxes of the classical extended functional calculus, *Matematicheskii sbornik*, 1938, No. 2, pp. 287-308 [in Russian].

11. V.B. Borshev, V.A. Brudno, M.V. Khomyakov, Algebraic Description of the Structure of Dependencies in a Database, *Nauchno-Tekhnicheskaya Informatsiya*, Ser. 2, 1977, No. 3, pp. 17-18 [in Russian].

12. A.P. Budunova, V.V. Poroikov, V.G. Blinova, and V.K. Finn, The JSM-method of hypothesis generation: Application for analysis of the relation "Structure - hepatoprotective detoxifying activity", Nauchno-Tekhnicheskaya Informatsiya, no. 7, pp.12-15, 1993 [in Russian].

13. Data Mining Cup (DMC), http://www.data-mining-cup.de

14. V.V. Dunaev, O.M. Polyakov, Methodological Aspects of Relational Classification Theory, *Nauchno-Tekhnicheskaya Informatsiya*, Ser. 2, 1987, No. 4, pp. 21-27 [in Russian].

15. E.F. Fabrikantova, Problems of computer modeling of metabolic transformations of xenobiotics in a human organism *Itogi Nauki i Tekhniki, Seriya Informatika*, **15**, 115-135, 1991 [in Russian].

16. S. Férré and O. Ridoux, The Use of Associative Concepts in the Incremental Building of a Logical Context in *Proc. 10th Int. Conf. on Conceptual Structures, ICCS'2002*, U. Priss, D. Corbet, G. Angelova, Eds., Lecture Notes in Artificial Intelligence, **2393**, 2002, 299-313.

17. V.K. Finn, On Machine-oriented Formalization of Plausible Reasoning in F.Bacon-J.S.Mill Style, *Semiotika i Informatika*, 1983, No.20, pp. 35-101 [in Russian].

18. V.K. Finn, Plausible Reasoning in Systems of JSM Type, *Itogi Nauki i Tekhniki, Seriya Informatika*, **15**, 54-101, 1991 [in Russian].

19. V.K. Finn, On a variant of argumentation logic, *Nauchno-Tekhnicheskaya Informatsiya*, Ser. 2, 1996, No. 5-6, pp. 3-19 [in Russian].

20. V.K. Finn, Synthesis of cognitive procedures and the problem of induction, *Nauchno-Tekhnicheskaya Informatsiya*, ser. 2, 1999, No. 1-2, pp. 8-44 [in Russian].

21. V.K Finn, M. A. Mikheyenkova, On logical means of conceptualization of opinion analysis, *Nauchno-Tekhnicheskaya Informatsiya*, ser. 2, 2002, No.6, pp. 4-21 [in Russian].

22. B.A. Galitsky, S.O. Kuznetsov, and M. V. Samokhin, Analyzing Conflicts with Concept-Based Learning, Proc. 11th Int. Conf. on Conceptual Structures, ICCS'04, F. Dau, M.-L. Mugnier, Eds., Lecture Notes in Artificial Intelligence (2005).

23. B.Ganter, P.A. Grigoriev, S.O. Kuznetsov, M.V. Samokhin, Concept-Based Data-Mining with Scaled Labeled Graphs, *Proc. 12th Int. Conf. on Conceptual Structures (ICCS'04)*, Lecture Notes in Artificial Intelligence, **3127**, 2004, pp. 94-108.

24. B. Ganter and S.O. Kuznetsov, Formalizing Hypotheses with Concepts, *Proc. 8th Int. Conf. on Conceptual Structures, ICCS'98*, G. Mineau and B. Ganter, Eds., Lecture Notes in Artificial Intelligence, **1867**, 2000, pp. 342-356.

25. B. Ganter and S. Kuznetsov, Pattern Structures and Their Projections, *Proc. 9th Int. Conf. on Conceptual Structures, ICCS'01*, G. Stumme and H. Delugach, Eds., Lecture Notes in Artificial Intelligence, **2120** 2001, pp. 129-142.

26. B. Ganter and S.O. Kuznetsov, Hypotheses and Version Spaces, *Proc. 10th Int. Conf. on Conceptual Structures, ICCS'01*, A.de Moor, W. Lex, and B.Ganter, Eds., Lecture Notes in Artificial Intelligence, **2746** 2003, pp. 83-95.

27. B. Ganter and R. Wille, *Formal Concept Analysis: Mathematical Foundations*, Springer, 1999.

28. T. Gergely, V.K. Finn, On Solver of "Plausible Inference + Deduction" Type, in Intelligent Information-Computing Systems, *Artif. Intel. IFAC Ser. No. 9*, London Pergamon Press, 1984.

29. P.A. Grigoriev, SWORD-systems or JSM-systems for strings employing statistical considerations, *Nauchno-Tekhnicheskaya Informatsiya*, Ser. 2, 1996, No. 5-6, pp. 45-51 [in Russian].

30. P.A. Grigoriev, On computer forecast of repeated hypophysis adenoma, *Nauchno-Tekhnicheskaya Informatsiya*, Ser. 2, 1999, No. 1-2, pp. 83-88 [in Russian].

31. P.A. Grigoriev, S.O. Kuznetsov, S.A. Obiedkov, S.A. Yevtushenko, On a Version of Mill's Method of Difference, *Proc. ECAI 2002 Int. Workshop on Advances in Formal Concept Analysis for Knowledge Discovery in Databases*, Lyon, 2002, pp. 26-31.

32. P.A. Grigoriev and S.A. Yevtushenko, Elements of an Agile Discovery Environment, *Proc. 6th International Conference on Discovery Science (DS 2003)*, G. Grieser, Y. Tanaka and A. Yamamoto, Eds., *Lecture Notes in Artificial Intelligence*, **2843**, 2003, pp. 309–316.

33. P.A. Grigoriev, S.A. Yevtushenko and G.Grieser, QuDA, a data miner's discovery enviornment, Tech. report, FG Intellektik, FB Informatik, Technische Universität Darmstadt, 2003, AIDA 03 06, http://www.intellektik.informatik.tu-darmstadt.de/~peter/QuDA.pdf.

34. V.Ya. Gusakov, S.M. Yakubovich (Gusakova), Galois Connection and Some Theorems on Representing Binary Relations, *Nauchno-Tekhnicheskaya Informatsiya*, Ser. 2, 1974, No. 7, pp. 3-6 [in Russian].

35. V.Ya. Gusakov, S.M. Yakubovich (Gusakova), On Classification Algorithms, *Nauchno-Tekhnicheskaya Informatsiya*, Ser. 2, 1976, No. 12, pp. 17-22 [in Russian].

36. S.M. Gusakova, V.K. Finn, On Formalization of Local and Global Similarities, *Nauchno-Tekhnicheskaya Informatsiya*, Ser. 2, 1986, No. 6, pp. 16-19 [in Russian].

37. S.M. Gusakova, Canonical Representation of Similarities, *Nauchno-Tekhnicheskaya Informatsiya*, Ser. 2, 1987, No. 9, pp. 19-22 [in Russian].

38. S.M. Gusakova, V.K. Finn, On New Means for Formalization of Local and Global Similarities, *Nauchno-Tekhnicheskaya Informatsiya*, Ser. 2, 1987, No. 10, pp. 14-22 [in Russian].

39. S.M. Gusakova, V.K. Finn, Similarity and Plausible Reasoning, *Izvestia Akademii Nauk (Tekhnicheskaya Kibernetika)*, 1987, No. 5, pp. 42-63 [in Russian].

40. S.M. Gusakova, M.A. Mikheenkova, V.K. Finn, On Logical Means for Automated Analysis of Opinions, *Nauchno-Tekhnicheskaya Informatsiya*, Ser. 2, 2001, No. 5, pp. 4-24 [in Russian].

41. S.M. Gusakova, Paleography with JSM-method. Technical Report, VINITI, 2001.

42. H. Hirsh, Generalizing Version Spaces, *Machine Learning* **17**, 5-46, 1994.

43. S.O. Kuznetsov, On the lattice on graph sets for graphs with ordered vertex labels, Proc. Workshop on Semiotical Aspects of Formalization of Intelligent Activity, Borzhomi (Georgia, USSR), 1988, vol. 1, pp. 204-207 [in Russian].

44. S.O. Kuznetsov, Similarity operation on hypergraphs as as a Basis of Plausible Inference, Proc. 1st Soviet Conference on Artificial Intelligence, 1988, vol. 1, 442-448 [in Russian].

45. S.O. Kuznetsov, Interpretation on Graphs and Complexity Characteristics of a Search for Specific Patterns, *Nauchn. Tekh. Inf., Ser. 2 (Automat. Document. Math. Linguist.)* no. 1, 23-27, 1989 [in Russian].

46. S.O. Kuznetsov, JSM-method as a machine learning method, *Itogi Nauki i Tekhniki, ser. Informatika*, vol. 15, pp.17-50, 1991 [in Russian].

47. S.O. Kuznetsov, Mathematical aspects of concept analysis, Journal of Mathematical Science, Ser. Contemporary Mathematics and Its Applications, **18**, pp. 1654-1698, 1996 [in Russian].

48. S.O. Kuznetsov, Learning of Simple Conceptual Graphs from Positive and Negative Examples. In: J. Zytkow, J. Rauch (eds.), Proc. *Principles of Data Mining and Knowledge Discovery, Third European Conference, PKDD'99*, Lecture Notes in Artificial Intelligence, **1704**, pp. 384-392, 1999.

49. S.O. Kuznetsov, On Computing the Size of a Lattice and Related Decision Problems, *Order*, 2001, **18**(4), pp. 313-321.

50. S.O. Kuznetsov, Machine Learning and Formal Concept Analysis, *Proc. 2nd Int. Conf. on Formal Concept Analysis (ICFCA'04)*, P. Eklund, Ed., Lecture Notes in Artificial Intelligence, **2961**, 2004, pp. 287-312.

51. S.O. Kuznetsov, Learning in Concept Lattices from Positive and Negative Examples, *Discrete Applied Mathematics*, 2004, **142**, pp. 111-125.

52. S.O. Kuznetsov and V.K. Finn, Extension of Expert Systems of JSM-type to Graphs, *Izvestia AN SSSR, ser. Tekhn. Kibern.*, 1988, No. 5, p. 4-11.

53. S.O. Kuznetsov and V.K. Finn, On a model of learning and classification based on similarity operation, *Obozrenie Prikladnoi i Promyshlennoi Matematiki* **3**, no. 1, pp. 66-90, 1996 [in Russian].

54. S.O. Kuznetsov, S.A. Obiedkov, Comparing performance of algorithms for generating concept lattices, *J. Exp. Theor. Artif. Intell.*, 2002, vol. 14, nos. 2-3, pp. 189-216.

55. S.O. Kuznetsov, M.V. Samokhin, Learning Closed Sets of Labeled Graphs for Chemical Applications, Proc. 15th Conference on Inductive Logic Programming, ILP'05, S. Kramer, B. Pfahringer, Eds., Lecture Notes in Artificial Intelligence (2005).

56. A.E. Leibov, Some methods of realization of the similarity operations for chemically oriented expert systems of JSM type, *Nauchno-Tekhnicheskaya Informatsiya*, Ser. 2, 1996, No. 5-6, pp. 20-32 [in Russian].

57. D. Maier, The Theory of Relational Databases, Computer Science Press, 1983.

58. A.A. Matveev, E.F. Fabrikantova, Algorithmic and Programming Means for Metabolism Forecasting, *Nauchno-Tekhnicheskaya Informatsiya*, Ser. 2, 2002, No. 6, pp. 26-34 [in Russian].

59. S.V. Meien, Yu.A. Shreider, Methodological Aspects of Classification Theory, *Problemy Filosofii*, 1976, No. 12, pp. 67-69 [in Russian].
60. D. van der Merwe, S. A. Obiedkov, D. G. Kourie, AddIntent: A New Incremental Algorithm for Constructing Concept Lattices, *Proc. 2nd Int. Conf. on Formal Concept Analysis (ICFCA'04)*, P. Eklund, Ed., Lecture Notes in Artificial Intelligence, **2961**, 2004, pp. 372-385.
61. M.A. Mikheenkova, V.V. Avidon, and S.A. Sukhanova, On Program Realization of the JSM-method of Automated Hypothesis Generation with Nonelement set of attributes, *Nauchno-Tekhnicheskaya Informatsiya*, Ser. 2, 1984, No. 11, pp. 20-26 [in Russian].
62. M.A. Mikheenkova, V.K. Finn, On a Class of Expert Systems with Incomplete Information, *Izvestia AN SSSR, ser. Tekhn. Kibern.*, 1986, No. 5, pp. 82-103 [in Russian].
63. J. S. Mill, *A System of Logic, Ratiocinative and Inductive*, London, 1843.
64. T. Mitchell, Generalization as Search, *Artificial Intelligence* **18**, no. 2, 1982.
65. T. Mitchell, *Machine Learning*, The McGraw-Hill Companies, 1997.
66. J.M. Nicolas, Mutual Dependencies and Some Results on Undecomposable Relations, *Proc. 4th Int. Conf. on Very Large Data Bases*, West Berlin, 1978, pp.360-376.
67. S. A. Obiedkov, Modal Logic for Evaluating Formulas in Incomplete Contexts, *Proc. 10th Int. Conf. on Conceptual Structures, ICCS'02*, Lecture Notes in Artificial Intelligence, **2393**, 2002, pp.314-325.
68. S.A. Obiedkov, V. Duquenne, Incremental Construction of the Canonical Implication Basis, Proc. International Conference Journee de l'Informatique Messine (JIM03), Metz (2003). To appear in Discrete Applied Mathematics (2005).
69. E.S. Pankratova, D.V. Pankratov, V.K. Finn, I.P. Shabalova, Application of the JSM-method for forecasting high pathogenicity viruses of human papilloma, *Nauchno-Tekhnicheskaya Informatsiya*, Ser. 2, 2002, No. 6, pp. 22-25 [in Russian].
70. N.S. Panova, Yu.A. Shreider, On Symbolic Nature of Classification, *Nauchno-Tekhnicheskaya Informatsiya*, Ser. 2, 1974, No. 12, pp. 3-10 [in Russian].
71. D.V. Popov, V.G. Blinova, E.S. Pankratova, Drug Design: JSM-method of Hypothesis Generation for Antitumor Activity and Toxic Effects Forecast with Respect to Plant Products, FECS 5th Int. Conf. Chem. and Biotechnol. Biologica. Act. Nat. Prod. Sept. 18-23, 1989, Varna, Bulgaria, pp. 437-440.
72. O.M. Polyakov, V.V. Dunaev, Classification Schemes: Synthesis through Relations, *Nauchno-Tekhnicheskaya Informatsiya*, Ser. 2, 1985, No. 6, pp. 15-21 [in Russian].
73. O.M. Polyakov, Classification Data Model, *Nauchno-Tekhnicheskaya Informatsiya*, Ser. 2, 1986, No. 9, pp. 13-20 [in Russian].
74. O.M. Polyakov, On Systematization, *Nauchno-Tekhnicheskaya Informatsiya*, Ser. 2, 1988, No. 12, pp. 21-28 [in Russian].
75. J.R. Quinlan, Induction on Decision Trees, *Machine Learning*, **1**, No. 1, pp. 81-106 (1986).
76. A.A. Raskina, I.S. Sidorov, and Yu.A. Shreider, Semantical Foundations of Object-Attribute Languages, *Nauchno-Tekhnicheskaya Informatsiya*, Ser. 2, 1976, No. 5, pp. 18-25 [in Russian].
77. Yu.A. Shreider, Mathematical Model of Classification Theory, VINITI, Moscow, 1968, pp.1-36 [in Russian].
78. Yu.A. Shreider, Equality, Similarity, Order, Moscow, Nauka, 1971 [in Russian].
79. Yu.A. Shreider, Logic of Classification, *Nauchno-Tekhnicheskaya Informatsiya*, Ser. 2, 1973, No. 5, pp. 3-7 [in Russian].

80. Yu.A. Shreider, Algebra of Classification, *Nauchno-Tekhnicheskaya Informatsiya*, Ser. 2, 1974, No. 9, pp. 3-6 [in Russian].

81. Yu.A. Shreider, Typology as a Base of Classification, *Nauchno-Tekhnicheskaya Informatsiya*, Ser. 2, 1981, No. 11, pp. 1-5 [in Russian].

82. D. Soergel, Mathematical Analysis of Documentation Systems, *Inf. Stor. Retr.*, 1967, No. 3, pp. 129-173.

83. M.Sh. Tsalenko, Semantical and Mathematical Models of Databases, *Itogi Nauki i Tekhniki*, Ser. Informatika, vol. 9, 1985, pp. 3-207 [in Russian].

84. M.Sh. Tsalenko, Canonical Representation of Irreducible Systems and Classification Schemes *Nauchno-Tekhnicheskaya Informatsiya*, Ser. 2, 1985, No. 1, pp. 30-34 [in Russian].

85. A. Tversky, Features of Similarity, *Psychological Review*, vol. 84, no. 4, pp. 327-352, 1977.

86. D.V. Vinogradov, Logical Programms for Quasi-Axiomatic Theories, *Nauchno-Tekhnicheskaya Informatsiya*, Ser. 2, 1999, No. 1-2, pp. 61-64 [in Russian].

87. D.V. Vinogradov, Formalization of Plausible Reasoning in FOPL, *Nauchno-Tekhnicheskaya Informatsiya*, Ser. 2, 2000, No. 11, pp. 17-20 [in Russian].

88. D.V. Vinogradov, Correct Logical Programms for Plausible Reasoning, *Nauchno-Tekhnicheskaya Informatsiya*, Ser. 2, 2001, No. 5, pp. 25-27 [in Russian].

89. R. Wille, Restructuring Lattice Theory: an Approach Based on Hierarchies of Concepts, In: *Ordered Sets* (I. Rival, ed.), Reidel, Dordrecht–Boston, 1982, pp. 445-470.

90. R. Wille, Conceptual Structures of multicontexts. In: P. W. Eklund, G. Ellis, G. Mann (eds.): *Conceptual Structure Representation as Interlingua*. Springer, Berlin-Heidelberg-New York, 1996, pp. 23-39.

91. S.M. Yakubovich (Gusakova), Axiomatic Theory of Similarity, *Nauchno-Tekhnicheskaya Informatsiya*, Ser. 2, 1968, No. 10, pp. 15-19 [in Russian].

92. S.M. Yakubovich (Gusakova), On Properties of Conjugated Tolerance Spaces, *Information Problems of Semiotics, Linguistics, and Machine Translation*, 1971, vol. 1, pp. 116-123 [in Russian].

93. S.M. Yakubovich (Gusakova), Decomposition of Tolerance Spaces, *Nauchno-Tekhnicheskaya Informatsiya*, Ser. 2, 1973, No. 3, pp. 26-28 [in Russian].

94. M.I. Zabezhailo, V.K. Finn, A.V. Avidon, V.G. Blinova et al., On experiments with a database using JSM-method of hypothesis generation, *Nauchno-Tekhnicheskaya Informatsiya*, Ser. 2, 1983, No. 2, pp. 28-32.

95. M.I. Zabezhailo, V.G. Ivashko, S.O. Kuznetsov, M.A. Mikheenkova, K.P. Khazanovskii, O.M. Anshakov, Algorithmic and Program Means of JSM-method Of Automatic Hypothesis Generation, *Nauchno-Tekhnicheskaya Informatsiya*, Ser. 2, 1987, No. 10, pp. 1-14.

96. M.I. Zabezhailo, On search problems arising in automatic hypothesis generation with JSM-method, *Nauchno-Tekhnicheskaya Informatsiya*, Ser. 2, 1988, No.1, pp. 28-31 [in Russian].

97. M.I. Zabezhailo, Formal models of reasoning in decision making: Applications of the JSM-method in systems of intelligent control and automation of scientific research, *Nauchno-Tekhnicheskaya Informatsiya*, Ser. 2, 1996, No. 5-6, pp. 20-32 [in Russian].

98. E.C. Zeeman, The Topology of Brain and Visual Perception, In *The Topology of 3-Manifolds and Related Topics*, K.M. Ford, ed., pp. 240-256, Prentice Hall, 1965.

Conceptual Knowledge Processing in the Field of Economics*

Rudolf Wille

Fachbereich Mathematik, Technische Universität Darmstadt
Schloßgartenstr. 7, D–64289 Darmstadt
wille@mathematik.tu-darmstadt.de

Abstract. *Conceptual Knowledge Processing* is obliged to a pragmatic understanding of knowledge according to which human knowledge is obtained and supported in a process of human thinking, reasoning, and communicating. It is based on a mathematical theory of concepts directed toward an interaction of formal and material thoughts. How this theoretical conception enables effects in the economic practice is explained in this paper, guided by the key processes of the organizational knowledge management. These key processes are *knowledge identification, knowledge acquisition, knowledge development, knowledge distribution and sharing, knowledge usage*, and *knowledge preservation*. For each key process, an example demonstrates the use of specific methods of Conceptual Knowledge Processing. Finally, objectives and evaluation of knowledge management are included into the discussion.

1 Conceptual Knowledge Processing

"Conceptual Knowledge Processing" is grounded on an understanding of human thinking based on concepts. According to this understanding, concepts are the basic units of thought containing both experiences and knowledge of the world. Human beings depend upon concepts to find their way in the world, to act in an adequate manner and to communicate with other human beings. As there is a great variety of structures in conceptual thinking, mathematical means may be used successfully to support it. Such means have been elaborated within *Formal Concept Analysis* [GW99] at Darmstadt University of Technology for more than twenty years by members of the *Research Group on Concept Analysis*. For applying Formal Concept Analysis, methods and procedures for Conceptual Knowledge Processing have been developed multifariously [WZ94],[SW00],[Wi00]. For performing these methods and procedures, software was established to match the theoretical and methodological developments. Especially, the TOSCANA software [KSVW94],[Vo96], was used in various ways in research projects at universities as well as in commercial projects outside universities (mainly by the firm NAVICON AG).

"Conceptual Knowledge Processing" refers to an understanding of knowledge according to which *ambitious knowledge* may only be obtained and supported by

---

* A German version of this paper has been presented in [Wi02]

B. Ganter et al. (Eds.): Formal Concept Analysis, LNAI 3626, pp. 226–249, 2005.

conscious reflection, discursive argumentation and human communication based on the existent understanding of life-world, social conventions, and personal experience [Wi00]. This understanding of knowledge being strictly related to humans is in accordance with the present ideas of knowledge management stating that knowledge (in contrast to data and information) is always linked to humans. The connection between data, information, and knowledge may therefore, according to [PRR99] and [De99], be defined formula-like as follows:

Data   = Signs + Syntax
Information = Data + Meaning
Knowledge = Internalized information + Ability to utilize the information

The component "Conceptual" in the term "Conceptual Knowledge Processing" shall render the constitutive role of the thinking, arguing, and communicating human being for knowledge and its processing (cf. [Wi04]).

"Processing" in knowledge processing refers to a process in which something is gained which may be knowledge or something else like a forecast, an opinion, a reason etc. To process knowledge, formal elements of language and formal processes are called upon to a large extent. This presupposes formal representations of knowledge and, in turn, knowledge must be constituted from such representations by humans. To understand this process better, the basic relationship between *form* and *content* must be clarified for Conceptual Knowledge Processing [Wi94]. A branch of philosophy concerned with this relationship is the *pragmatic philosophy* founded by Ch. S. Peirce which is presently continued among others in the discourse philosophy of K.-O. Apel and J. Habermas. According to the pragmatic philosophy knowledge is formed in an unlimited process of human thinking, arguing, and communicating. In this process, reflecting on the effects of conceptions is significant and real experience give causes to iterative rethinking. Form and content are related in this process so closely that they may not be separated without loss (see [Wi95]).

As Conceptual Knowledge Processing is theoretically based on Formal Concept Analysis, the basic terms of Formal Concept Analysis shall be explained by an example [GZ90][1]: A data table as in Fig.1 is mathematically understood to be a *formal context* consisting of a set of *objects* (in the example: companies interviewed during the fair "Kontakta '89" in Darmstadt) and a set of attributes (in the example: criteria for hiring employees), and a *relation* (indicated in the example by the crosses). The relation states which object has which attribute (in the example: which company considers which criterion to be particularly important). Certain formal concepts belong to a given formal context. They form a so-called *concept lattice* of the formal context with respect to the subconcept-superconcept-ordering. A *formal concept* consists of a set of objects, its extent, and a set of attributes, its intent. The extent contains all the objects of the context having all the attributes of the intent. And the intent contains all the attributes of the context common to all the objects of the extent. A formal con-

---

[1] Those who are already familiar with Formal Concept Analysis may skip the rest of this section.

	Diplomnote	Fächerkombination	Studiendauer	Zusatzqualifikation	Auslandsaufenthalte	außeruniv. Aktivitäten	vorherige Ausbildung
AEG(N)	X	X					
AKZO(N)	X	X		X			
Audi		X			X		
Black & Decker(I)	X						
Degussa	X	X					
Dow(I)		X					
Effem	X				X		
Mannesmann		X					
INA(I)		X					
Merck(N)			X				
JPMorgan					X	X	
NCR		X			X		
Opel				X			
KPMG(I)	X		X				
KPMG(N)	X		X	X	X		X
Procter & Gamble(I)					X	X	
Rütgers(N)	X			X			
SEL(I)		X					
VW	X	X	X				

Fig. 1. Formal context of firms and hiring criteria for graduates from university

cept is a *subconcept* to another formal concept if its extent is contained in the extent of the other concept or – which is equivalent – if its intent contains the intent of the other concept.

To make conceptual connections in data tables more transparent, it proved effective to represent concept lattices by line diagrams as in Fig.2. The small circles of a line diagram represent the formal concepts of the given formal context and the ascending pathes of line segments represent the subconcept-superconcept-ordering. For instance, in Fig.2 the small circle labelled by "VW" represents a subconcept to the formal concept which is represented by the small circle labelled by "duration of studies". This indicates that the criterion "duration of studies" is particularly important to the company "VW". In general, the extent and the intent of a formal concept may be read from a line diagram as follows: The extent of the concept consists of all the objects whose labels are to be found on the paths descending from the circle representing the concept. And the intent of the concept consists of all the attributes whose labels are to be found on paths ascending from the circle representing the concept. In Fig.2 the small circle marked "Effem" represents – according to the aforementioned – the formal concept the extent of which consists of the companies "Effem" and "KPMG(N)" and the intent of which consists of the criteria for hiring "diplomagrade" and "study abroad". What was previously said has as consequence the fact that the original context can be reconstructed from the line diagram, i.e., no data is lost when the concept lattice and its corresponding line diagram is established. For arbitrary data tables concept lattices may be constructed even if numbers or

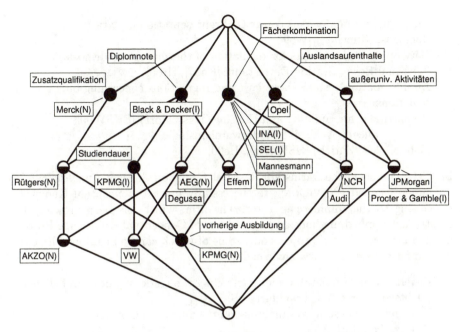

Fig. 2. Concept lattice of the formal context in Fig.1

other symbols are used instead of crosses; for such cases, the method used is called *conceptual scaling* (see [GW99]) which allows one in general to deduce formal contexts and their respective concept lattices from arbitrary data tables (databases).

2   Applications in the Field of Economics

Conceptual Knowledge Processing is applied in the field of economics in a multitude of cases. Members of the Research Group on Concept Analysis have been in charge of and have carried out the following projects (among others):

- An information system for the configuration of personal computers
  (IBM Wiesbaden) [Wi89]
- Conceptual optimization of industrial production processes
  (IBM Mainz) [WS93]
- Conceptual control of industrial production processes
  (DEGUSSA, Frankfurt) [Wo94]
- A conceptual system for designing pipelines
  (ABB Mannheim) [Vg95]
- Conceptual quality control in distillation columns
  (DEGUSSA, Frankfurt) [Wo95]
- Conceptual analysis of data on flight movements
  (Flughafen Frankfurt/M AG) [Kf96]

- Three-dimensional versus conceptual representation of data
  (Siemens, München) [Wo96]
- Development of concepts for regulation of thermic refuse disposal
  (Thermal Power Station Darmstadt) [Ka97],[Hr00]
- An information system about laws and regulations concerning building con-
  structions
  (Department for Building and Housing, Düsseldorf) [EKSW00]
- Formal Concept Analysis in data warehousing
  (Jelmoli AG, Zürich) [HSWW00],[He00]

In 1994 members of the Darmstadt Research Group on Concept Analysis founded
a company, called "NaviCon", to successfully market Conceptual Knowledge
Processing to the commercial sector. The firm, which has been growing constantly, has applied methods and procedures of Conceptual Knowledge Processing in a large number of projects only some of which can be mentioned here (for
more information see: www.navicon.de):

- Development of a database for controlling the trade of stocks and shares
  (Deutsche Börse AG, Frankfurt)
- Drawing up a system for analysis and documentation of data
  (Eurex Frankfurt AG)
- Development of an information system for IT security management
  (Bank Julius Bär & Co. AG, Zürich)
- Development of a system to support quality function deployment
  (Daimler-Chrysler AG, Department for Research and Development,
  Frankfurt)
- Development of a TOSCANA-system for analysis and exploration of data
  (Max-Planck-Institute for Aeronomy, Kaltenburg-Lindau)

How can the wide range of applicability of Conceptual Knowledge Processing,
which is a mathematically founded method, be explained? The main reason is the
sound mathematization of concepts and concept systems, the basic structures
of human knowledge [Se01],[Wi01a]. In this way, interrelations in thought may
be rendered transparent and intelligible which effectively supports the creation
and processing of knowledge in human thinking. By showing concept systems
graphically, the user easily grasps connections, activating background knowledge
for creating and processing knowledge. This becomes obvious when users, who
are experts in their respective fields, often notice mistakes in the data contexts
when examining the corresponding line diagrams. In this sense, the program
TOSCANA which renders a conceptual collective view of large data contexts possible is understood as a tool for navigating "conceptual landscapes of knowledge"
[Wi99]. In general, Conceptual Knowledge Processing supports acts of thinking
such as exploring, searching, recognizing, identifying, investigating, analyzing,
making conscious, deciding, improving, restructuring, remembering, informing
etc.; these issues are fully explained in [Wi00].

To clarify the multitude of possibilities to use Conceptual Knowledge Processing in the field of economics, different aspects of potential applications are

discussed on the basis of an appropriate structure of *organizational knowledge management*. Such a structure is convincingly presented in the book *"Wissen managen: wie Unternehmen ihre wertvollste Resource nutzen"*; this book has been translated to English under the title *"Managing Knowledge: Building Blocks for Success"* [PRR99]. The book stresses the following key processes of knowledge management: *knowledge identification, knowledge acquisition, knowledge development, knowledge distribution and sharing, knowledge usage,* and *knowledge preservation*. All of these key processes are more or less tightly bound with each other. For a better embedding in the corporate strategy, the book recommends including the fields *knowledge objectives* and *knowledge evaluation*. The key processes are discussed in great detail in [PRR99]. They will be explained here to the point where the possibilities of the respective application of Conceptual Knowledge Processing may be shown by an example for one of these applications.

2.1   Identifying Knowledge

How can transparency about existing knowledge be obtained internally and externally? Companies often complain about the absence of transparency of the existing knowledge store and also about the rising flood of information which causes further lack of transparency as a means of selection are absent. The objective is to remedy this matter by using methods for *knowledge identification*. According to [PRR99] (p.106) these methods produce transparency of knowledge enabling the individual to improve his orientation and find better access to the internal and external sphere of knowledge. In so doing, synergy may be produced, cooperations may be formed, and valuable contacts made. This results in more efficient use of internal and external resources and faster responses for the company.

In [PRR99], many ideas are brought up to make the internal and external identification of knowledge easier. Drawing up suitable *"knowledge maps"* proved to be an efficient and relatively easy method to improve the knowledge identification. According to [Ep97], knowledge maps are generally graphical indices of knowledge support, knowledge stores, knowledge sources, knowledge structures, or knowledge applications:

- *Knowledge holder maps* mainly fulfill the function to show which types of knowledge and which content are present in which knowledge holder.
- *Knowledge source maps* give information about which people or other aids within the team, within the organisation or the external environment, may supply important knowledge for the task in question.
- *Knowledge store maps* show where and how certain stores of knowledge are located.
- *Knowledge structure maps* depict units of knowledge as far as significant structures of content and interrelations, respectively, are concerned.
- *Knowledge application maps* structure knowledge as far as uses and applicational relations are concerned.

The *line diagrams of concept lattices* may be understood as informative knowledge structure maps which may be employed as *conceptual query structures* in a flexible way using the program TOSCANA. How this may be effected will be explained by using the project *"Conceptual analysis of data on flight movements"* [Kf96] as an example (see also [SWW98]). The objective of this project was to investigate to what extent conceptual knowledge processing may contribute to the task of extracting useful information from the large amount of continuously accumulating data on flight movements at the airport in Frankfurt. The project was based on all available data about flight movements in June 1996.

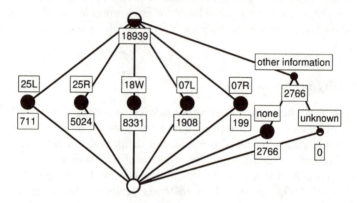

Fig. 3. Query structure "G13 Runway [RWY]"

Generally, in a TOSCANA-system, the *objects* of a field of investigation are stored in a *relational database* so that, by means of an SQL-query, they may be activated to update the conceptual structures of the investigation. The line diagrams which represent the query structures first indicate the *numbers* (or *percentages*) of the objects which represent the respective size of the extent of the concept shown. In the query structure "G13 Runway [RWY]" in Fig.3, the number 8331, which is attached from below to the circle labelled by "18W", stands for the 8331 takeoffs in June 1996 from the runway called Runway West. More specific information about the objects is given if a mouse click is made on one of the numbers which produces a list of the *names of the objects* of the respective extent of the concept. Clicking on one of the names of the objects in turn releases further information on the object.

The strength of TOSCANA-systems lies in the fact that query structures may be gradually refined by further query structures which render possible a flexible and informative *navigation* through large stores of data. If, for example, the query structure "G24 code for range and size of positions [SWC]" in Fig.4 is zoomed into the circle "18W" of the query structure "G13 Runway [RWY]", the second line diagram in Fig.4 results. The percentages in the second line diagram show that a major part of the 8331 takeoffs on Runway West were undertaken by large aircraft (the size of positions 4 and 5 constitute a proportion

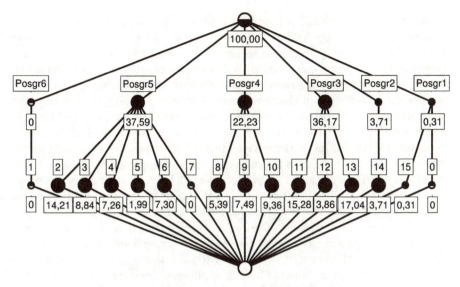

Fig. 4. Query structure "G24 code for range and size of positions [SWC]"

of 59.82 percent). If the query structure "G26 noise classification of the aircraft according to the ICAO annex 16 [A16]" is zoomed into the circles "Posgr4" and "Posgr5" of the concept lattice in Fig.4, the amount of the noise pollution of the people living near the airport becomes apparent. The fact that more than 95 percent of the aircrafts of class 3 belong to the so called "aircrafts with low noise levels" may immediately be discerned. This example was to explain that, by using TOSCANA-systems, the demands (connected with knowledge identification) for more transparency, better orientation, easier access, improved possibilities of usage, and faster responses in the knowledge environment are fulfilled to a large extent.

2.2   Acquiring Knowledge

How is external knowledge acquired? According to [PRR99], companies cover a considerable part of their demand for knowledge from sources outside of the company. The following activities may be distinguished:

- Acquisition of knowledge from external knowledge holders,
- Acquisition of knowledge from other companies,
- Acquisition of knowledge from stakeholders,
- Acquisition of knowledge products.

To get a hold of knowledge and faculties of people who are valuable to the company as *knowledge holders*, the company usually tries to recruit the people in question. For the desired knowledge, some distinctive factors must be observed. These factors are mainly whether the knowledge is of a general or a specific kind and whether it is potential knowledge or readily useful knowledge. Instead of hiring individual knowledge holders or external knowledge experts for a given span of time, the company may choose to secure access to *knowledge bases of other companies* by forming a cooperation. A form of cooperation which is frequently discussed and formed is a strategic alliance where the cooperating partners set up common objectives and thus compensate at least partially their weaknesses by the strengths of the other partners and vice versa. *The potential and the store of knowledge of stakeholders* are very important for a company. Stakeholders in that sense are all groups in the environment of the company who have special interests or demands as far as the activities of the company are concerned. The following are often mentioned as important stakeholders: customers, suppliers, owners, employees, representatives of the association of workers, politicians, the media and opinion leaders, the financial world, and the public. The purchase of knowledge in the form of *knowledge products* independent from the individual, for example software or CD-ROMs, may well be a good means of effective knowledge management. However, it is necessary to take care that these products can be integrated appropriately in the field of application they are intended for. How this may be achieved by using methods of conceptual knowledge processing shall be explained by an example.

In the course of the project *"A conceptual system for designing pipelines"* [Vg95] which was conducted in cooperation with Dipl.-Ing. F. Meinl (an engineer at ABB Mannheim), the task was to find out how the *German industrial norms (DIN)*, relevant for the construction of pipelines, could be made available to the engineer in the development department as an external *product of knowledge*. The variety of standardized parts used in mechanical engineering is so enormous that even experts have great difficulties to find suitable standardized parts for a given task. Therefore a prototype of a *TOSCANA-system* was designed to adequately integrate all information on relevant norms according to DIN into the line of thought and action of an engineer of a development department in order to support that engineer in *acquiring the necessary knowledge* for his work. For the prototype the following standardized parts were chosen as objects:

(1) seemless steel pipes according to DIN 24481,
(2) bow pipes according to DIN 2605-1&2, T-pieces according to DIN 2615-2, reducing pieces according to DIN 2616-2,
(3) welding flanges according to DIN 2631 (Nd6) and DIN 2632 (Nd10), loose flanges according to DIN 2641 (N6) and DIN 2642 (Nd10).

This resulted in a (many-valued) data context of nearly 4,000 objects and 54 (many-valued) attributes like external diameter, thickness, material, permissible pressure at 100 degrees Celsius and so on. The finished prototype has 33 query structures at disposal.

Fig.5 shows the query structure "Selection of pipe parts" which combines the part types "bow pipes" ("Rohrbögen"), "reducing pieces" ("Reduzierstücke"), and "flanges" ("Flansche") with their respective DIN norms as attributes of that structure. This is a useful overall view which makes a comparative search for a suitable part easier. The query structure "Thickness", shown in Fig.6,

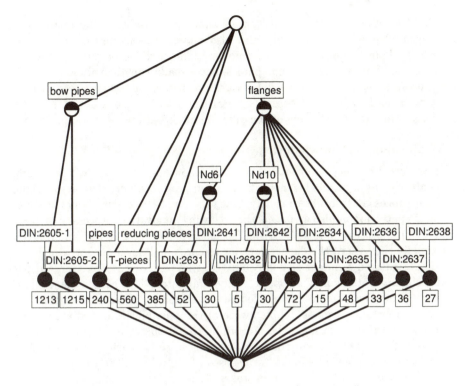

Fig. 5. Query structure "Selection of pipe parts"

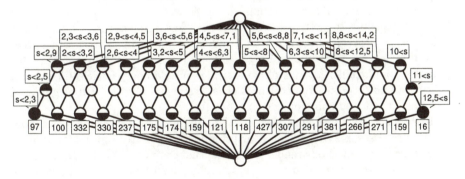

Fig. 6. Quey structure "Thickness of pipes"

consists of a conceptual pattern where small intervals of the thickness s supply the attributes. This structure is justified by the fact that there might be no flanges or bows to be found matching the thickness of some of the pipes; in turn this results in the necessity of reworking the pipe, flange or bow (therefore one interval overlaps with up to four other intervals). These two examples are presented to make clear that external products of knowledge, like the DIN norms, must be *integrated by adding integrational factors* into the thinking and acting on the spot to make these products fully usable.

Besides supporting the acquisition of knowledge products, the methods of Conceptual Knowledge Processing may also support the acquisition of external knowledge, knowledge bases, and stakeholders. The introductory example in Section 1 on the hiring criteria of companies may already convincingly explain this for *procedures for hiring new employees.* As a matter of fact, concept lattices have been used in hiring processes several times to include a better base for discussion and an overall view of the differing qualifications of the applicants.

2.3   Developing Knowledge

How is new knowledge internally developed? Knowledge development in a company aims for new knowledge, new abilities and products, better ideas and more efficient processes. In order to clarify, how this may be brought about, it is recommended to distinguish between the following innovative processes: *product innovations, process innovations, and social innovations.* Product innovations are usually the task of the department for research and development, which is – of course – often dependent on competent external partners. Process innovations and social innovations may not be so easily pinpointed. This causes the need for knowledge management to create adequate attention for these innovations.

It is important to make people aware of the possibilities of individual and collective knowledge development. According to [GP95], creativity in *individual knowledge development* is supported by working in small, easily comprehensible units, by company wide mobility, by a sense of family, by suitable objectives, by friendly treatment in the event of mistakes, by a long time frame, which creates leisure, and by a culture of fair dispute. The key spheres in *collective knowledge development* are interaction, communication as well as transparency and integration. According to [PRR99], the distinctive key competences of companies today are the processes in which a multitude of members of the respective organization take part. Instruments for collective knowledge development are – for example – think tanks (groups of experts), learning arenas (nuclei of learning), lessons learned (reports of experiences given systematically) as well as giving of scenarios (structured communication processes).

To develop new knowledge, new abilities and products, better ideas and more efficient processes, may be achieved by using methods of Conceptual Knowledge Processing. These methods render the conceptual connections between form and contents more transparent in many ways and – by so doing – activate creative thinking. During a research project carried out by the Research Group on Concept Analysis from Darmstadt and the Swiss retail combine Jelmoli AG (Zürich), TOSCANA was integrated into a data warehouse in order to analyze the shopping habits of the customers and, by so doing, strengthen the *database marketing activities* of the retailer [HSWW00],[He00]. The results of the project were received in a very positive manner as the heads of the department for marketing were able to discuss the data in a sophisticated way with their business partners based on the presentation of the data according to the methods of Formal Concept Analysis. New insight into the shopping habits of the customers were gained and new ideas for marketing acitivities were developed based on the resulting new knowledge.

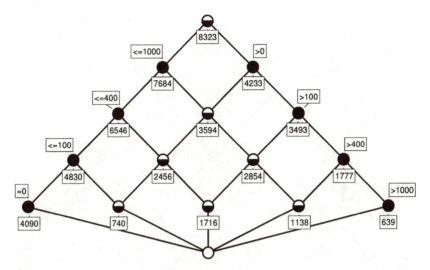

Fig. 7. Concept lattice for analysing purchases in the depatment "women's wear"

The labelled line diagram in Fig.7 shows how the spending of 8323 female customers in the department "women's wear" is distributed (for example: 1716 customers spent an amount higher than 100 and lower than 400 Swiss francs and 1777 customers spent more than 400 Swiss francs). The line diagram in Fig.8 combines two *aspects of purchasing activities*: The line diagram restricted to the black circles represents the distribution according to the number of departments which were visited during the shopping session, and the inserted rhombs with a black circle on top differentiate the respective number (at the top) further as to the departments "household goods" (at the left side), "interior decoration" (at the right side), and both of these departments (at the bottom). For the marketing department the nested diagram proved interesting in many ways. For instance, of the 6546 female customers who spent little on clothing (400 Swiss francs at the most according to Fig.7), the line diagram shows their distribution according to the number of departments where they purchased and according to their purchases in the department "household goods" and "interior decoration". As target group for *direct mailing* which aims at stimulating purchases in the department "women's wear", the 2001 customers who have purchased in 5 up to 12 departments, especially in the departments "household goods" and "interior decoration", could be selected, for example.

Once again, the Jelmoli project proved that the materialization of conceptual interrelations by means of line diagrams prompts the *individual and collective formation of new knowledge*. As mentioned earlier, line diagrams showing contents activate the background knowledge of their viewers and thereby trigger new insights and ideas. As line diagrams are also an effective means of communication for conceptual interrelations, they particularly support collective development of ideas and knowledge.

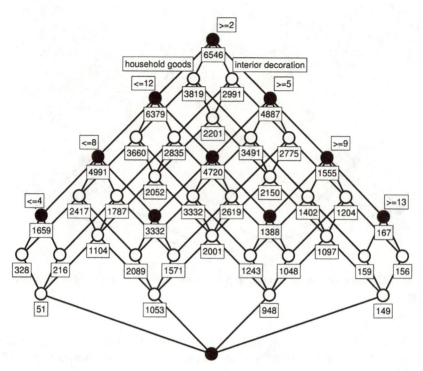

Fig. 8. Concept lattice for analysing purchases in the departments "household goods" and "interior decoration"

2.4   Distributing and Sharing Knowledge

How can we distribute and share knowledge appropriately? Distributing and sharing knowledge and experience within an organisation is a compelling prerequisite to render knowledge in isolated existence usable to the organisation as a whole. These activities do not end in the mechanical allocation and transport of packages of knowledge, as knowledge is an item which often can only be transferred in personal exchange between individuals. According to [PRR99], the way to distribute knowledge may be subdivided into three tasks:

– multiplying knowledge by fast transmission to a multitude of employees,
– securing and sharing previous experiences,
– simultaneous exchange of knowledge which leads to the development of new knowledge.

Multiplying knowledge describes a centrally controlled operation which aims to spread certain stores of knowledge quickly to a large number of employees. The multiplication of knowledge is to promote continual education and foster the socialization of the employees. According to [PRR99], socialization may be understood as getting acquainted with the values and norms common in the organisation on the one hand and communicating basic behavior patterns and expected role models on the other. In short: it may be understood as the bedding down in the corporate culture. For *securing and sharing experiences and knowledge* the latest developments in information and communication technology (particularly in the field of computer networking) open a multitude of possibilities, but criteria for the sensible use of these possibilities are mostly still missing. In order to effect *productive exchange of knowledge* under a suitable technological infra-structure, multiple individual and cultural barriers that prevent knowledge sharing must be overcome. Individual barriers may be the absence of ability or the willingness to share. Cultural barriers may be missing elements in the culture of the company which exercise an authorizing and supporting influence on the knowledge distribution.

How methods of Conceptual Knowledge Processing may be used successfully to distribute and share knowledge will be explained for a field of management which continually gains in importance: the field of IT security management. The companies NAVICON (Frankfurt) and r^3 ag (Zürich) developed a TOSCANA based *Information System for IT Security Management*, which serves the following operational tasks [BSWZ00]:

– The system offers a data model and means for its use for analysis of requirements (i.e. settlement of IT units and registration of risk) and for taking measures (i.e. assignments of demands, directives, and checklists).
– The structure of the basic IT management system is open to extension and change in the event of revisions.
– A graphical interface is at the user's disposal for the collection and maintenance of data and for the analysis and control of the respective standard of security.

This information system provides both the people responsible and the people concerned with the necessary knowledge about risk in the IT field and about countermeasures because, as the query structures indicate the contents, a well-directed question for the sought information is possible. In this way knowledge about IT-security can be distributed, but it also allows to share knowledge up to the discussion of improving the IT management system.

If (for example) a new answering machine is to be examined for possible risks with the aim of taking counteraction, then it is advisable to open the query structure "answering machine" which is represented in Fig.9. The data context upon which the query structure is based is taken from the Basic IT Safety Manual of the German Federal Office of Safety in Information Technology [So98]. In this data context, the risk of answering machines determines the "objects" while the countermeasures make up the "attributes". If (on the other hand) all objects concerned and possible countermeasures against a certain risk like *variation in voltage* are to be considered, the query structure "variation in voltage", shown in Fig.10, may be opened. In this case the modem, TC-equipment, and PC-net are the "objects" and the measures against present risk are the "attributes".

In the information system of NAVICON and r^3 – on top of being imbedded in the mentioned query structures – the measures are captured in *directives*

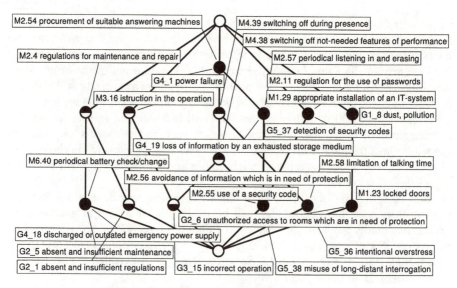

Fig. 9. Query structure "Answering machine"

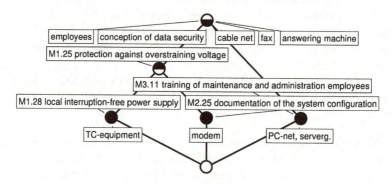

Fig. 10. Query structure "Voltage fluctuations"

and checklists that may be activated. In these directives and checklists the specific needs of the respective company may be taken into account more effectively. Altogether, the TOSCANA based information system achieves the desired distribution of the respectively necessary knowledge about risks and countermeasures in the IT-sector of a company to the responsible co-workers.

2.5   Using Knowledge

How is existing knowledge put into active use? Knowledge usage, meaning the productive use of organisational knowledge for the benefit of the company, is, according to [PRR99], *objective and purpose of knowledge management*. A working environment must be created which supports the application of the acquired knowledge and promotes the willingness to put it to use on an individual and a collective basis. On an *individual* basis, the preparedness to continually question the existing states and structures should be stimulated. Asking questions has to be considered, not as a sign of a lack of competence, but as the readiness to learn and to change. On a *collective* basis, knowledge should be considered a resource which must be employed for the common good of the organization. In this context it is of no importance which source the knowledge comes from, but it is crucial how this knowledge may be put into use most effectively for the organization.

To activate existing knowledge it is important to supply a *user-friendly infrastructure*. Essential criteria which must be fulfilled are the following elements: "easy-to-use", "just-in-time" as well as "ready-to-connect". User-friendliness comprises an *arrangement* of the material and situation of work starting with the preparation of documents up to the form of the workplace that is *beneficial to the user*. According to [PRR99], user-friendliness must be central in all respects of knowledge management. This means that the *needs of the users of the knowledge* should be considered in all measures taken by knowledge management.

An extensive project, the purpose of which was to use existing knowledge, was carried out by members of the Research Group on Concept Analysis from Darmstadt together with members of the Department for Building and Housing of the State of Nordrhein-Westfalen. The objective was to build a prototype of a TOSCANA *information system about laws and regulations concerning building construction* integrating the specialized knowledge that existed in the department [EKSW00]. The main purpose of the TOSCANA-system was defined to be a support for the planning department and building control office as well as for people that are entitled to present building projects to the office in order to enable these groups to consider the laws and technical regulations in planning, controlling, and implementing building projects. An extensive data context was elaborated for the system. The objects of this context were the relevant *paragraphs or units of text* of the bulding laws and regulations and technical instructions, while the attributes were *building components and requirements for building components* put in terms that are used as search words for the respective paragraphs or units of text. As it is typical for TOSCANA-systems, numerous conceptual query structures were deduced from the basic data context and represented by line diagrams in order to use them as *conceptual search structures*.

The methodology of Conceptual Knowledge Processing substantially contributed to the *useful activation* of existing specialized knowledge about building laws and regulations and structural engineering for the desired information system. After several attempts of the building experts to structure the specialized knowledge for the information system, the question of the suitable "objects" finally led to a breakthrough. When the staff involved had reached the agreement that the relevant paragraphs and units of text, respectively, should be the "objects" and the building components and requirements should be the "attributes", the basic data context (which finally extended to more than 50,000 pieces of information) was quickly established. Nevertheless, the later deduced query structures still caused several revisions of the data.

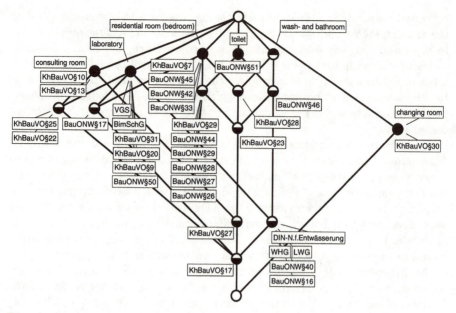

Fig. 11. Query structure "functional rooms in a hospital" of a TOSCANA information system about laws and regulations concerning building construction

An *instructive case of one of the mentioned revisions* came up for the query structure "functional rooms", shown in Fig.11, which is relevant in the building of hospitals: In order to test the legibility of the corresponding diagrams, a secretary was included in the discussion during one of the working sessions in the department. The secretary was quite astonished that in the version of the diagram "functional rooms" as discussed, §51 of the building regulations of Nordrhein-Westfalen ("BauONW§51") demanding building according to the needs of the disabled was attached to the circle labelled "toilet" and therefore was only applicable to the construction of toilets. She was unable to understand that, for instance, "wash- and bathroom" did not have to be built according to the needs of the disabled. The experts were surprised too, but an additional reading of §51 confirmed that only the toilet were part of that regulation. Only after extensive discussion (*activation of collective knowledge!*) did the group come to the opinion that – according to higher legal considerations – an application of §51 to "wash- and bathroom" had to be called for. Finally, by similar reasons, the "consulting room" and the "residential room" were also included so that, in the line diagram of Fig.11, the label "BauONW§51" moved down to the circle with the label "KhBauVO§27" (cf. [Wi95]).

2.6   Preserving Knowledge

How can we avoid knowledge loss? The well-directed preservation of experience, information, and documents requires management to make an effort. In order not to relinquish valuable expert knowledge, processes of selecting knowledge worth retaining, knowledge saving in an adequate way, and knowledge updating on a regular basis must be shaped deliberately. The objective of this selection is to separate valuable experience from valueless experience and to transfer valuable data, information and experience to organizational systems useful to the whole company. As a rule, organizations are not able to manage all processes necessary for the selection of knowledge, therefore it is advisable to concentrate the selection on knowledge which will prove useful in the future.

When it comes to saving knowledge, three forms of preservation are to be distinguished: the individual, the collective, and the electronic saving of organizational knowledge. The *individual knowledge* inside the heads of single employees can be made available (partly) but this requires considerable effort. The *collective knowledge* is more than the summing up of individual knowledge, for knowledge is created in the interaction of individuals also. For this reason the locus of preservation of the collective knowledge has been called the *collective memory* of the respective social group. The saving of knowledge in the *electronic memory* of the company becomes increasingly significant. The great advantages of digitalized memory media are easy editing, reusability, and little expense when being distributed via networks. Selecting and saving must be structured in a manner which supports the updating of the saved knowledge adequately, i.e. so that the user may retrieve the wanted information in good quality.

As early as the mid '80s, before even the thought of developing TOSCANA software was born, the Research Group on Concept Analysis in Darmstadt worked on a project application the objectives of which were selecting, saving, and updating of knowledge on a large scale. The task of the project was to develop an information system which was to support the configuration of personal computers with parts manufactured by the IBM corporation. Starting the project the participants were surprised to find out that the necessary information on the existing personal computer parts could not be called up anywhere at IBM as a whole. Therefore the information had to be gathered from the different sources which proved to be an enormous task. The methodological approach of Formal Concept Analysis to base all conceptual systems on formal contexts was helpful in so far as the purposeful selection of the data for parts as "objects" and their characteristics as "attributes" of a comprehensive data context has rendered working simpler and more efficient. The finally resulting data context of personal computer parts of IBM was treasured as a highly valuable store of knowledge. We were under the impression that the data context was the most valuable result of the project to our IBM-partners.

Nevertheless, the following development of a system to update and actively use the knowledge thus stored was very interesting too. For the various types of parts, such as CPUs, screens etc., partial contexts were extracted from the comprehensive data context and for each of these contexts the respective concept lattice was determined. Fig.12 shows the concept lattice of the context having as objects the screens 5151, 5272, 5279, 5379, 4861 und 5175 and as attributes screen sizes D12, D13, D14, resolutions 640×200, 640×350, 640×480, 720×512,

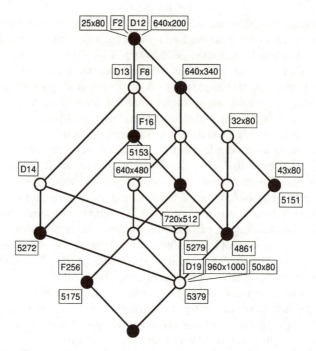

Fig. 12. Concept lattice for inspecting PC screens and their attributes

960 × 1000, numbers of colours F2, F8, F16, F256, and text presentations 25 × 80, 32 × 80, 43 × 80, 50 × 80. The blackening of certain circles is a consequence of the preceding decision to choose an SDL-CPU as part of the configuration.

3   Knowledge Objectives and Evaluation

The key processes of knowledge management discussed are concerned with operational problems which may arise when handling the resource knowledge. As a strategy for corporations, these processes do, however, need a framework for orientation and coordination for which objectives of knowledge and evaluation of knowledge are fields of basic importance. According to [PRR99], *objectives of knowledge* establish what abilities are to be built up on which levels. Normative objectives of knowledge point to the creation of a knowledge conscious corporate culture. Strategic objectives of knowledge define the core knowledge of the organization and thereby describe the need for future competence. Operational objectives of knowledge translate knowledge management into action and secure that normative and strategic objectives are definitely converted. According to [PRR99], *evaluation of knowledge* is to measure whether the objectives of knowlege were adequately defined and the measures of knowledge management successfully put into effect. Promising approaches to do so are an understanding of cause-effect-mechanisms and indirect evaluation via indicators of knowledge. "Knowledge control" is to be based upon the evaluation of knowledge. Using "knowledge control" as an aid, the manifold activities can be directed towards the vision and strategy of the corporation in relation to knowledge. Concerning these characterizations of knowledge objectives and evaluation, one has to consider that knowledge objectives and knowledge evaluation, because of the process-like discursive nature of human knowledge, influence each other: the knowledge objectives stipulate knowledge evaluation and, conversely, knowledge evaluation causes modifications of knowledge objectives; concerning those influences, the key processes of knowledge management have their effects too.

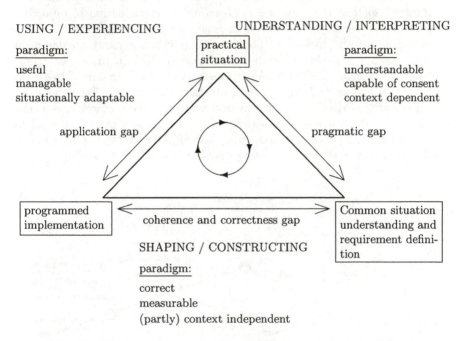

Fig. 13. The iterative process of system development in informatics

In [An97], this interrelationship is discussed for system developments in software engineering in a threefold relation *definition of reqirements – implementation – practical situation* (see Fig.13). The analysis of requirements establishes the purpose and the objective and (by so doing) determines the type of technical implementation which – in turn – must prove to be worthwhile and is evaluated according to its performance. There are to be bridged the "pragmatic gap" between practical situation and definition of requirements, the "coherence and correctness gap" between definition of requirements and implementation and finally the "application gap" between implementation and practical situation.

– Bridging the *pragmatic gap* may be supported by a social process of understanding the validity and the adequacy of the intended system development with respect to the given life-world context.

- Bridging the *coherence and correctness gap* may be supported by a process of understanding about the coherence between requirements and implementation.
- Bridging the *application gap* may be supported by a process of understanding about the usage, the reliability and the acception of the implemented system with regard to the respective practical situations.

As these processes of understanding are mutually dependent, the whole process of system development, and also the processes of knowledge management, must be designed so that the three fields of action – determination of objectives, performance, and evaluation – are to be gone through several times.

In [An97], a framework of methods to discursively analyze the requirements is given which was evaluated with several case studies. In these case studies methods of Conceptual Knowledge Processing were used. In one of these *case studies* which was carried through at Darmstadt University of Technology as a funded research project, the point was to develop adequate information technological means; the purpose of these means was to support the relevant official control and hearing procedures concerning technical devices and installations by simulations and virtual realities before they have even been produced.

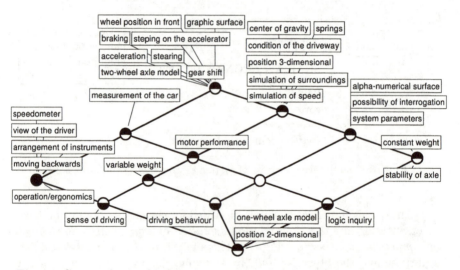

Fig. 14. Concept lattice for determining a conceptual model for car development

As an example, the framework of methods was applied in the field of car manufacture (cf. [An97], p.259ff.). After the *contextual frame* for the discursive analysis of the requirements had been negotiated, the required modelling aspects for the desired functionalities were formally conceptualized, i.e. a conceptual model was developed; and in doing so, the methods of Formal Concept Analysis have been proven valuable. The line diagram in Fig.14 presents, by extract, the respective characteristics and special functions which are considered to be important for the following functionalities: operation/ergonomics, sense of driving, driving behaviour, logical inquiry, and stability of the axles. After establishing the conceptual model, the modelling scheduled for the given task was discursively validated against the conceptual model developed by reasons of contents. In the course of the discussion about validation, the representations by line diagrams were helpful when it came to detecting incorrect specifications, underlying misunderstandings, and wrong interpretations of the importance of individual characteristics. Altogether the methods of Conceptual Knowledge Processing contributed substantially to the negotiations for *common interpretation and understanding* of complex situations and tasks.

References

[An97]   U. Andelfinger: *Diskursive Anforderungsanalyse. Ein Beitrag zum Reduktionsproblem bei Systementwicklungen in der Informatik.* Peter Lang, Frankfurt 1997.

[BSWZ00] K. Becker, G. Stumme, R. Wille, U. Wille, M. Zickwolff: Conceptual information systems discussed through an IT-security tool. In: R. Dieng, O. Corby (eds.): *Knowledge Engineering and Knowledge Management: Methods, Models, and Tools.* LNAI **1937**. Springer, Heidelberg 2000, 352-365.

[De99]   K. Devlin: *Infosense. Turning Information into Knowledge.* Freeman, New York 1999.

[EKSW00] D. Eschenfelder, W. Kollewe, M. Skorsky, R. Wille: Ein Erkundungssystem zum Baurecht: Methoden der Entwicklung eines TOSCANA-Systems. In: [SW00], 254-272.

[Ep97]   M. J. Eppler: Praktische Instrumente des Wissensmanagements – Wissenskarten: Führer durch den "Wissensdschungel". *Gablers Magazin* **8** (1997), 10-13.

[GW99]   B. Ganter, R. Wille: *Formal Concept Analysis: mathematical foundations.* Springer, Heidelberg 1999.

[GZ90]   B. Ganter, M. Zickwolff: Nach welchen Kriterien wählen Firmen Hochschulabsolventen aus? FB4-Preprint Nr. 1343, TU Darmstadt 1990.

[GP95]   P. Gomez, G. J. B. Probst: Die Praxis ganzheitlichen Problemlösens – Vernetzt denken – Unternehmerisch handeln – Persönlich überzeugen. Haupt, Bern/Stuttgart/Wien 1995.

[He00]   J. Hereth: *Formale Begriffsanalyse im Data Warehousing.* Diplomarbeit. FB Mathematik, TU Darmstadt 2000.

[HSWW00] J. Hereth, G. Stumme, R. Wille, U. Wille: Conceptual knowledge discovery and data analysis. In: B. Ganter, G. W. Mineau (eds.): *Conceptual Structures: Logical, Linguistic, and Computational Issues.* LNAI **1867**. Springer, Heidelberg, 2000, 421-437.

[Hr00]   C. Herr: *Innovative Analyse und primärseitige Prozessführungsoptimierung thermischer Abfallbehandlungsprozesse – am Beispiel der Mülleingangsklassifizierung bei der Rostfeuerung.* Schriftenreihe WAR 119, TU Darmstadt 2000.

[Ka97]   E. Kalix: *Entwicklung von Regelungskonzepten für thermische Abfallbehandlungsanlagen.* Diplomarbeit, FB13, TU Darmstadt, 1997.

[Kf96]   U. Kaufmann: *Begriffliche Analyse von Daten über Flugereignisse – Implementierung eines Erkundungs- und Analysesystems mit TOSCANA*. Diplomarbeit, FB4, TU Darmstadt, 1996.

[KSVW94]   W. Kollewe, M. Skorsky, F. Vogt, R. Wille: TOSCANA – ein Werkzeug zur begrifflichen Analyse und Erkundung von Daten. In: [WZ94], 267-288.

[PRR99]   G. J. B. Probst, S. Raub, K. Romhardt: *Wissen managen: wie Unternehmen ihre wertvollste Resource optimal nutzen*. 3. Aufl. Gabler, Wiesbaden 1999; English version: *Managing knowledge: building blocks for success*. Wiley, New York 1999.

[Se01]   Th. B. Seiler: *Begreifen und Verstehen: Eine Buch über Begriffe und Bedeutungen*. Verlag Allgemeine Wissenschaft, Mühltal 2001.

[So98]   H. Söll: *Begriffliche Analyse triadischer Daten: Das IT-Grundschutzhandbuch des Bundesamts für Sicherheit in der Informationstechnik*. Diplomarbeit, FB4, TU Darmstadt 1998.

[SW00]   G. Stumme, R. Wille (Hrsg.): *Begriffliche Wissensverarbeitung: Methoden und Anwendungen*. Springer, Heidelberg 2000.

[SWW98]   G. Stumme, R. Wille, U. Wille: Conceptual knowledge processing in databases using formal concept analysis methods. In: J. M. Zytkov, M. Quafofou (eds.): *Principles of data mining and knowledge discovery*. LNAI **1310**. Spinger, Heidelberg 1998, 450-458.

[Vg95]   N. Vogel: *Ein begriffliches Erkundungssystem für Rohrleitungen*. Diplomarbeit, FB4, TU Darmstadt, 1995.

[Vo96]   F. Vogt: *Formale Begriffsanalyse mit C++: Datenstrukturen und Algorithmen*. Springer, Heidelberg 1996.

[Wi89]   R. Wille: Lattices in data analysis: how to draw them with a computer. In: I. Rival (ed.): *Algorithms and Order*. Kluwer, Dordrecht 1989, 33-58.

[Wi94]   R. Wille: Plädoyer für eine philosophische Grundlegung der Begrifflichen Wissensverarbeitung. In: [WZ94], 11-25.

[Wi95]   R. Wille: Begriffsdenken: Von der griechischen Philosophie bis zur künstlichen Intelligenz heute. *Dilthey-Kastanie*, Ludwig-Georgs-Gymnasium Darmstadt 1995, 77-109.

[Wi99]   R. Wille: Conceptual landscapes of knowledge: a pragmatic paradigm for knowledge processing. In: W. Gaul, H. Locarek-Junge (eds.): *Classification in the Information Age*. Springer, Heidelberg 1999, 344-356.

[Wi00]   R. Wille: Begriffliche Wissensverarbeitung: Theorie und Praxis. *Informatik Spektrum* **23** (2000), 357-369; gekürzte Version in: *Thema Forschung: Information, Wissen, Kompetenz* (TU Darmstadt), Heft 2/2000, 128-140.

[Wi01a]   R. Wille: Mensch und Mathematik: Logisches und mathematisches Denken. In: K. Lengnink, S. Prediger, F. Siebel (Hrsg.): *Mathematik und Mensch: Sichtweisen der Allgemeinen Mathematik*. Verlag Allgemeine Wissenschaft, Mühltal 2001, 139-158.

[Wi01b]   R. Wille: Boolean Judgment Logic. In: H. Delugach, G. Stumme (eds.): *Conceptual structures: broadening the base*. LNAI **2120**. Springer, Heidelberg 2001, 115-128.

[Wi02]   R. Wille: Begriffliche Wissensverarbeitung in der Wirtschaft. *Information – Wissenschaft und Praxis* (Organ der Deutschen Gesellschaft für Informationswissenschaft und Informationspraxis e.V.) **53** (2002), 149–160.

[Wi04]   R. Wille: Formal Concept Analysis as mathematical theory of concepts and concept hierarchies. This volume.

[WZ94]   R. Wille, M. Zickwolff (Hrsg.): *Begriffliche Wissensverarbeitung: Grundfragen und Aufgaben.* B.I.-Wissenschaftsverlag, Mannheim 1994.

[Wo94]   K. E. Wolff: Conceptual control of complex industrial production processes. *Advances in Knowledge Organization* **4** (1994), 294.

[Wo95]   K. E. Wolff: Conceptual quality control in chemical destillation columns. In: J. Janssen, S. McClean (eds.): *Applied Stochastic Models and Data Analysis.* University of Ulster 1995, 652-654.

[Wo96]   K. E. Wolff: Comparison of graphical data analysis methods. In: F. Faulbaum, W. Bandilla (eds.): *SoftStat '95. Advances in Statistical Software.* Lucius & Lucius, Stuttgart 1996, 139-151.

[WS93]   K. E. Wolff, M. Stellwagen: Conceptual optimization in the production of chips. In: J. Janssen, C. H. Skiadas (eds.): *Applied Stochastic Models and Data Analysis.* Vol.II. World Scientific Publ. Comp. 1993, 1054-1064.

A Survey of Formal Concept Analysis Support for Software Engineering Activities

Thomas Tilley[1], Richard Cole[1], Peter Becker[1], and Peter Eklund[2]

[1] School of Information Technology and Electrical Engineering
University of Queensland, Brisbane, Australia
{tilley,rcole,pbecker}@itee.uq.edu.au
[2] School of Information Technology and Computer Science
The University of Wollongong, Wollongong, Australia
peklund@uow.edu.au

Abstract. Formal Concept Analysis (FCA) has typically been applied in the field of software engineering to support software maintenance and object-oriented class identification tasks. This paper presents a broader overview by describing and classifying academic papers that report the application of FCA to software engineering. The papers are classified using a framework based on the activities defined in the ISO12207 Software Engineering standard. Two alternate classification schemes based on the programming language under analysis and target application size are also discussed. In addition, the authors work to support agile methods and formal specification via FCA is introduced.

1 Introduction

In the domain of software engineering, Formal Concept Analysis (FCA) has typically been applied to support software maintenance activities – the refactoring or modification of existing code – and to the identification of object-oriented (OO) structures. There is also a body of literature reporting the application of FCA to the identification and maintenance of class hierarchies in database schemata [14, 48, 49]. While a database system typically forms the backbone of most Computer Assisted Software Engineering (CASE) tools the discussion of database related applications is beyond the scope of this paper.

Beyond the identification of classes, FCA has also been applied to other areas of software engineering including requirements analysis and component retrieval. The aim of this paper is to provide a broad overview of the area by describing and classifying academic papers that report the application of FCA to a range of software engineering activities. These papers are classified using a framework based on the activities defined in the ISO12207 Software Engineering standard [31]. Two alternate classification schemes based on the programming language applicability and target application size are also presented along with a brief analysis of authorship and citation patterns within the survey literature.

The next section of the paper introduces the framework used to classify the different reported approaches based on their applicability to well-defined software engineering activities. Section 3 then presents the classified papers to provide an overview of FCA support for software engineering. Approaches related to the early phases of software

B. Ganter et al. (Eds.): Formal Concept Analysis, LNAI 3626, pp. 250–271, 2005.

development and an application to formal specification are discussed in Section 4. Section 5 reports work related to software maintenance before Section 6 concludes the paper.

2 A Software Engineering Life Cycle Framework

To understand how software engineering can be supported by FCA we must first have some understanding of what software engineering is or at least of the processes involved. This section of the paper sets out a framework that will be used to classify papers from a software engineering perspective.

The development of software has traditionally been described by life-cycle models. These models grew out of a need to better understand and manage the software engineering process which has been characterised by failed, late, and bug-laden projects. Royce [46] proposed the classic "waterfall" model which consists of seven *steps* or *phases* that proceed in a linear fashion: System Requirements, Software Requirements, Analysis, Program Design, Coding, Testing, and Operations.

The waterfall model focuses heavily on the documentation produced during each implementation phase and there may be some iteration between successive steps. Royce realised that sometimes iterations happen across non-consecutive steps which is undesirable. To address this he proposed some extensions to alleviate the "risk" which largely focused on the production of additional documentation.

The "V" model [45] is a variant of the waterfall model where each step down the left hand side of the "V" has a corresponding validation or verification step on the right hand side. This model presents the opportunity for more "formal" development where documents from the left hand-side feed into the validation activities on the right. The spiral model [8] is another alternative life-cycle that directly incorporates risk analysis as one of four major activities that also includes: planning, engineering and customer evaluation. Starting in the centre of a spiral the developers work through a planning phase, followed by risk analysis, the engineering of a prototype system and then customer evaluation. The cycle then repeats and each move around the spiral progresses outwards towards the final system in an evolutionary fashion.

In addition to these three examples a number of other life-cycle models exist and the most appropriate model to use for a given project may depend on a number of factors including the type of project, the development style and the organisational maturity of both the developers and the customer. An alternative to the classic life-cycle approaches is to use a meta-model that defines common software engineering activities independently of a particular life-cycle model. Developers can then choose the most-appropriate life-cycle for their project and the activities can be mapped onto the chosen model.

2.1 ISO 12207

The ISO 12207 Software Engineering Standard describes such a meta-model for software engineering life-cycle processes that consists of thirteen activities that can be mapped onto a chosen life-cycle model [32]. The first activity, "process implementation", is related to starting the methodology itself, while another four of the activities are system related: "System requirements analysis", "System architectural design", "System integration" and "System qualification". The remaining eight are related to the software itself and the standard notes that "these activities and tasks may overlap or interact and may be performed iteratively or recursively". Short descriptions of these eight remaining activites obtained from the IEEE Standard Glossary of Software Engineering Terminology [30] are:

- **requirements analysis**, the process of studying user needs to arrive at a definition of system, hardware, or software requirements.
- **architectural design**, the process of defining a collection of hardware and software components and their interfaces to establish the framework for the development of a computer system.
- **detailed design**, the process of refining and expanding the preliminary design of a system or component to the extent that the design is sufficiently complete to be implemented.
- **coding and testing**, where *coding* is defined as "... the process of expressing a computer program in a programming language" and *testing* is "the process of analyzing a software item to detect the differences between existing and required conditions (that is, bugs) and to evaluate the features of the software items".
- **integration**, the process of combining software components, hardware components, or both into an overall system.
- **qualification testing**, testing conducted to determine whether a system or component is suitable for operational use.
- **installation**, the period of time in the software cycle during which a software product is integrated into its operational environment and tested in this environment to ensure that it performs as required.
- **acceptance support**, formal testing conducted to determine whether or not a system satisfies its acceptance criteria and to enable the customer to determine whether or not to accept the system.

2.2   Software Maintenance

In addition to the eight activities defined above an understanding of software maintenance is also required. The process of software maintenance requires iteration through some or all of the previously defined activities and in terms of the waterfall model it could be thought of as a feedback loop to previous stages. The IEEE Standard Glossary of Software Engineering Terminology defines it as:

- **software maintenance**, the process of modifying a software system or component after delivery to correct faults, improve performance or other attributes, or adapt to a changed environment.

The next section of the paper uses these nine activities within a framework to classify academic papers reporting the application of FCA to software engineering activities. The intention is not to classify a paper according to a single activity but to record all of the activities that are supported by a reported approach.

3 FCA Support for Software Engineering Activities

This section of the paper presents an overview of FCA support for software engineering via a paper survey. The nine software engineering activities defined in Section 2 are used to classify 47 academic papers and theses published between 1992 and 2003. The papers report the application of FCA to software engineering and the approaches they describe are summarised in Sections 4 and 5. Note that this paper assumes some familiarity with FCA and the reader is pointed to Ganter and Wille's classic text for further details [24].

The papers within the survey were analysed using FCA. A many-valued context was constructed that considers the papers as a set of formal objects. Part of the context appears in Table 1. References to papers included in the survey use the naming format adopted by the Research Index (formerly known as "CiteSeer") digital library[1]. Paper names are composed of the first author's surname, the last two digits of the year of publication, and the first word of the title (excluding words like "an", "the", "a", etc.). For example the paper "On The Inference of Configuration Structures from Source Code" published by Krone and Snelting in 1994 appears as *Krone94inference* [33].

A first classification of the papers, derived from the many-valued context by considering the ISO activities as scale attributes, appears in Figure 1. From the line diagram it can be seen that 27 of the papers describe applications to both *software maintenance* and *detailed design*. Note that only the object count has been included for this concept to aid the readability of the diagram. These papers are typically reporting the use of FCA to identify class candidates in legacy code or the maintenance of class hierarchies[2]. Considering that concept lattices are hierarchies this is an obvious application. An emerging body of literature related to *requirements analysis* can also be seen although typically a number of the papers are reporting the same example. It is also worth noting that there are only two papers describing applications to *testing* and no papers explicitly reporting application to *software integration, qualification testing, acceptance support* or *coding*, and thus these areas present an opportunity to FCA researchers.

3.1 Alternate Classification Schemes

In addition to the ISO activity categorisation a number of other attributes were used to categorise the papers in the survey. In total 133 attributes were identified including: the names of the authors, citations of other papers in the survey, the year of publication, inputs, outputs, target application languages (e.g. C++, Java) and the "size" of any reported application target.

The context in Table 1 represents the application of the techniques described in a paper to a particular programming or design language. The attributes here are the programming languages: C, C++, COBOL, Fortran, Java, Modula-2, Smalltalk, and the design or specification languages: OMT, UML and Z. Both procedural and OO languages are represented. The attribute values record the size of any reported target application in KLOC ("thousand lines of code) – for example, 106 KLOC represents an

---

[1] See http://citeseer.nj.nec.com/

[2] To avoid confusion the terms "class" or "class candidate" will typically be used to refer to Object-oriented objects as opposed to formal objects in FCA

Table 1. A Formal Context showing reported application languages for the 47 papers in the survey. The attribute values represent the size of the application in KLOC ("thousand Lines Of Code"). A KLOC value of "0" indicates that the paper reported application to a particular language but no size was quoted.

	C	C++	COBOL	FORTRAN	Java	Modula-2	OMT	Smalltalk	UML	Z
Ammons03debugging [1]	0									
Andelfinger97diskursive [2]										
Arevalo03understanding-a [3]								0		
Arevalo03understanding-b [4]								0		
Ball99concept [5]	0									
Boettger01reconciling [10]										
Bojic00reverse [9]		0								
Canfora99case [11]			200							
Dekel02applications [13]					0					
Duwel98identifying [16]										
Duwel99enhancing [15]										
Duwel00bridging [17]										
Eisenbarth01aiding [18]	0									
Eisenbarth01feature [19]	76									
Eisenbarth03locating [20]	1,200									
Fischer98specification [21]										
Funk95algorithms [23]	1.6									
Godin93building [27]								0		
Godin95applying [26]										
Godin98design [25]										
Huchard99from [28]					0					
Huchard02when [29]									0	
Krone94inference [33]	1.6									
Kuipers00types [34]			100							
Leblanc99environment [35]		0			0		0		0	
Lindig95concept [36]										
Lindig97assessing [37]			5	106				1.5		
Richards02assisting [39]										
Richards02controlled [40]										
Richards02recocase [41]										
Richards02representing [38]										
Richards02using [42]										
Sahraoui97applying [47]	47									
Schupp02right [50]		0								
Siff97identifying [51]	28									
Snelting96reengineering [52]	1.6									
Snelting98concept [53]	1.6		0	106				1.5		
Snelting98reengineering [55]		0								
Snelting99reengineering [56]	0				9					
Snelting00software [54]	1.6	0	0	106	9			1.5		
Snelting00understanding [57]		0			12					
Streckenbach99understanding [59]		0			12					
Tilley03software [63]										0
Tilley03towards [62]										0
Tonella99object [65]		21								
Tonella99object [64]	249									
vanDeursen98identifying [66]			100							

application containing 106,000 lines of source code. KLOC is also sometimes referred to as SKLOC (Source Thousands Lines of Code) and is a metric that is often reported to indicate project size in software engineering. While KLOC is not necessarily a consis-

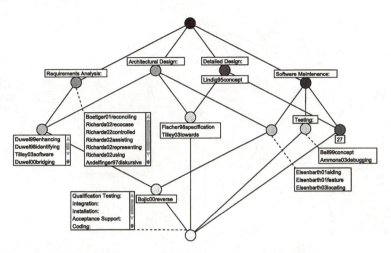

Fig. 1. A concept lattice showing the 47 papers categorised according to the ISO software engineering activities they support.

tent or meaningful metric it gives a raw indication of the size of applications to which the technique has been applied and may imply tool support.

A number of the papers report application to a specific language but do not report the size of a particular application and the KLOC value for these papers appears as "0" in the context. It is also interesting to note that where a non-zero value repeats in the context it typically refers to the same example being reported in a number of papers. For example, the 1.6 KLOC C application appears in the papers *Funk95algorithms, Krone94inference, Snelting96reengineering, Snelting98concept* and *Snelting00software*. Similar patterns can also be seen for the 106 KLOC FORTRAN, 100 KLOC COBOL and 1.5 KLOC Modula-2 applications.

Figure 2 presents a concept lattice that treats Table 1 as a simple one-valued context where any KLOC value including 0 relates the object to the attribute. It can be seen that 14 of the 47 papers do not report any application to a particular programming or design language. Also of note is the paper *Snelting00software* [54] which reports applications to all of the programming languages except Smalltalk. This is a paper by Snelting that surveys earlier results from a number of papers he has either authored or co-authored. The lattice shows that Smalltalk and Z have only been described in isolation within papers, while in contrast C, C++, Java, Modula-2, and Fortran have been analysed in connection with other languages.

The line diagram in Figure 3 summarises the same context but only considers the maximum reported KLOC across all languages for each paper. The figure gives an indication of which papers fall into the different KLOC ranges. It can be seen that there are eight papers in the survey reporting application to systems of 100 KLOC or more, however, these actually refer to only five different examples. The analysis of a 106 KLOC FORTRAN system is discussed in the three papers: *Lindig97assessing, Snelting98concept* and *Snelting00software*. In addition the 100 KLOC COBOL examples reported by Kuipers and Moonen in *Kuipers00types* and Van Duersen and Kuipers in *vanDuersen98identifying* also describe the same application example.

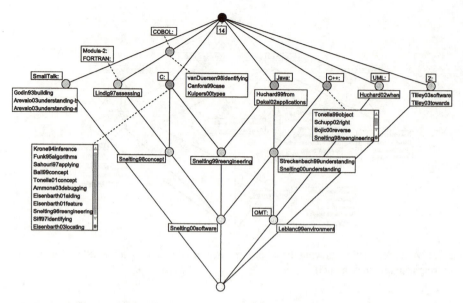

Fig. 2. Lattice based on Table 1 showing reported application languages ignoring size.

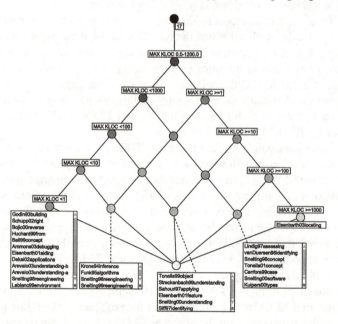

Fig. 3. An inter-ordinal scale showing the maximum application size across all languages in each paper. Note that papers reporting UML, OMT and Z are ignored in this scale because KLOC is not a suitable metric for design/specification languages.

The largest application in the survey describes the analysis of a 1,200 KLOC semiconductor testing tool written in C. The work by Eisenbarth, Koschke and Simon in

Eisenbarth03locating [20] is an order of magnitude larger than any of the other examples and demonstrates that FCA-based software analysis tools are capable of handling real-world projects.

A summary of collaboration between the authors within the survey papers is presented in Figure 4. This concept lattice represents those authors who have collaborated on papers with different authors. There are 13 papers at the top of the lattice whose authors only appear once across the 47 papers or who always work with the same co-authors. Only a count of the number of papers is shown and the structures below Godin, Boettger and Snelting are of particular interest. These structures represent collaboration on multiple papers with a reasonably large number of different authors. Snelting's collaboration with others is the likely reason for the large number of language applications described in *Snelting00software*. The fact that the diagram can be horizontally decomposed into eight sublattices indicates that research has been performed rather independently within these research groups the largest of which are led by Snelting, Hesse, Boettger and Richards, Huchard, and Godin; there are no joint publications across these groups.

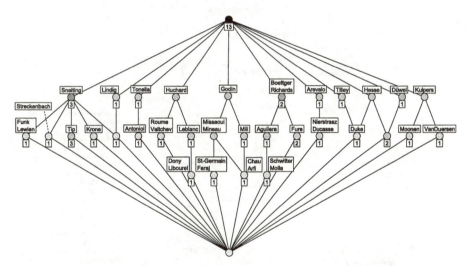

Fig. 4. Lattice showing collaboration between authors within the set of survey papers. Note that only papers where the authors have worked with different co-authors (within the collection of papers examined) are listed.

Finally in this section, Figure 5 presents a line diagram showing the transitive closure of citations within the set of survey papers. For example, if paper *B* cites paper *C*, and paper *A* cites paper *B*, then *A* transitively cites *C*. At the top of the line diagram there are 9 papers listed and these papers have not cited any of the other literature within the survey set. There are a number of explanations for the location of these papers. The earliest papers in the survey, *Godin93building* and *Krone94inference*, by definition have no earlier work to cite within the survey collection. *Andelfinger97diskursive* is a paper written in German and the list of citations was unavailable. The remaining papers all

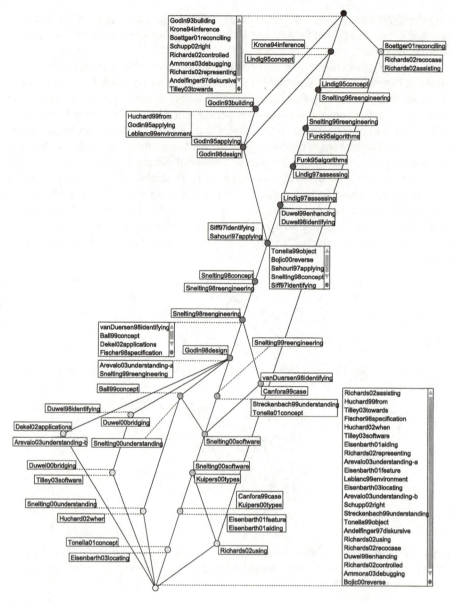

Fig. 5. Concept lattice representing the transitive closure of citations within the set of survey papers.

cite the work of Wille as FCA background but do not build directly on any of the work described in the other papers.

At the bottom of the diagram are 20 papers which are not cited within the survey collection. Papers with earlier publication years appear to have been ignored by the community while more recent papers may not have been around long enough to be cited yet. *Andelfinger97diskursive* also appears in this list and this could be because it is the only German language paper in an otherwise English language set.

Papers with the most impact appear at the top of the diagram with long chains (i.e. containing many concepts) beneath them. It can be observed that most of the work is nearly linear in terms of citations which reflects some coherence within the community. For example, Snelting and Godin cite each others work before large forks appear in the structure.

The structure down the right hand-side of the line diagram is also interesting. The papers by Böttger and Richards et al. either contained no citations within the survey set or they cited their own work in *Boettger01reconciling*. *Richards02using*, however, cites *Snelting00software* which connects their work back into the main "trunk".

4   Early Phase Activities

Software engineering activities that occur before the commencement of coding can be considered as "early-phase" activities. This section summarises the papers shown in Figure 1 that support the early-phase software engineering activities. This section is organised according to the specific activities described in these papers. To see how we placed each paper with respect to the software engineering activities please refer back to Figure 1. An overview of papers describing FCA support for software maintenance activities appears in Section 5.

4.1   Requirements Analysis

Andelfinger's thesis [2] describes a discoursive environment for requirements gathering based on Habermas' theory of "communicative rationality". Habermas described a perhaps somewhat idealistic discoursive environment that attempts to make any agendas obvious during negotiation and considers all viewpoints equally.

Within the thesis FCA is used as a question answering and discussion tool. The value of unrealised concepts is highlighted as it promotes questions about things that are missing which may be indicative of incomplete requirements. While the three case studies presented are not directly related to software engineering they parallel typical problems encountered in requirements gathering. It is interesting to note that one of the case studies (Beta) describes the gathering requirements for an FCA based retrieval system for the library of the "Center of Interdisciplinary Technology Research"(ZIT) [44].

Use Cases are a tool used in requirements gathering and analysis where a task is described from a certain perspective or role. Typically these descriptions are written in natural language although sometimes controlled vocabularies are used. The work of Düwel [15] and Düwel and Hesse [16, 17] attempts to identify class candidates in use case descriptions. The use cases themselves are considered as objects in a formal context and nouns identified within the text are considered as formal attributes. The structure of the corresponding concept lattice is then considered as a starting point for a class hierarchy[3]. In [63] Tilley et al. present a case study applying Düwel's approach to an Object-Z specification.

---

[3] See also Hesse's contribution to this volume

Böttger and Richards et al. [10, 38–43] also apply FCA to use cases in an attempt to reconcile descriptions written by different stake-holders using a controlled vocabulary and grammar. Their tool RECOCASE (RECOnciling CASE tool) exploits a Prolog answering system called ExtrAns and LinkGrammar – an English language parser using link grammar theory – to translate sentences into unambiguous flat logical forms. The formal nature of this controlled language facilitates the analysis of use cases to identify misunderstandings, inconsistencies and conflicts. Furthermore, a context can be produced where the sentences are formal objects and the flat logical forms are broken into word phrases which are then treated as formal attributes. Similar concepts and differences in terminology can then be identified from the resulting concept lattice.

4.2   Component Retrieval for Software Reuse

Lindig [36] describes a retrieval system that could be used for retrieving software components from a library indexed by keywords. An example using keywords from man pages describing Unix commands is presented where the keywords are considered as formal attributes and the Unix commands as formal objects. In this case the commands represent the components to be retrieved. The retrieval system provides a query-by-refinement interface in which a boolean query, B, is mapped to the formal concept, (B', B''), whose lower cover (within the concept lattice) is offered as a set of possible refinements to the user.

4.3   Formal Specification

Fischer [21] builds on the component retrieval work of Lindig, however, instead of keywords a formal specification that captures the behaviour of a software component is used. By exploiting the power of Formal Methods, components can be retrieved based on explicit properties required for a component selection or on implicit similarity with other components. The underlying structure is a concept lattice which is computed "offline" and it facilitates the browsing of software component libraries as well as retrieval.

The component specifications consist of axioms describing pre-conditions and post-conditions. These specifications are used as both the objects and the attributes of a formal context. Functions or partial functions within the specifications are also considered as formal attributes. In addition to library navigation and component retrieval, the resulting lattice can also be used to improve the library. Unlabelled concepts and extents containing intuitively "unexpected" components may indicate missing attributes or features that can then be added to the library.

4.4   Visualizing Z Specifications via FCA

Z is a state based formal method that exploits set theory and first order predicate logic. Specifications in Z are composed of named *schema* boxes that describe operations by their input/output behaviour. Models are constructed by specifying and composing a series of schemas which can be further refined to reflect the desired level of system abstraction. An example schema from Spivey's "BirthdayBook" specification [58] appears in Figure 6.

$$\underline{\quad AddBirthday \quad}$$

$\Delta\, BirthdayBook$
$name? : NAME$
$date? : DATE$

$name? \notin known$
$birthday' = birthday \cup \{name? \mapsto date?\}$

Fig. 6. The AddBirthday schema from the BirthdayBook specification in Z.

As a result of the mathematical nature of the notation most Z tools are comprised of at least a formatting package for LaTeX and a type-checker. There is a continued call for formal methods tool support and in particular tools that support the ability to view specifications at different levels of abstraction. This section provides a brief overview of the authors work to create a tool for interactively exploring Z specifications based on ZML [60], an XML representation of Z, and the open-source, cross-platform FCA tool ToscanaJ. This work has recently been reported in [62] and an overview of the process is presented in Figure 7.

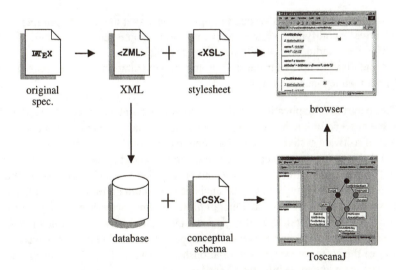

Fig. 7. Overview of the specification browsing system.

While Z specifications are traditionally written in LaTeX and viewed as postscript or PDF documents, a more recent alternative for "marking-up" specifications is ZML. Specifications written in ZML are easily parsed and have the advantage that they can be transformed using XSL stylesheets into HTML which can be rendered in a Web browser. Specifications can then be easily accessed in an on-line form complete with HTML anchors and hyperlinks for navigation.

ToscanaJ is a FCA tool for rendering concept lattices that supports extent/intent highlighting, conceptual scaling, nesting, and zooming [7]. A formal context representing the static structure of a Z specification can be constructed by considering each schema as a formal object and the individual mark-up elements as attributes. This context can be stored in a relational database and accessed using ToscanaJ. Scales can either be constructed by the user or pre-defined to "query" the specification and reveal properties of interest based on, for example, Z language features or data-types within a specification. ToscanaJ's extensible view interface can then be exploited so users can click on a schema name within a line-diagram and a web browser will be launched displaying the relevant part of the specification rendered using ZML. In this way users can conceptually navigate and explore a Z specification using FCA and retrieve the relevant parts of the original specification as desired.

5   Software Maintenance

This section deals with the use of FCA to suggest modifications to existing software programs or systems. This activity may be performed within different contexts and with different objectives. The term *software maintenance* usually refers to the modification of a software system that has already been deployed to the customer. Four types of software maintenance are identified [61]:

1. corrective – modifying a system to improve the way it meets its requirements.
2. adaptive – modifying a system to operate correctly in a new environment.
3. perfective – adding new functionality to the system.
4. preventative – improving the design and implementation to better accommodate future maintenance activities.

The spectrum of applications of FCA to software maintenance covers each of these types, but has a common thread – extracting understandable structures that organise the artifacts of software systems. The various applications of FCA to software maintenance vary on their inputs, the concept lattices they create, and the use to which they put the concept lattices. In an effort to organise and present these different approaches they are grouped into the following categories: analysis of dynamic information, application to legacy systems, and review of class hierarchies. We summarise these categories briefly; they are discussed in depth in Snelting's contribution to this volume. Following this we will present a framework for merging these approaches using a knowledge base storing artifacts of the software system and relationships between them and a mechanism for deriving concept lattices using graph based queries.

5.1   Dynamic Analysis

Approaches for the analysis of dynamic aspects of software systems have been reported by Ammons et al. [1], Ball [5], Eisenbarth et al. [18–20], and Bojic et al. [9]. Ammons et al. and Ball examine test coverage while Bojic and Eisenbarth recover software architecture related to use cases.

Dynamic information is typically extracted from programs by executing them with a profiler. The profiler records which software artifacts such as procedures and variables were accessed during the run of the program and in what order.

A computer program essentially consists of a large number of instructions. Each instruction is identified by its position within the program. For example we may consider the instruction at memory location $0x00f1$. A run of a computer program produces a *trace*. A trace is the sequence of instructions that were run. Two notions concerning instructions, with respect to a collection of traces, are important: dominance and pre-domination. An instruction x *dominates* another instruction y if any trace prefix that ends on y contains x. In other words x dominates y if the only way to execute y is to have already executed x. Similarly x *post-dominates* y if any trace postfix starting with y also contains x. In other words x post-dominates y if any execution of y requires that x will subsequently be executed.

While there is not a direct correspondence between dominance, post-dominance and the lattice structure, dominance and post-dominance lead to implications in the lattice. If we consider program traces as formal objects, and instructions as formal attributes, then if x either dominates or post-dominates y, there will be an implication in the lattice of the form $y \rightarrow x$.

Recently, Ammons, Mandelin, Bodik, and Larus [1] have incorporated FCA and Formal Methods in their work to debug temporal specifications. While very small specifications can be debugged by inspection, larger specifications are verified using tools that check the specification against a number of programs. There may be hundreds or thousands of execution traces from these checks and these are used as the formal objects in their analysis. Each of the execution traces must be classified by an expert who decides if they are correct or erroneous. By considering transitions within the finite automata that represent the specifications as the formal attributes, a concept lattice can be produced that clusters similar traces together. An expert can then classify clusters of traces rather than classifying them all individually.

Ball examines test coverage by comparing the implicational logic in the concept lattice generated from traces extracted from test programs with dominance and post-dominance relationships extracted by static code analysers. Any additional implications in the concept lattice are considered to see if they can be removed by the introduction of a new test.

Eisenbarth et al. describe a technique for locating the computational units within software that actually implement a feature or functionality of interest. For example, they are interested in locating the portion of the web-browser code related to the use of the browser history. They combines both static and dynamic analysis and of particular note is the application of their technique to the 1,200 KLOC example discussed in Section 3.1. A number of test cases or "scenarios" are constructed which cover the use cases of interest and these are treated as the formal objects in their analysis. The computational units executed during runs of the program are then considered as the formal attributes. The attribute contingents of object concepts in the resulting lattice are of particular interest since they contain the program artifacts introduced by specific scenarios.

Bojic et al. report a similar approach, but they additionally arrange the artifacts within the attribute contingents as UML diagrams using a UML reverse engineering tool. In this way the specific parts of the software architecture related to use cases can be extracted and viewed. This capability is particularly useful in the preparation of traceability in the software engineering process whereby aspects of the system architecture can be traced back to requirements.

5.2   Application to Legacy Systems

FCA has been applied to extract structure from legacy systems and has been compared with hierarchical clustering as a technique for organising the artifacts of legacy systems. Snelting [23, 33, 52] used FCA to analyse the preprocessor commands in legacy C programs in order to examine the configuration structure. Formal objects are the code fragments included by the preprocessor commands, while, the formal attributes are disjunctive expressions governing the inclusion of the code fragments. The concept lattice is constructed and the notion of an interference is introduced. An *interference* is a meet-reducible concept with a non empty extent. Two types of undesirable interference are identified, those corresponding to illegal configurations – for example an interference between XWINDOWS and DOS – and those corresponding to orthogonal attributes – for example an interference between a variable related to the graphics subsystem and one related to the operating system. If there are no interferences then the concept lattice can be horizontally decomposed.

Experiments with legacy systems revealed that few configuration lattices can be directly decomposed into a horizontal sum of disjoint sub-lattices. In order to simplify the configuration structure the notion of a k-*interference* is introduced. A k-interference (see [23], p.8 for a formal definition) is a collection of k meet-reducible incomparable concepts whose downset removal yields a lattice that is decomposable into a disjoint horizontal sum of k terms. An additional constraint on a k-interference is that no subset of the concepts is a $k - 1$ interference. The concepts involved in such k-interferences are of particular interest since they are most likely interferences between orthogonal aspects of the system configuration.

Other techniques to simplify the concept lattice include limiting the nesting depth of preprocessor commands considered and merging rows which differ by fewer than k elements. These techniques are of use when the objective is to get an overview of the configuration structure present in a software program.

Legacy programs written in languages where access to common data structures is the norm, e.g. FORTRAN and COBOL, have been considered by van Deursen et al. [66], Kuipers et al. [34], Lindig et al. [37] and Canfora [11]. Van Deursen and Kuipers compare the use of formal concept analysis for grouping fields within a large legacy COBOL program to that of hierarchical clustering. Hierarchical clustering involved defining a distance metric between COBOL procedures, extending the metric to sets of procedures, starting with every procedure in its own cluster and then repeatedly merging the two closest clusters to produce a binary tree of clusters. Hierarchical clustering is generally criticised because it can yield different inputs for the same data as in some instances several clusters are equidistant, leading to an arbitrary choice of the next clustering step generally based on the input order, and very different clustering results for slightly different distance metrics are often obtained. The results produced in contrast by FCA are always the same, not dependent on the definition of a distance metric, and were much closer to that produced by software engineers familiar with the legacy system. Since the objective was to focus on domain specific procedures rather than on those performing system functions, procedures having a high degree of fan-in[4] were judged as being system procedures and were discarded. This judgment was controlled by an operator-set threshold.

Canfora et al. follow a similar approach but are interested in organising legacy COBOL systems into components suitable for distribution via CORBA. They considered programs and their use of files (relational tables). The formal context was pruned by removing objects and attributes in isolated concepts – concepts that are directly below the top concept and directly below the bottom concept and therefore don't have any intent or extent intersection with other concepts[5]. Files, i.e. relation tables, having the same structure were also merged. This case arises when several files are used to perform some operation on a table – like sorting it. Also programs that used only a single file were removed. Canfora et al. apply their rules until no more formal objects and formal attributes can be removed. The result was a concept lattice that was almost horizontally decomposable into four domain areas, except for a number of interferences corresponding to operations involving more than one domain area.

The task of deriving object oriented models from legacy systems written in C has also been considered by Sahraoui et al. [47], Siff and Reps [51], and Tonella [64]. The general approach is to consider C functions as formal objects and the attributes as either commonly accessed data structures or fields within commonly used structures. Both Siff and Reps, and Tonella are concerned with re-organising the functions into a different, perhaps more fine grained, module structure based on the access of functions to either (i) common data structures [64], or (ii) fields within commonly accessed data types [51].

Starting with a formal context based on access to common data structures by functions, Tonella [64] seeks a partitioning of the objects of the formal context (functions) into software modules. To do this Tonella's method is to seek a partitioning of the objects from the formal context. Tonella's method searches for an optimal partitioning in the following way: (i) choose a set of concepts, X, whose extents partition the objects, (ii) assign each attribute, m, to the concept in X that has maximum overlap with the extent of m, (iii) measure both the number of concepts in X, and the number of concepts in X that have no attributes assigned. One set of concepts X_1 is better than another, X_2, if X_1 has more concepts and fewer unlabelled concepts than X_2. The search produces a number of optimal X, and each X is considered as a candidate for placing the functions into software modules described by concept intents and containing functions in concept extents.

5.3 Reengineering Class Hierarchies

Snelting [54, 55, 57] explains a mechanism to re-organise class hierarchies using FCA. Variables in C++ are taken as formal objects, and methods and fields of the objects to which the variables refer are taken as formal attributes. A variable is associated with a field or method if that variable is used to access the method or field. A number of rules are employed to account for assignment between variables and conservatively account

---

[4] Fan-in refers to the number of other procedures that call a particular procedure

[5] Other than the top and bottom concepts

for dynamic dispatch. The real objective of investigation are the objects existing during a run of a program. Snelting and Tip access these via static analysis through the medium of variables. See Snelting's contribution to this volume for more details.

Schupp et al. consider class hierarchies in the C++ standard template library (STL). They have classes as formal objects and documented properties of the classes as formal attributes. They introduce the notions of "well abstracting", "lacking orthogonality" and "lacking refinement" to describe class libraries. Rather than inspecting various aspects of the structure they attempt to construct the whole concept lattice, render it and draw conclusions. Inspection of aspects of the STL as shown, for example, in Figure 8 reveal a very regular structure. The example shows three complementary pairs of attributes: unique and multiple associative, sorted and hashed, and pair and simple associative. Complementary attributes are related by *exclusive or* – in other words all objects have exactly one of the two attributes.

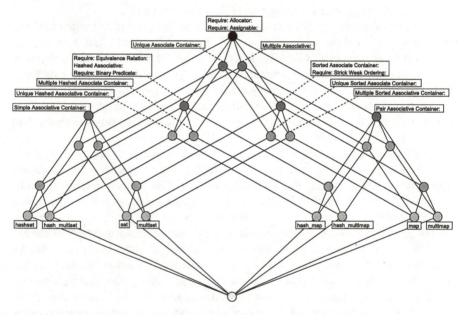

Fig. 8. Concept lattice of set and map classes in the STL.

Godin et al. [27] consider a context where the formal objects are messages (methods in SMALLTALK) and formal attributes are classes. They consider concepts having an empty attribute contingent, i.e. those not labeled by a class as new class candidates. After approval by the designer these class candidates may be added to the software. Godin et. al.'s approach is discussed in more detail in Godin and Valtchev's contribution to this volume. We applied a similar technique to an analysis of the collection classes in JAVA and discovered a new inheritance link not present in the collection. Our approach is briefly outlined in Section 5.4.

While Leblanc, Dony, Huchard and Libourel [35] describe an environment for re-engineering class hierarchies, Huchard and Leblanc [28] consider a concept lattice generated with classes as formal objects and attributes derived from method signatures. Their approach thereby includes information about parameter types and return values. Again each concept is considered as a candidate for a Java interface.

Huchard, Roume and Valtchev [29] address the problem of representing and analysing data via FCA where relationships exist between the objects. The binary inter-object relationships are represented by a *relational context family*. Their approach is applied to UML class diagrams representing both classes and association relationships between classes where the classes are considered as the formal objects and the variables and methods as attributes.

Tonella and Antoniol [65] attempt to recover design patterns in C++ source code using a formal context whose objects are triples each consisting of three C++ classes. The attributes of the formal context are triples of the form (i, j, r) where i and j are indexes into the object triple and r is a relation type. For example, an object (A, B, C), being associated with an attribute $(1, 2, derived - from)$ would indicate that A is derived from B. Tonella and Antoniol discover as one of the concepts in the concept lattice the well known "adapter pattern".

The work of Arévalo [3], and Arévalo, Ducass and Nierstrasz [4] is also concerned with detecting patterns in software via FCA. While their work is similar to that of Tonella and Antoniol they apply the approach to Smalltalk and also take into account behavioral information related to the derivation of subclasses.

On another track, Dekel's proposes a mechanism to provide a suggested reading order for Java classes [13]. The idea is to propose a reading order to a human reader in which constructs are first encountered in the class where they are introduced, rather than in classes where they are used.

5.4   Conceptual Analysis of Software Structure

The process of software design and implementation often contains many arbitrary decisions, such as the name of methods or variables or indeed the structure of a class hierarchy. As the design proceeds these decisions need to be reviewed in order to achieve consistent, orthogonal and simple designs. Agile methods, and extreme programming (XP) in particular, advocate regular refactoring activities undertaken to regularise and revise the software structure [6, 22]. Our tool, Conceptual Analysis of Software Structure (CASS) [12], attempts to address these requirements. CASS exploits source code analysers and profilers to extract information which is stored in a knowledge base as a large collection of triples of the form *(subject, predicate, object)*. A rule based system is then used to extend the knowledge base with new relationships and artifacts. Graph based queries can then be used to define an aspect of the code to be explored and these are used to generate result sets that are visualised using concept lattices. These hypotheses or questions may then be investigated either by generating new lattices displaying new aspects of the software structure, or by navigating back to the source artifacts within the software or its documentation. Since each concept lattice is generated from a query graph, a natural refinement ordering allows general views to be elaborated and made more specific. Thus the user is able to progress from a general view to a more specific view, or vice versa. In addition, the theory of FCA allows two or more aspects of the software structure to be combined coherently in nested diagrams.

6 Conclusion

This paper has provided a snapshot of where FCA has been applied to support software engineering activities using the activities defined in the ISO 12207 standard and a number of other criteria. The power of a line-diagram to convey and summarise large amounts of information has been demonstrated on real-world examples via a number of approaches and tools. The paper has also provided an insight into authorship groups and citation patterns within the survey literature.

The majority of the reported work has been in the areas of detailed design and software maintenance where FCA has been applied to OO re-engineering and class identification tasks. While these approaches could be seen as obvious applications because of the specialisation/generalisation relationship between the concepts in a concept lattice, the range of different formal objects ranging from compiled code through to use-cases is surprising. Other novel applications have included support for test-coverage analysis and perhaps in the future tools and examples will emerge that support coding, integration and qualification testing.

References

1. G. Ammons, D. Mandelin, R. Bodik, and J.R. Larus. Debugging temporal specifications with concept analysis. In *Proceedings of the Conference on Programming Language Design and Implementation PLDI'03*. ACM, June 2003.
2. U. Andelfinger. *Diskursive Anforderungsanalyse. Ein Beitrag zum Reduktionsproblem bei Systementwicklungen in der Informatik*. Peter Lang, Frankfurt, 1997.
3. G. Arévalo. Understanding behavioral dependencies in class hierarchies using concept analysis. In *Proceedings of LMO 2003 (Langages et Modéles á Object)*, Paris (France), February 2003. Hermes.
4. G. Arévalo, S. Ducass, and O. Nierstrasz. Understanding classes using x-ray views. In *MASPEGHI 2003, MAnaging SPEcialization/Generalization HIerarchies (MASPEGHI) Workshop at ASE 2003*, Montreal, Canada, 2003. Preliminary Version.
5. T. Ball. The concept of dynamic analysis. In *Proceedings of ACM SIGSOFT Symposium on the Foundations of Software Engineering*, pages 216–234, September 1999.
6. K. Beck. *Extreme Programming Explained: Embrace Change*. Addison-Wesley, 2000.
7. P. Becker and J. Hereth Correia. The ToscanaJ suite for implementing conceptual information systems. In *This volume*.
8. B.W. Boehm. A spiral model of software development and enhancement. In R.H. Thayer, editor, *Tutorial: Software Engineering Project Management*, pages 128–142. IEEE Computer Society, Washington, 1987.
9. D. Bojic and D. Velasevic. Reverse engineering of use case realizations in UML. In *Symposium on Applied Computing - SAC2000*. ACM, 2000.
10. K. Böttger, R. Schwitter, D. Richards, O. Aguilera, and D. Mollá. Reconciling use cases via controlled language and graphical models. In *INAP'2001 - Proceedings of the 14th International Conference on Applications of Prolog*, pages 20–22, Japan, October 2001. University of Tokyo.
11. G. Canfora, A. Cimitile, A. De Lucia, and G.A. Di Lucca. A case study of applying an eclectic approach to identify objects in code. In *Workshop on Program Comprehension*, pages 136–143. IEEE, 1999.
12. R. Cole and T. Tilley. Conceptual analysis of software structure. In *Proceedings of Fifteenth International Conference on Software Engineering and Knowledge Engineering, SEKE'03*, pages 726–733, USA, June 2003. Knowledge Systems Institute.

13. U. Dekel. Applications of concept lattices to code inspection and review. In *The Israeli Workshop on Programming Languages and Development Environments*, chapter 6. IBM Haifa Research Lab, IBM HRL, Haifa University, Israel, July 2002.

14. H. Dicky, C. Dony, M. Huchard, and T. Libourel. ARES, adding a class and restructuring inheritance hierarchy. In *BDA : Onzièmes Journées Bases de Données Avancées*, pages 25–42, 1995.

15. S. Düwel. Enhancing system analysis by means of formal concept analysis. In *Conference on Advanced Information Systems Engineering 6th Doctoral Consortium*, Heidelberg, Germany, June 1999.

16. S. Düwel and W. Hesse. Identifying candidate objects during system analysis. In *Proceedings of CAiSE'98/IFIP 8.1 Third International Workshop on Evaluation of Modelling Methods in System Analysis and Design (EMMSAD'98)*, Pisa, 1998.

17. S. Düwel and W. Hesse. Bridging the gap between use case analysis and class structure design by formal concept analysis. In J. Ebert and U. Frank, editors, *Modelle und Modellierungssprachen in Informatik und Wirtschaftsinformatik. Proceedings "Modellierung 2000"*, pages 27–40, Koblenz, 2000. Fölbach-Verlag.

18. T. Eisenbarth, R. Koschke, and D. Simon. Aiding program comprehension by static and dynamic feature analysis. In *Proceedings of ICSM2001 - The International Conference on Software Maintenance*, pages 602–611. IEEE Computer Society Press, 2001.

19. T. Eisenbarth, R. Koschke, and D. Simon. Feature-driven program understanding using concept analysis of execution traces. In *9th Int'l Workshop on Program Comprehension*, pages 300–309. IEEE, 2001.

20. T. Eisenbarth, R. Koschke, and D. Simon. Locating features in source code. *IEEE Transactions on Software Engineering*, 29(3):195–209, March 2003.

21. B. Fischer. Specification-based browsing of software component libraries. In *Automated Software Engineering*, pages 74–83, 1998.

22. M. Fowler. *Refactoring, Improving the Design of Existing Code*. Addison Wesley, 1999.

23. P. Funk, A. Lewien, and G. Snelting. Algorithms for concept lattice decomposition and their applications. Technical Report 95-09, TU Braunschweig, December 1995.

24. B. Ganter and R. Wille. *Formal Concept Analysis: Mathematical Foundations*. Springer-Verlag, Berlin, 1999.

25. R. Godin, H. Mili, G. W. Mineau, R. Missaoui, A. Arfi, and T.-T. Chau. Design of class hierarchies based on concept (Galois) lattices. *Theory and Application of Object Systems (TAPOS)*, 4(2):117–134, 1998.

26. R. Godin, G. Mineau, R. Missaoui, M. St-Germain, and N. Faraj. Applying concept formation methods to software reuse. *International Journal of Knowledge Engineering and Software Engineering*, 5(1):119–142, 1995.

27. Robert Godin and Hafedh Mili. Building and maintaining analysis-level class hierarchies using Galois lattices. In *Proceedings of the OOPSLA'93 Conference on Object-oriented Programming Systems, Languages and Applications*, pages 394–410, 1993.

28. M. Huchard and H. Leblanc. From Java classes to Java interfaces through galois lattices. In *Actes de ORDAL'99: 3rd International Conference on Orders, Algorithms and Applications*, pages 211–216, Montpellier, 1999.

29. M. Huchard, C. Roume, and P. Valtchev. When concepts point at other concepts: the case of UML diagram reconstruction. In *Advances in Formal Concept Analysis for Knowledge Discovery in Databases, FCAKDD 2002*, pages 32–43, 2002.

30. IEEE. *IEEE Std 610.12-1990 — IEEE Standard Glossary of Software Engineering Terminology*. IEEE, New York, September 1990.

31. IEEE. *IEEE/EIA 12207.0-1996 — Standard for Information Technology – Software life cycle processes*. IEEE, New York, March 1998.

32. ISO. *ISO/IEC 12207:1995 — Standard for Information Technology – Software life cycle processes*. ISO, New York, March 1995.

33. Maren Krone and Gregor Snelting. On the inference of configuration structures from source code. In *Proceedings of the International Conference on Software Engineering (ICSE 1994)*, pages 49–57, 1994.

34. T. Kuipers and L. Moonen. Types and concept analysis for legacy systems. Technical Report SEN-R0017, Centrum voor Wiskunde en Informatica, July 2000.

35. H. Leblanc, C. Dony, M. Huchard, and T. Libourel. An environment for building and maintaining class hierarchies. In A. Moreira and S. Demeyer, editors, *ECOOP'99: Workshop "Object-Oriented Architectural Evolution"*, number 1743 in Lecture Notes in Computer Science, Heidelburg, 1999. Springer-Verlag.

36. C. Lindig. Concept-based component retrieval. In J. Köhler, F., Giunchiglia, C. Green, and C. Walther, editors, *Working Notes of the IJCAI-95 Workshop: Formal Approaches to the Reuse of Plans, Proofs, and Programs*, pages 21–25, August 1995.

37. C. Lindig and G. Snelting. Assessing modular structure of legacy code based on mathematical concept analysis. In *Proceedings of the International Conference on Software Engineering (ICSE 97)*, pages 349–359, Boston, 1997.

38. D. Richards and K. Boettger. Representing requirements in natural language as concept lattices. In *22nd Annual International Conference of the British Computer Society's Specialist Group on Artificial Intelligence (SGES), (ES2002)*, Cambridge, December 2002.

39. D. Richards and K. Boettger. Using RECOCASE to compare use cases from multiple viewpoints. In *Proceedings of ACIS2002*, 2002.

40. D. Richards, K. Boettger, and O. Aguilera. A controlled language to assist conversion of use case descriptions into concept lattices. In *Proceedings of 15th Australian Joint Conference on Artificial Intelligence*, 2002.

41. D. Richards, K. Boettger, and A. Fure. RECOCASE-tool: A CASE tool for RECOnciling requirements viewpoints. In *Proceedings of the 7th Australian Workshop on Requirements Engineering, AWRE'2002*, 2002.

42. D. Richards, K. Boettger, and A. Fure. Using RECOCASE to compare use cases from multiple viewpoints. In *Proceedings of the 13th Australasian Conference on Information Systems ACIS 2002*, Melbourne, December 2002.

43. D. Richards and P. Compton. Combining formal concept analysis and ripple down rules to support reuse. In *Proceedings of Software Engineering Knowledge Engineering SEKE'97*, Madrid, June 1997. Springer-Verlag.

44. T. Rock and R. Wille. Ein TOSCANA-Erkundungssystem zur Literatursuche. In G. Stumme and R. Wille, editors, *Begriffliche Wissensverarbeitung: Methoden und Anwendungen*, pages 239–253, Berlin-Heidelberg, 2000. Springer-Verlag.

45. P. Rook. Controlling software projects. *Software Engineering Journal*, 1(1):7–16, January 1996.

46. W. W. Royce. Managing the development of large software systems. In R.H. Thayer, editor, *Tutorial: Software Engineering Project Management*, pages 118–127. IEEE Computer Society, Washington, 1987. Originally published in Proceedings of WESCON'97.

47. H.A. Sahraoui, W. Melo, H. Lounis, and F. Dumont. Applying concept formation methods to object identification in procedural code. In *Proceedings of International Conference on Automated Software Engineering (ASE '97)*, pages 210–218. IEEE, November 1997.

48. I. Schmitt and S. Conrad. Restructuring object-oriented database schemata by concept analysis. In T. Polle, T. Ripke, and K.-D. Schewe, editors, *Fundamentals of Information Systems (Post-Proceedings 7th International Workshop on Foundations of Models and Languages for Data and Objects FoMLaDO'98)*, pages 177–185, Boston, 1999. Kluwer Academic Publishers.

49. I. Schmitt and G. Saake. Merging inheritance hierarchies for database integration. In *Proceedings of the 3rd International Conference on Cooperative Information Systems (CoopIS'98)*, New York, August 1998.

50. S. Schupp, M. Krishnamoorthy, M. Zalewski, and J. Kilbride. The "right" level of abstraction - assessing reusable software with formal concept analysis. In G. Angelova, D. Corbett, and U. Priss, editors, *Foundations and Applications of Conceptual Structures - Contributions to ICCS 2002*, pages 74–91. Bulgarian Academy of Sciences, 2002.

51. M. Siff and T. Reps. Identifying modules via concept analysis. In *Proceedings of the International Conference on Software Maintenance*, pages 170–179. IEEE Computer Society Press, 1997.

52. G. Snelting. Reengineering of configurations based on mathematical concept analysis. *ACM Transactions on Software Engineering and Methodology*, 5(2):146–189, April 1996.

53. G. Snelting. Concept analysis — a new framework for program understanding. In *SIGPLAN/SIGSOFT Workshop on Program Analysis for Software Tools and Engineering (PASTE)*, pages 1–10, Montreal, Canada, June 1998.

54. G. Snelting. Software reengineering based on concept lattices. In *Proceedings 4th European Conference on Software Maintenance and Reengineeering*, pages 3–12. IEEE, 2000.

55. G. Snelting and F. Tip. Reengineering class hierarchies using concept analysis. Technical Report RC 21164(94592)24APR97, IBM T.J. Watson Research Center, IBM T.J. Watson Research Center, P.O. Box 704, Yorktown Heights, NY 10598, USA, 1997.

56. G. Snelting and F. Tip. Reengineering class hierarchies using concept analysis. In *Proceedings of ACMSIGSOFT Symposium on the Foundations of Software Engineering*, pages 99–110, November 1998.

57. G. Snelting and F. Tip. Understanding class hierarchies using concept analysis. *ACM Transactions on Programming Languages and Systems*, pages 540–582, May 2000.

58. J.M. Spivey. An introduction to Z and formal specifications. *Software Engineering Journal*, 4(1):40–50, January 1989.

59. M. Streckenbach and G. Snelting. Understanding class hierarchies with KABA. In *Workshop on Object-Oriented Reengineering - WOOR'99*, Toulouse, France, September 1999.

60. J. Sun, J.S. Dong, J. Lui, and H. Wang. Object-Z web environment and projections to UML. In *WWW10 10th International World Wide Web Conference*, pages 725–734, New York, 2001. ACM.

61. E. B. Swanson. The dimensions of maintenance. In *Proceedings of the 2nd International Conference on Software Engineering*, pages 492–497. IEEE Computer Society Press, 1976.

62. T. Tilley. Towards an FCA based tool for visualising formal specifications. In B. Ganter and A. de Moor, editors, *Using Conceptual Structures: Contributions to ICCS 2003*, pages 227–240. Shaker Verlag, 2003.

63. T. Tilley, W. Hesse, and R. Duke. A software modelling exercise using FCA. In B. Ganter and A. de Moor, editors, *Using Conceptual Structures: Contributions to ICCS 2003*, pages 213–226. Shaker Verlag, 2003.

64. P. Tonella. Concept analysis for module restructuring. *IEEE Transactions on Software Engineering*, 27(4):351–363, April 2001.

65. P. Tonella and G. Antoniol. Object-oriented design pattern inference. In *Proceedings of CSM 1999*, pages 230–240, 1999.

66. A van Deursen and T. Kuipers. Identifying objects using cluster and concept analysis. In *Proceedings of the 21st International Conference on Software Engineering, ICSE-99*, pages 246–255. ACM, 1999.

Concept Lattices in Software Analysis

Gregor Snelting

Universität Passau

Abstract. About ten years ago, the first serious applications of concept lattices in software analysis were published. Today, a wide range of applications of concept lattices in static and dynamic analysis of software artefacts is known. This overview summarizes important papers from the last ten years, and presents three methods in some detail: 1. methods to extract classes and modules from legacy software; 2. the Snelting/Tip algorithm for application-specific, semantics-preserving refactoring of class hierarchies; 3. Ball's method for infering dynamic dominators and control flow regions from program traces. We conclude with some perpectives on further uses of concept lattices in software technology.

1 Overview

Concept lattices were already introduced more than 50 years ago in Birkhoff's first book on lattice theory[1]. More than 20 years ago, Ganter and Wille started to expand the theory considerably and investigated serious applications of concept analysis e.g. in the social sciences. But only 10 years ago, a few researchers started to explore the possibilities of concept lattices for computer science, in particular software technology. Godin, Mili and their coworkers in Montreal applied concept analysis to software design, in particular object-oriented design; this line of reasearch is described in Godin's contribution to the current book. The current author and his group, then in Braunschweig, came up with the first applications of concept lattices in software analysis. Meanwhile, a wealth of results is available, and it is the goal of this article to present important uses of concept lattices and their structure theory for static and dynamic analysis of software artefacts.

It is very natural to apply concept lattices for software analysis, as every software artefact contains an abundance of relations between "objects" and "attributes". To explore hidden structure in such relations is a natural task whenever one wants to understand old software artefacts, or reengineer legacy systems. As a result, a wave of concept lattice applications in software technology was proposed. Some of the applications were well-motivated and based in a thorough understanding of the underlying theory, while others just generated lattices from "yet another relation", without validating the resulting structures. In the following we will concentrate on some substantial contributions; some

---

[1] We will not give references to general literature on concept analysis or software technology, but restrict ourselfes to citations of specific papers which utilize concept lattices in software analysis.

B. Ganter et al. (Eds.): Formal Concept Analysis, LNAI 3626, pp. 272–287, 2005.

of the latest papers are covered as well as "classics". Thus in the intention of the author, the current article also serves as a successor to the earlier overview articles [Sne98, Sne00].

This paper starts out with summaries of concept lattice applications in program understanding, software reengineering, component retrieval, refactoring, and dynamic analysis; it also presents some data on the scientific impact of concept analysis in software engineering. The rest of the paper describes three applications in more detail: extraction of modules or classes from legacy code; automatic refactoring; and dynamic analysis. We conclude with some remarks on future applications of concept lattices in software technology.

1.1 Infering Configuration Structures from Source Code

One of the very first nontrivial applications of concept lattices for software analysis was Krone's and Snelting's work on the inference of configuration structures from source code. The paper was presented in 1994 at the International Conference on Software Engineering (ICSE) [KS94], and an expanded version later appeared in the ACM Transactions on Software Engineering and Methodology [Sne96]. The authors analysed the relationship between code pieces and preprocessor variables in Unix system software, as the preprocessor is typically used for configuration management in older Unix programs. Not only did implications and interferences between configurations become visible in the lattice; the structure theory of concept lattices (irreducible elements and implication base) allowed for a restructuring of the preprocessor variables, and the configuration space could be modularized according to algebraic decompositions of the lattice.

1.2 Identifying Modules and Classes in Legacy Software

"Modularization" was also the keyword for a whole series of papers which came out in the following years; triggered by the Y2K problem and its corresponding reengineering challenges. Old legacy systems typically have been developed without modern software technology; in particular, there is no explicit modularization. Identifying modules or classes in legacy code therefore is an important task in order to make such systems survive ("software geriatry" [Parnas]), and concept analysis turned out to be quite helpful.

The author's approach to modularize old Fortran systems, which was presented at ICSE 1997 [LS97], will be described later in detail. Generally speaking, it explores the relationship between program variables and procedures in order to identify modules. At the same time, Siff and Reps presented a similar approach to the restructuring of C programs [SR97]. Van Deursen and Kuipers applied basically the same idea to Cobol legacy programs and published it at ICSE 1999 [vDK99]. All three papers have shown that it is not enough to just compute the lattice, but that background knowledge has to be exploited for a careful selection of "objects" and "attributes", and that the lattice must be simplified, decomposed and interpreted by experts. Other authors have stepped into the footsteps of these three publications, but not with the same success and impact.

1.3 Software Component Retrieval

A side line of the work on modularization resulted in support for software component retrieval. Godin et al. were probably the first authors to apply concept lattices for component retrieval [GMA93]. Lindig's dissertation [Lin99] went a considerable step further: it carefully engineered concept-based component retrieval and validated its effectiveness for interactive retrieval. Later Fischer combined Lindig's approach with formal specifications, where match relations between specifications are checked beforehand by a theorem prover in order to obtain the initial table from which the lattice is generated; this work won the best Paper Award at the Conference on Automated Software Engineering 1998 [Fis98]. Other authors have proposed similar approaches, but did not really improve on Lindig's or Fischer's work.

1.4 Refactoring Class Hierarchies

As concept lattices are natural inheritance structures, a natural application field is class hierarchies for object-oriented languages. The work by Godin as well as Hesse's approach to requirements engineering aim at the construction of a class hierarchy from some requirements and are described elsewhere in this book. But in practice, evolution of existing systems is more important than the construction of new systems, and in the object-oriented world, refactoring of class hierarchies is the method of choice. Refactoring applies a sequence of (hopefully) semantics-preserving transformations to a class hierarchy such as moving methods to other classes, splitting classes, or extracting new methods from statements. The overall goal is to improve the hierarchy according to software engineering principles such as high cohesion and low coupling, or to identify design patterns in existing code.

One approach was proposed by Tonella, who used concept lattices to identify design pattern in existing code, and reports some success for small examples [TA99]. The semantics-preserving refactoring method by Snelting and Tip was first published at the 1998 Symposium on Foundations of Software Engineering (FSE) [ST98]; a much more detailed version appeared later in the ACM Transactions on Programming Languages and Systems [ST00]. It is based on a fine-grained analysis of a given hierarchy together with a set of applications, and will be explained later in more detail. Work on refactoring using concept lattices is still ongoing, and we will see more results in the future.

1.5 Dynamic Analysis

Only a few years ago, the program analysis community started to pay attention not only to static program analysis, but also to dynamic program analysis. Static analysis relies on the source text alone, and precise static analysis is often expensive, if not undecidable. Dynamic analysis uses a set of execution traces in addition; of course the analysis results are valid only for a set of specific inputs or program runs, but are much cheaper to compute and in practice often quite sufficient. In the last years, dynamic analysis has thus become a very active research topic.

It is therefore not surprising that concept lattices were used for dynamic analysis as well. The first paper presenting such an approach was Ball's reconstruction of control flow graphs from program traces; it was published at FSE 1999 [Bal99] and will be explained later in this article. A more general (but also less precise) approach was presented by Koschke in 2001 and received the Best Paper Award at the International Conference on Software Maintenance [EKS01].

The latest article in this line of research was presented by Ammons, Mandelin, Bodik and Larus at the Conference on Programming Language Design and Implementation (PLDI) [AMBL03]. The authors use concept lattices to debug specifications in temporal logic. The idea is to analyse execution traces (e.g. counterexamples generated by a model checker) and group similar traces into "concepts"; this reduces the debugging work. The paper is not only remarkable due to its high-tech combination of temporal specifications, model checkers, specification extractors, and concept lattices, but also due to the fact that two authors are academics, one is from Microsoft, and one is from IBM.

1.6   Impact

These days, citation databases are gaining influence, and we therefore browsed the CiteSeer data base, which contains most publications and citations in computer science. The above overview contains only the most important publications; CiteSeer lists about 40 papers on "concept analysis" or "concept lattices", and more are coming out. The articles sketched above have all been published at very selective and influential conferences and journals (for example, PLDI is according to CiteSeer the most cited of all computer science conferences), and as a consequence CiteSeer lists more than 400 citations of concept analysis papers in computer science. Ganter and Wille made it into CiteSeer's list of the 10000 most cited computer scientists, even though they are mathematicians.

This success would not have been possible without readily available implementations. In Germany, the concept lattice software from Wille's group is quite well-known, but on an international scale the most popular software is Lindig's implementation of Ganter's algorithm for concept lattice generation [Lin], as it is very efficient, robust, and usable as a background tool without own GUI. The software has been installed at ca. 50 sites worldwide.

2   Modularization

Let us now describe our work on modularization of old Fortran programs in some detail. While the above-mentioned later papers by Siff/Reps and van Deursen/Kuipers were more successful from a practical viewpoint, our work introduced the basic idea. The project was based on a cooperation with a national research institution, who aimed at reengineering their aerodynamics software written in Fortran. Several approaches to modularize the system had failed, so it was decided to try concept lattices.

The fundamental idea is to investigate the relation between global variables and procedures. If a set of variables V and a set of procedures P can be identified,

```
SUBROUTINE R1(...)          SUBROUTINE R3(...)
COMMON /C1/ V1,V2           COMMON /C2/ V3,V4
...                         COMMON /C4/ V6,V7,V8
END                         ...
                            END


SUBROUTINE R2(...)          SUBROUTINE R4(...)
COMMON /C2/ V3,V4           COMMON /C2/ V3,V4
COMMON /C3/ V5              COMMON /C3/ V5
...                         COMMON /C4/ V6,V7,V8
END                         ...
                            END
```

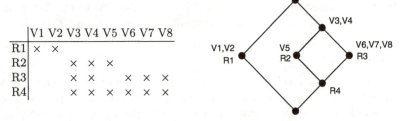

	V1	V2	V3	V4	V5	V6	V7	V8
R1	×	×						
R2			×	×	×			
R3			×	×		×	×	×
R4			×	×	×	×	×	×

Fig. 1. A Fortran fragment, its context table, and its concept lattice

where all procedures in P use only variables in V, and all variables in V are only used by procedures in P, then P together with V is definitely a module candidate. The reason is that modules implement information hiding, hence a module's variables may only be accessed through its interface procedures. Figure 1 presents a small example of four procedures, acting on various global variables which are organized in several "Common" blocks. The goal was to identify modules as described, and restructure the "Common" blocks such that there is one "Common" block per module. Figure 1 also presents the context table extracted from the source code, and the corresponding concept lattice.

The general situation is depicted in figure 2. Modules correspond to rectangle *shapes* in the context table (remember that in the context table, row and column permutations do not matter!), but need not be completely filled rectangles, as not every procedure accesses all module variables. The corresponding lattice is horizontally decomposable, and every rectangle shape in the table corresponds to one horizontal summand. Figures 1 and 2 both present horizontally decomposable lattices, hence a modularization is possible. In case horizontal summands are connected by a few additional infima, these are called interferences. Interferences prevent modularization (as the information hiding principle is violated), but can usually be removed by some small behavior-preserving transformations of the source code.

Now let us come back to our project with the national research institution. We analysed a 106 KLOC Fortran program, which was 25 years old and had

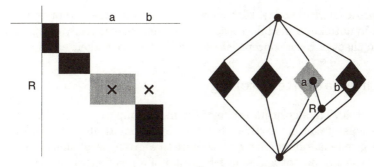

Fig. 2. Horizontally decomposable lattice and an interference

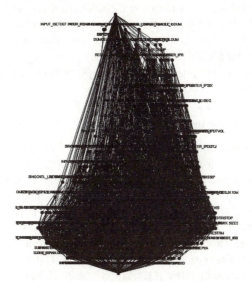

Fig. 3. Concept lattice for Fortran aerodynamics program

undergone countless modifications. 317 procedures were acting on 492 global variables, distributed over 40 "Common" blocks. After extraction of the context table, the lattice was computed and layoutet. The result can be seen in figure 3. The lattice has more than 2000 elements, is definitely not decomposable, but consists basically of interferences. A modularization based on a repartitioning of the global variables is therefore not possible. The national institution decided to cancel the reengineering project and develop a new system.

Let us add that the basic method can be extended in various ways: Siff/Reps not only used variables and procedures, but also types, and they explicitly coded the fact that $p \in P$ does *not* use variable $v \in V$ or type t. van Deursen/Kuipers preprocessed the variables, in order to distingish temporary variables from those relevant to modules. For the Fortran analysis, Wegman proposed to transform the program into static single assignment form first, as Fortran programs often misuse the same variable for different purposes. All this will improve the results of modularization. But today, with a few years distance, the author does not really believe in automatic modularization any more, because really old programs are just too chaotic. Even concept lattices will not prevent their entropy death.

3   Automatic Refactoring

The Snelting/Tip algorithm is one of the most complex, but also most powerful applications of concept lattices. It serves to automatically restructure ("refactor") a given class hierarchy with respect to a given set of client programs. As clients typically do not access every feature of a given hierarchy, the result is a refactored hierarchy which is "specialised" or "taylored" to the specific clients. In particular, all objects will contain only members and methods they really need (with respect to client behaviour). The method combines program analysis, type constraints and concept analysis to compute the most fine-grained refactoring which is still preserving client behaviour.

In this section, we recapitulate the basic properties of this algorithm. Full details can be found in [ST00].

3.1   Collecting Member Accesses

The algorithm is based on a fine-grained analysis of object access patterns. For all objects or object references o, it determines whether o does access member m from class C. The result is a binary relation, coded in form of a table T.

As an example, consider the program fragment in figure 4. B, being a subclass of A, redefines $f()$ and accesses the inherited fields x, y. The main program creates two objects of type A and two objects of type B, and performs some field accesses and method calls. Table T for this example consists of rows labelled with object references $a1, a2, b1, b2, A.f.this, B.f.this, B.g.this, B.h.this$ as well as object creation sites $A1, A2, B1, B2$. Columns are labelled with fields and methods $A.x, A.y, A.z, A.f(), B.f(), B.g(), B.h()$. For methods, there is an additional distinction between declarations and definitions ($dcl(C.f())$ vs. $def(C.f())$), which makes the analysis much more precise.

Now let $Type(o) = C$ be the static type of an object reference o, and let member accesses $o.m$ resp. $o.f()$ be given. Table T will contain entries $(o, C.m)$ resp. $(o, dcl(C.f()))$. Furthermore, points-to analysis is used to determine for o to which object creation sites it might point to at runtime; this set is denoted $pt(o) = \{O1, O2, \ldots\}$. $pt(o)$ may be too big (i.e. unprecise), but never too small (i.e. pt is a conservative approximation); in the example we have e.g. $pt(a1) = \{A1\}$, $pt(a2) = \{A2, B2\}$. Additional entries $(O_i, def(D_i.f()))$ are created for all $O_i \in pt(o)$ where $D_i = StaticLookup(Type(O_i), f)$. For the above example, the resulting table is shown in figure 5.

3.2   Type Constraints

In a second step, a set of type constraints is extracted from the program, which are necessary for preservation of behaviour. The refactoring algorithm computes

```
class A {
  int x, y, z;
  void f() {
    y = x;
  }
}


class B extends A {
  void f() {
    y++;
  }
  void g() {
    x++;
    f();
  }
  void h() {
    f();
    x--;
  }
}
```

```
class Client {
    public static void
    main(String[] args) {
        A a1 = new A();   // A1
        A a2 = new A();   // A2
        B b1 = new B();   // B1
        B b2 = new B();   // B2

        a1.x = 17;
        a2.x = 42;
        if (...) { a2 = b2; }
        a2.f();
        b1.g();
        b2.h();
    }
}
```

Fig. 4. A small Java class hierarchy

	A.x	A.y	A.z	dcl(A.f)	def(A.f)	dcl(B.f)	def(B.f)	dcl(B.g)	def(B.g)	dcl(B.h)	def(B.h)
a1	×										
a2	×			×							
b1								×			
b2						×				×	
A1											
A2					×						
B1							×		×		
B2							×				×
A.f.this	×	×			×						
B.f.this		×					×				
B.g.this	×					×			×		
B.h.this	×					×					×

Fig. 5. Member access table for figure 4

a new type (i.e. class) for every variable or class-typed member field, and a new "home" class for every member. Therefore, constraints for a variable or field x are expressed over the (to be determined) new type of x in the refactored hierarchy, $type(x)$; constraints for a member or method $C.m$ are expressed over its (to be determined) new "home class", $def(C.m)$.

There are basically two kinds of type constraints:

1. Any (explicit or implicit) assignment x = y; in the program text gives rise to a type constraint $type(y) \leq type(x)$. Such constraints are called assignment constraints.

2. If subclass B of A redefines a member or method m, and some object x accesses both $A.m$ and $B.m$ (that is, $\exists x : (x, def(A.m)) \in T \wedge (x, def(B.m)) \in T$), then $def(B.m) < def(A.m)$ must be retained in order to avoid ambiguous access to m from x. Such constraints are called dominance constraints. A more obvious, similar dominance constraint requires that for all methods $C.f$, $def(C.f) \leq dcl(C.f)$.

Once all type constraints have been extracted, they are incorporated into table T. To achieve this, we exploit the fact that a constraint can be seen as an *implication* between table rows resp. columns, and that there is an algorithm to incorporate any given set of implications into a table. First we observe that even in the refactored hierarchy, a subtype inherits all members from its supertype. Therefore $type(y) \leq type(x)$ emforces that any table entry for x must also be present for y; that is $\forall m : (x, m) \in T \Rightarrow (y, m) \in T$, or $x \rightarrow y$ for short. Second, $def(B.m) < def(A.m)$ enforces that any table entry for $def(B.m)$ must also be present for $def(A.m)$, which is written as $def(B.m) \rightarrow def(A.m)$.

Reconsidering figure 4, the following assignment constraints are collected in form of implications:

$$A.y \rightarrow A.x, A.f.this \rightarrow a2, B.f.this \rightarrow a2, B.g.this \rightarrow b1,$$
$$B.h.this \rightarrow b2, a1 \rightarrow A1, a2 \rightarrow A2, b1 \rightarrow B1, b2 \rightarrow B2, a2 \rightarrow b2$$

Furthermore, the following obvious dominance constraints are collected:

$$def(A.f) \rightarrow dcl(A.f), def(B.f) \rightarrow dcl(B.f),$$
$$def(B.g) \rightarrow dcl(B.g), def(B.h) \rightarrow dcl(B.h)$$

as well as the non-obvious dominance constraints

$$def(B.f) \rightarrow def(A.f), dcl(B.f) \rightarrow dcl(A.f)$$

These implications are incorporated into the initial table by copying row entries from row y to row x resp. column entries from column $def(A.f)$ to column $def(B.f)$ etc. Note that in general there may be cyclic and mutual dependences between row and/or column implications, thus a fixpoint iteration is required to incorporate all constraints into the table. The final table for figure 4 is presented in figure 6.

3.3   The Refactored Hierarchy as a Concept Lattice

In a final step, concept analysis is used to construct the refactored hierarchy from the final table. Concept lattices can naturally be interpreted as inheritance hierarchies. The concept lattice for figure 4, as constructed from the final table (figure 6), is given in figure 7. Every lattice *element* represents a *class* in

	A.x	A.y	A.z	dcl(A.f)	def(A.f)	dcl(B.f)	def(B.f)	dcl(B.g)	def(B.g)	dcl(B.h)	def(B.h)
a1	×										
a2	×			×							
b1								×			
b2	×			×	×					×	
A1	×										
A2	×	×		×	×						
B1	×	×		×		×	×	×	×		
B2	×	×		×		×	×			×	×
A.f.this	×	×		×	×						
B.f.this		×		×		×	×				
B.g.this	×	×		×		×	×	×	×		
B.h.this	×	×		×		×	×			×	×

Fig. 6. Table after incorporating type constraints for figure 4

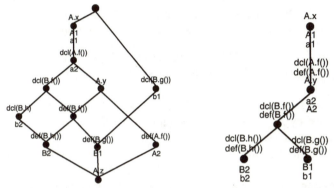

Fig. 7. Concept lattice and its simplified version for figure 6

the refactored hierarchy. Method or field names *above* an element represent the *members* of this class. Objects or pointers *below* an element will have that element (i.e. class) as its new *type*. In particular, all objects now have a new type which contains only the members the object really accesses.

Typically, original classes are split and new subclasses are introduced by the previous steps. This is particularly true for figure 7 (left), where the raw lattice introduces 12 refactored classes instead of the original two classes. These new classes represent object behaviour patterns: $a1$ and $A1$ use $A.x$ but nothing else, which is clearly visible in the lattice. $a2$ additionally calls $a.f()$ and therefore needs the declaration of this method. $b1$ calls $B.g()$ and nothing else; $b2$ calls $B.h(), B.f()$ plus anything called by $a2$. The "real objects" $A2, B2, B1$ are located far down in the hierarchy and use various subsets of the original hierarchies' members. $B2$ in particular not only accesses everything accessed by $b2$, but also calls $B.f()$, which leads to a multiple inheritance in the lattice. Note that the raw lattice clearly distinguishes between a class and its interface: several new classes contain only $dcl(...)$ entries, but no $def(...)$ entries or fields, meaning that they are interfaces.

As the lattice respects not only the member accesses, but also the type constraints, it guarantees preservation of behaviour for all clients. The lattice is rather fine-grained, and in its raw form represents the most fine-grained refactoring which respects the behaviour of all clients. But from a software engineering viewpoint, the lattice must be simplified in order to be useful. Some simplifications are quite obvious: "empty" elements (i.e. new classes without own members) such as the top and bottom element in figure 6 (middle) can be removed; multiple inheritance can in many cases be eliminated (e.g. by moving members upward in the hierarchy), and lattice elements can be merged according to certain (behaviour-preserving) rules [SS03]. In particular, the distinction between a class and its interface can be removed by merging lattice elements. The final result is in general not a lattice anymore, only a partial order – but for object-oriented programming, this is fine.

Figure 7 (right) presents a simplified version of figure 7 (left), which can be generated automatically. Now the empty elements and the interfaces are gone, and the different access patterns for the objects are visible even better:

- The two objects of original type B have different behaviour, as one calls g and the other calls h. Therefore, the original B class is split into two unrelated classes.
- The two objects of original type A have related behaviour, as $A2$ accesses everything accessed by $A1$, plus $A.f()$. Therefore, the original A class is split into a class and a subclass.
- $A1$ does only contain $A.x$ and not $A.y$. $A.z$ is dead, as it appears at the bottom element in the lattice. Thus objects become smaller in general, as unused members are physically absent in objects of the new hierarchy.

One might think of simplifying even further by merging the two topmost elements in figure 7 (right), but that would make $A1$ bigger than necessary by including $A.y$ as a member. It is the refactorer's decision whether this disadvantage is outweighted by a simpler structure of the refactored hierarchy. If so, the refactoring editor must guarantee that behaviour of all clients is still preserved after simplification.

3.4   The KABA System

KABA (KlassenAnalyse mit BegriffsAnalyse) is an implementation of the Snelting/Tip method for Java. KABA consists of four parts: a static analysis, a dynamic analysis, a graphical class hierarchy editor and a bytecode transformation tool.

KABA will display the (original or simplified) lattice, and offers browsing as well as back links to the original hierarchy. But the true value of the KABA hierarchy editor lies in its ability to manipulate the (refactored or original) hierarchy – where of course preservation of behaviour is always guaranteed. For example, classes can be merged or methods can be moved to neighbour classes. Eventually, Java code can again be generated. Note that all original statements

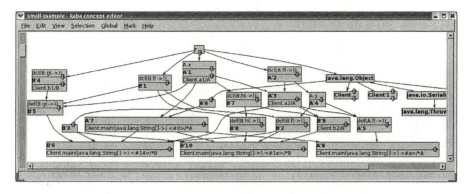

Fig. 8. Kaba screenshot for figure 4

remain unchanged – only the hierarchy and the declarations of variables, fields and methods change, as the classes from the new hierarchy have to be used.

Figure 8 shows a KABA screenshot. The reader should be aware that the implementation of the Snelting/Tip method for KABA and its application to real Java programs is much more complex than described above: the full Java language must be handled as well as libraries; questions of scale-up do matter, etc. Some case studies using KABA on real-world programs can be found in [SS03].

4   Dynamic Analysis

Ball [Bal99] was the first one to use concept lattices for dynamic analysis. His scenario assumes that no source code is given, just an executable program, and the task is to reconstruct the control flow graph (CFG) and its dominator relations. The starting point is an execution profile which for every test run says which statements or functions have been executed.

In order to understand Ball's method more fully, we will have to introduce a few definitions. CFG's are well known, and figure 9 (left) presents a small example. In CFG's, the definition of a dominator is very important. Statement x is a *predominator* of statement y, if x must always be executed before y: every path from the CFG start to y must pass through x. In the standard example, x is a while loop entry point, and y is a statement in the loop body; obviously y can never be executed unless x has been executed. Statement y is a *postdominator* of x if y must always be executed after x: every CFG path from x to the CFG exit must pass through y.

Figure 9 (right) presents a few examples for these definitions. Note that e.g. B is not a predominator for E since E can be reached via D, but B is a predominator for C as there is no way to reach C except via B. The relations x *predom* y and y *postdom* x are partial orders, and in fact pre- resp. postdominators can always be arranged in a tree.

If x *predom* y and y *postdom* x, then x and y are said to be in the same control flow *region*. Statements in the same region are always executed together

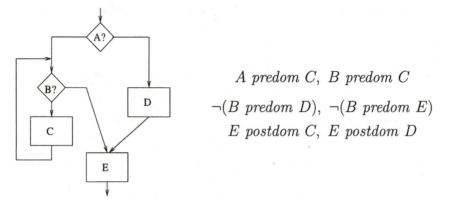

$$A \text{ predom } C, \; B \text{ predom } C$$

$$\neg(B \text{ predom } D), \; \neg(B \text{ predom } E)$$

$$E \text{ postdom } C, \; E \text{ postdom } D$$

Fig. 9. A simple control flow graph and some pre- and postdominator relationships

	add	rotate	rem	Min	Succ	DelFix
t1	X		X			X
t2	X	X	X			X
t3	X	X	X	X		
t4	X	X	X	X	X	X
t5	X	X	X	X	X	X

Fig. 10. A trace table and its concept lattice

or not at all; it is easy to see that these regions form an equivalence relation on the CFG nodes. In the example, A and E are in the same region as A *predom* E and E *postdom* A, but other non-trivial regions do not exist (B, C, D are all in their own singleton region).

As defined above, dominators are static relations: they are valid for every possible execution. If source code is missing, static dominators cannot be determined. All that can be said is that for the given test runs, x was always executed before (or after) y. Such *dynamic* dominators are valid only for some specific set of executions, but not for all program executions. The more executions are run, the more likely it is that a dynamic dominator is in fact a static dominator. Thus dynamic dominators are good candidates for static dominators and perhaps allow a reconstruction of the CFG! This was Ball's idea, together with his insight that dynamic dominators can be determined by computing concept lattices from program traces.

Let us consider an example. In figure 10 we see a table summarizing the results of profiling five test runs. For every test run (i.e. row) we see which functions among six functions were executed. The concept lattice for this table is shown in the same figure. What is the interpretation of this lattice? First of all, the concepts are *dynamic regions*: all functions in a concept's intent are executed together or not at all. Furthermore, upward arcs are implications: any test that executes min and succ also executes rotate. Therefore, rotate is a dynamic dominator of both `min` and `succ`! (Unfortunately, we cannot tell whether it is a pre- or postdominator, as the trace table does not say anything about the *temporal order* of function executions.)

Next, suprema resp. infima correspond to forks of control flow ("if-statements"): `add` dominates both `Min` and `DelFix`, but there are tests which distinguish execution of both. Let us assume that `add` is in fact a dynamic *predominator* of `min` and `succ`, then there must be a case distinction at `add` leading to either `min` or `succ` (and the case distinction cannot be earlier in the execution, since `add` is at the supremum!). In this situation, the infimum of `Min` and `DelFix` correponds to the "join point" (dynamic postdominator) in the CFG, where the two branches of the "if" merge again.

But note that the situation could be the other way round, that is the infimum could correspond to the "if" predominating the two branches, and the supremum could be the join point[2]. In any case, the lattice is an order-preserving image of the CFG according to the following equations:

$$x \; predom \; y \implies \mu(x) \geq \mu(y)$$
$$x \; postdom \; y \implies \mu(x) \leq \mu(y)$$

If the trace table can be enriched with information saying which function was executed earlier (and this should usually be easy!), the lattice can definitely distinguish (dynamic) pre- and postdominators. The more test cases are used, the more fine-grained this lattice will become, and in the limit case of an infinite number of tests covering all CFG paths, the CFG can be order-embedded into the lattice. Note that the CFG is only a quasi-order as it usually contains cycles.

Summarizing this preliminary discussion, we see that the concept lattice allows to uncover the control flow graph and its regions and dominators from test cases. This is a very useful method for reengineering old executables where the source code has been lost – a situation which occurs in practice. Let us hope that Ball will proceed to work out the details and apply it to real-world examples.[SS03]

5   Conclusion

This overview article centered around applications of concept lattices in software analysis. Several other applications of concept analysis in software technology are described elsewhere and have been left out due to space restrictions. Examining the applications we have discussed, one can clearly distinguish two different "historical" phases: early applications of concept lattices in software technology centered on design and static analysis, while later applications are based on program transformation and dynamic analysis.

It is kind of surprising that all these applications stick to the basic theory of concept lattices and their corrresponding implication base, but do not apply

---

[2] Ball for some reason assumed that suprema always correspond to predominators, but Ganter pointed out that the dual situation could also be the case.

more advanced results, such as the structure theory of concept lattices or fuzzy contexts. In fact the author believed for a while that these advanced techniques can improve applications in software technology. But today we know that this is not true. The reason is that realistic lattices do not have the properties required for the advanced techniques. For examples, typical lattices in software technology have neither congruences nor block relations ("weak congruences"); the reason is that congruences have nonlocal effects on the lattice which have no counterpart in the world of software. Similarly, subdirect or subtensorial decompositions could not be found in our various applications. Future work in lattice theory must show whether the structure theory of concept lattices can be extended in such a way that typical "local" situations occuring in software analysis can be handled.

Nevertheless, concept lattices have received a huge wave of attention by software technology researchers in the last ten years, and proved to be a very helpful instrument. We will see many more concept lattices in software technology in the next ten years!

References

[AMBL03] Glenn Ammons, David Mandelin, Rastislav Bodik, and James Larus. Debugging temporal specifications with concept analysis. In *ACM SIGPLAN Conference on Programming Language Design and Implementation*, pages 182–193, 2003.

[Bal99] Thomas Ball. The concept of dynamic analysis. In *ESEC / SIGSOFT FSE*, pages 216–234, 1999.

[EKS01] Thomas Eisenbarth, Rainer Koschke, and Daniel Simon. Feature-driven program understanding using concept analysis of execution trace. In *Proc. Ninth International Workshop on Program Comprehension (IWPC'01)*, May 2001.

[Fis98] B. Fischer. Specification-based browsing of software component libraries. In *Automated Software Engineering*, pages 74–83, 1998.

[GMA93] R. Godin, R. Missaoui, and A. April. Experimental comparison of navigation in a galois lattice with conventional information retrieval methods. *International Journal of Man-Machine Studies*, 38, 1993.

[KS94] Maren Krone and Gregor Snelting. On the inference of configuration structures from source code. In *Proceedings of the 16th international conference on Software engineering*, pages 49–57. IEEE Computer Society Press, 1994.

[Lin] Christian Lindig. Concepts: a program for concept lattices. http://www.st.cs.uni-sb.de/~lindig/src/concepts.html.

[Lin99] Christian Lindig. *Algorithmen zur Begriffsanalyse und ihre Anwendung bei Softwarebibliotheken*. PhD thesis, Technische Universität Braunschweig, 1999.

[LS97] Christian Lindig and Gregor Snelting. Assessing modular structure of legacy code based on mathematical concept analysis. In *Proceedings of the 19th International Conference on Software Engineering*, pages 349–359. ACM Press, 1997.

[Sne96] Gregor Snelting. Reengineering of configurations based on mathematical concept analysis. *ACM Transactions on Software Engineering and Methodology (TOSEM)*, 5(2):146–189, 1996.

[Sne98]   Gregor Snelting. Concept analysis - a new framework for program un-
           derstanding. In *Proc. ACM SIGPLAN/SIGSOFT Workshop on Program
           Analysis for Software Tools and Engineering (PASTE)*, pages 1–10, 1998.
           Invited contribution.

[Sne00]   Gregor Snelting. Software reengineering based on concept lattices. In *Proc.
           4th European Conference on Software Maintenance and Reengineeering*,
           pages 3–12, 2000. Invited contribution.

[SR97]   Michael Siff and Thomas Reps. Identifying modules via concept analysis. In
           Proc. International Conference on Software Maintenance, pages 170–179.
           IEEE Computer Society Press, 1997.

[SS03]   Mirko Streckenbach and Gregor Snelting. Behaviour-preserving refactoring
           with KABA. August 2003. Submitted for publication.

[ST98]   Gregor Snelting and Frank Tip. Reengineering class hierarchies using con-
           cept analysis. In *Proc. ACM SIGSOFT Symposium on the Foundations of
           Software Engineering*, pages 99–110, Orlando, FL, November 1998.

[ST00]   Gregor Snelting and Frank Tip. Understanding class hierarchies using con-
           cept analysis. *ACM Transactions on Programming Languages and Systems*,
           pages 540–582, May 2000.

[TA99]   Paolo Tonella and Giuliano Antoniol. Object-oriented design pattern infer-
           ence. In *International Conference on Software Maintenance*, pages 230–,
           1999.

[vDK99]   Arie van Deursen and Tobias Kuipers. Identifying objects using cluster
           and concept analysis. In *Proceedings of the 21st international conference
           on Software engineering*, pages 246–255. IEEE Computer Society Press,
           1999.

Formal Concept Analysis
Used for Software Analysis and Modelling

Wolfgang Hesse[1] and Thomas Tilley[2]

[1] FB Mathematik und Informatik
Philipps-Universität
Marburg, Germany
hesse@mathematik.uni-marburg.de
[2] School of Information Technology and Electrical Engineering
University of Queensland
Brisbane, Australia
tilley@itee.uq.edu.au

Abstract. Formal Concept Analysis (FCA) has shown its benefits in many application areas – including the field of Software Engineering. In general, FCA can successfully be used in almost all phases of the software life cycle. Several applications deal with software architecture, modularisation, program/code and configuration analysis while the early phases of the software life cycle – including requirements analysis, domain and system modelling – have not been considered to the same extent so far.

This article focuses on the use of FCA during the early phases of software development. First of all, the software life cycle and the importance of concepts – in particular for object oriented modelling (OOM) – are discussed. In principle, FCA can be used wherever concepts play a significant role in the software process. Reported work in this area focuses on requirements engineering (RE), use case analysis (UCA), object-oriented modelling, the analysis of class/object hierarchies and component retrieval.

As a typical application, the task of finding or deriving class candidates from a given use case description is considered in more detail. FCA offers valuable support to bridge this well-known gap existing in almost all OO methods. FCA allows a "crossing of perspectives" – between the functional view represented by the use cases and the data view implied by the "things" occurring there. Finally, future perspectives for using FCA as an encompassing tool supporting major parts of the software process are discussed. Such an approach might open a new vision on a thorough concept-based software engineering process.

1 Introduction: The Role of Concepts in Software Development

Software Engineering (SE) is the field of Informatics dealing with the analysis, conception, implementation and validation of middle-sized up to large software

B. Ganter et al. (Eds.): Formal Concept Analysis, LNAI 3626, pp. 288–303, 2005.

systems. Such systems are mental artefacts of humans – usually called *software analysts, designers and engineers*. One preferred way of structuring the SE field is along the time axis of the *software development process* resulting in *process sections* or *phases*. In our own EOS model, we argue that any kind of software development process can be structured by four *phases* (cf. fig. 1 and [15]):

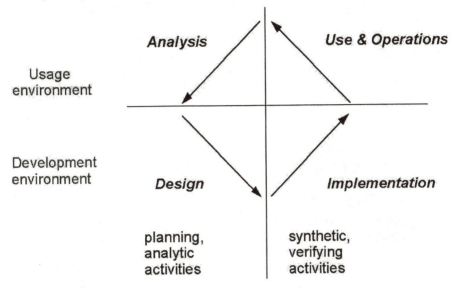

Fig. 1. Phases of a software development process

Analysis regarding the requirements on the piece of software to be developed, including the analysis and modelling of its application domain;

Design of a given piece of software including its specification and construction;

Implementation of the building block under consideration – including its function test;

Use and Operations of the building block under consideration, in its simulated or target environment.

The construction of middle-sized to large software systems implies their decomposition into smaller *building blocks* named – according to their size – *subsystems, components, modules* or *classes*. Many authors (including well-experienced practitioners) consider the tasks of analysis, modelling and design the most decisive and challenging ones. In most cases, the structure, the model elements and building blocks of a software system do not "fall from heaven" nor are they a "natural" or "straightforward map" of some piece of "reality" – but they are the result of highly complex conception and communication process involving many stakeholders – not to forget the software owners and users [17].

As various authors (e.g. Booch [4], Martin and Odell [23]) have pointed out, *concepts* play a central role during the whole software development process. If customers start to explain their problems to the software developers or the developers want to present their results to the customers, usually the problem of using different languages arises. The developers have to learn the language of their customers and understand the concepts of the application domain. Databases are designed according to conceptual models of the application domain. Consequently, many authors (including the cited ones) start with (application domain specific) 'concepts' when they describe the analysis and design phases of OO software development. Later on in the software life cycle, more software-specific concepts like *system, component, module, program, procedure, test case, test object, subsystem* etc. come into play. Data dictionaries, class libraries and software repositories are built and structured along concepts and searched for concepts – in short: concepts are (almost) everywhere around in the SE landscape (cf. fig. 2)!

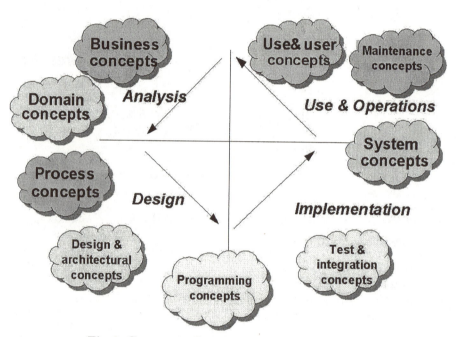

Fig. 2. Concepts in the software development process

Recently, *ontologies* have been advocated as a means for gathering and formalising application domain knowledge in order to make it available for human analysts as well as for automated knowledge processors. According to T. Gruber's definition, an ontology is "an explicit specification of a shared conceptualisation" [14]. Again, concepts are the key anchor point for understanding knowledge domains and building application software systems for them.

All this may be sufficiently motivating to focus on concepts during software development and to use formal methods and tools for supporting this approach. *Formal Concept Analysis* (FCA) is a method capable of uncovering and analysing the conceptual structure of any arbitrary application domain. In particular, the field of *already existing* software is such a domain offering manifold kinds of analyses for various purposes ranging from software understanding, code analysis, quality improvement and modularisation of legacy code to software maintenance, component retrieval and reuse. Another group of applications deal with software modelling, in particular with existing OO models and an analysis of their class hierarchies. These domains have been investigated by several authors, in particular by the research groups of R. Godin and G. Snelting. Recently, good surveys of this work have been presented by the aforementioned authors and by P. Eklund and his group (cf. sections 4 and 5).

In general, software modelling has a Janus-like, double-faced character: Models are either abstractions of some already existing original (*descriptive models*) or of some future "original" to be constructed (*prescriptive models*) [22, 31]. Accordingly, another important (but so far much less investigated) application area of concept analysis is the domain of *prescriptive software models*, i.e. those prescribing the structural units and dependencies of the software to *be developed*.

In OO modelling, the outstanding structural units are classes, objects and their associations. FCA can be used to analyse software requirements (e.g. in the form of use case descriptions) and to derive and compare possible class and object structures which are meant to reflect these requirements. This is the starting point of the BASE approach of Düwel and Hesse [7, 8] and a few related approaches which will be sketched in the following two sections.

2   The Early Phases: Requirements Engineering and Use Case Analysis

Every software project starts with *requirements*: These are the demands, wishes, vague ideas of the people who order (and normally pay for) a piece of software – thereafter called the (envisaged) *software system*. The addressees of requirements are the software builders: analysts, designers, implementers, testers etc. Some authors distinguish *system requirements* and *software requirements* (cf. e.g. [3]) where the former encompass the latter and refer to the whole target system including the hardware and the organisational environment for the software to be developed.

The predominant way to formulate requirements is writing natural language (NL) text. In most cases this is the adequate form since finding, formulating and fixing requirements is a social process linking the various people involved. FCA can support this process in a natural way as a document analysis and a question posing and answering tool. This is, for example, the focus of its proposed application in U. Andelfinger's work. In his Ph. D. thesis on "discursive requirements analysis" the author conducted various case studies in order to develop a "theory of discursive activity" [2]. FCA was used to characterise requirements (= formal objects) by categories of relevance or urgency (= formal attributes).

Other forms for requirements specifications comprise dedicated languages like PSL/PSA, tabular representations of requirements glossaries during software "predesign" [24], graphical representations of activities and data flows (like the SADT and data flow diagram techniques) and use case descriptions. In the SE community, the latter have become quite popular through the work of I. Jacobson and the dissemination of the Unified Modelling Language (UML) where use case diagrams and use case descriptions are incorporated [18, 36].

Jacobson and other recent OO methodologies based on UML recommend a *"use casedriven analysis"*: Functional requirements on the envisaged system are to be grouped and formulated as use cases, where a single *use case* is defined as *"the specification of a sequence of actions, including variants, that a system (or other entity) can perform, interacting with actors of the system."* [36]. Thus a use case represents a piece of functionality offered or to be offered by a technical and/or organisational system.

An obvious way to deal with use cases in FCA is to treat them as documents and thus as formal objects while selected keywords are taken as formal attributes in the FCA sense. D. Richards and K. Böttger follow a similar but finer grained approach. Their RECOCASE (RECOnciling CASE) tool considers sentences within use cases as the formal objects. The aim is to support the merging of multiple descriptions of the same use case by different individuals. The sentences are further broken down into word phrases into so called flat logical form and these become the formal attributes. Line diagrams can then be used to explore the overlap between different descriptions of the same use case for the purposes of merging as well as coverage checking in the differences [26].

3   Crossing Perspectives: From Use Cases to Class and Object Structures

3.1   The Use Case / OO Modelling Gap

Use cases are not only favoured as a structuring and documentation principle for software requirements but also as a starting point for *object-oriented analysis and modelling*, i.e. for building *class and object models*. Any such model reflects a particular view on the entities and associations of the application domain of the system to be built. An outstanding characteristic of the OO approach is to specify classes and objects as capsules of data together with their controlling and access operations. Thus the *structural* and *behavioural* aspects of a software unit are already combined in its specification. Another peculiarity of the class and object model is the way it functions as the "backbone" of the software process: It is defined rather early in the analysis phase, refined and enhanced during the design phase, further extended and transformed to running code during the implementation phase and maintained for further alternations, extensions and derivations in the succeeding operating phases or development cycles.

The OO approach has produced considerable improvements of software quality, and, in particular, has led to better testable, modifiable, extensible, maintainable and reusable software. However, the "genesis" of classes and objects, i.e. the way to derive or construct them out of a given requirements specification needs much more clarification. While many authors and methodologies recommend use cases as a necessary prerequisite to find a good class and object structure, almost no guidelines are available on how to bridge this gap in practice. As a representative example, we cite B. Meyer from his famous book *Object oriented software construction* [25]:

> "... object-oriented design is a natural approach: the world being modeled is made of objects – sensors, devices, airplanes, employees, paychecks, tax returns – and it is appropriate to organize the model around computer representations of these objects. This is why object-oriented designers usually do not spend their time in academic discussions of methods to find the objects: in the physical or abstract reality being modelled, **the objects are there just for the picking!** The software objects will simply reflect these external objects."

In his *Objectory* method, I. Jacobson recommends that designers build and maintain object lists during use case analysis and then to take the members of these lists as possible class and object candidates. Again, apart from appealing to intuition, experience and good modelling skills there is virtually no hint as to what qualifies an object list member for a class or object in the OO model nor any guideline for identifying or constructing them. Moreover, the decision whether to treat such a member as a class, as an object, as an attribute or anyhow is more or less left to the intuition of the designer.

3.2   FCA Used to Bridge the Gap

To support software analysts and designers in this difficult and decisive phase of their work by FCA is the central goal of the BASE system developed by St. Düwel at the University of Marburg [6]. BASE stands (in German) for *concept-based analysis during software development*. It starts with a requirements specification consisting of use cases and takes these as formal attributes in a formal context of FCA. All relevant "things" occurring in the use case descriptions are taken as formal objects in the FCA sense. This way, a concept lattice is generated which allows a *"crossing of perspectives"* combining the functional view (represented by the use cases) and the data view (represented by the "things") on a software application domain (cf. fig. 3). Each formal concept comprises data elements (its extent) together with corresponding functional elements (its intent) – this is exactly what we expect from classes and objects in OO design. Thus the lattice diagram can be used to visualise data/function dependencies and to identify class candidates for an OO model.

As is clear from its definition, FCA is a dualistic theory where formal objects and formals attributes are exchangeable. For example, use cases might also have been identified with formal objects and the occurring items with formal attributes – a quite "natural" approach which in fact had been considered for a while as a serious alternative. Two arguments have led to the eventual decision documented in this article:

(a) to identify formal "objects" with "things" in the use cases seemed to be obvious from a linguistic point of view,

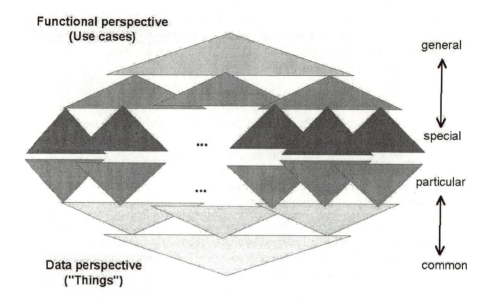

Fig. 3. Crossing of perspectives in OO modelling via FCA

(b) the resulting line diagram (in the shape of fig. 3) resembles the typical layer structure of many software architecture diagrams: Upper layers represent functional components while the lower ones stand for common services often associated with data clusters.

Some practice and experiences with this approach show that one cannot expect that all nodes of the resulting concept lattice will automatically correspond to classes in the resulting OO model. However, the approach reveals its benefits while working *interactively* with a given set of use cases and an evolving class and object structure:

- In a first step, the requirements specification is (re-)structured und use cases are (re-)formulated (if not already given).
- The use cases are examined and involved "things" are marked. Thus the first formal context is built – with the "things" as formal objects and the use cases as formal attributes.
- The corresponding concept lattice is generated and examined. Use cases should show up higher in the diagram the more general they are (i.e. the more "things" they involve), whereas "things" should show up lower in the diagram the more common they are (i.e. the more use cases they are involved in).
- The nodes of the lattice are checked for their suitability as good class candidates. "Things" in their extent are candidates for OO attributes while use cases (or their parts) in their intent are candidates for OO operations.
- Discussion of the lattice structure, its labelling, the class candidates with their possible attributes and operations leads to questions whether cer-

tain "things" should be explicitly mentioned in particular use cases or not, whether formal concepts represent good data capsules in the OO sense etc.
- As a result of the discussion, use cases are reformulated, the marking of "things" is reworked and a new formal context is built leading to a new lattice diagram.
- This process is iterated until a stable situation is achieved and those nodes which represent appropriate data capsules are taken as a basis for an OO class model.

3.3   A Brief Example

In order to briefly illustrate the approach, the business of a wine trading company called JWI is considered as an example. A more detailed version of this example can be found in [7]. Originally, the business has been described in the form of 28 statements. These have being grouped and (re-) reformulated resulting in 8 use cases. Two of them look as follows:

Use case "Receive order"

- The centre receives *orders* from *customers* by phone from 9:00 a.m. to 5:00 p.m. A received order is recorded on a form. An order may consist of many *detailed items*. Detailed items refer to single *products*. Each detailed item is recorded in a line of the form.

Use case "Create delivery instructions"

- The centre produces a *delivery instruction* ticket for each *delivery truck* by gathering the *detailed ordered items* in the *'assigned ordered items'* file, considering the *destinations* and the total amount of the orders for each item.

The example shows that certain decisions – e.g. separation of things, normalisation of flexion forms, leaving out irrelevant things etc. – have already been taken by the analyst. Of course, the marking of things might be more automated leaving these decision steps for the first iteration cycle done by the analyst.

The resulting formal context is depicted in fig. 4. The corresponding line diagram (concept lattice) is shown in fig. 5. It shows the dependencies between the data and functional view of our domain. Looking at the diagram down from the top element we can follow the refinement of the use cases representing the system functionality. Considering the diagram from the bottom element yields the data view.

"Things" that show up in a low position in the resulting line diagram are relevant for many use cases. These are "first class" class candidates. In our case the preferred class candidates of the JWI system *Detailed ordered item* and *Product* are in the lowest positions of the diagram. All use cases use at least one of them. Use cases occurring in their neighbourhood are the first candidates for class operations. This is the case for *determine inventory stock* and *Define minimal and maximal stock quantity*. A similar argument applies to things occurring in

	Receive order	Process order	Order missing products	Determine inventory stock	Create delivery instructions	Define maximal and minimal stock quantity	Process incoming deliveries	Process delivery results
Customer order	×	×						
Customer	×							
Detailed order item	×	×			×		×	×
Product	×	×	×	×		×	×	×
Stock quantity		×	×	×			×	
Assigned orders item file		×			×		×	×
Waiting order items file		×					×	
Missing quantity			×					
Order to supplier			×					
Supplier			×				×	
Minimal stock quantity			×			×		
Maximal stock quantity			×			×		
Delivery instruction					×			×
Delivery truck					×			
Destination					×			
Delivery result								×

Fig. 4. Formal context of use cases

the neighbourhood: At least if we assume that attribute candidates are always mentioned in the use cases together with the class candidates they belong to, we can expect them in their upper neighbourhood. This is the case for the attribute candidates *Stock quantity, Minimal* and *Maximal stock quantity* of the class candidate *Product*.

Of course, it is not sufficient to examine just the lowest level of the line diagram. For example, class candidates that are only interesting for single use cases show up at the same node marked by the use case. Intermediate nodes (including those without any labels like the one above *Detailed ordered item* and *Product*) are other interesting class candidates.

The lattice diagram can also successfully be used as a tool for checking use case descriptions for completeness and for inherent (i.e. not explicitly stated) assumptions. For example, *Customer* shows up higher in the diagram than *Customer order* – in contrast to our intuition. The reason is an implicit assumption of the *Customer* being involved in the use case *Process order*. The use case can easily be corrected from this observation in the line diagram.

3.4 Further Examinations Using Implications and Block Relations

Another sort of examination offered by the BASE tool is based on the idea of implications. It addresses the granularity of use cases and the resulting OO

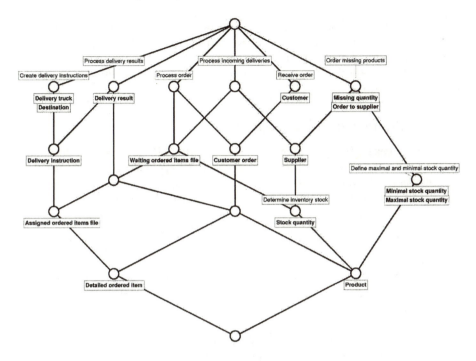

Fig. 5. Line diagram for the Wine trading company example

operations. In our application of FCA, an *implication* $A \rightarrow B$ with two sets A, B of "things" (= formal objects) holds if all operations which use all "things" in A also use all "things" in B. A basis of such implications can be computed. The implications of this basis are presented to the analysts in the form of questions and they have to decide if this implication really holds within the application domain. If not, the BASE tool suggests to refine the use case descriptions by introducing an additional operation which separates the "things" of A and B.

For example in fig. 4 the implication *Delivery instruction* $\rightarrow$ *Assigned ordered items file* holds. Examination by the analysts shows that this implication does not correctly reflect the "real world" situation. Therefore, the use case *Create delivery instructions* is refined leading to three sub-use cases *Create delivery instruction, Insert detailed ordered item,* and *Attach delivery instructions*. By this refinement the "unnatural" implication is removed and three potential class operations have been found [7].

Another sort of examinations is based on the idea of *block relations* first published by C. Lindig and G. Snelting [21]). In the BASE application of FCA, the incidence relation of a formal context reflects which data is used in which operation. Lindig and Snelting already examined this situation while looking for a modular structure of existing systems As module candidates they considered not only formal concepts, but also so-called *"blocks"*. A block can be generated from a pair (A,B) of formal object/formal attribute sets with incomplete incidence relation by filling in all the missing incidences. Similarly to Lindig and Snelting's approach this kind of analysis can be used to obtain suggestions for a modularisation of the system, i.e. its possible decomposition into components and modules. More details and an example of this application of FCA can be found in [7] and in [8].

3.5   Attribute and Object Exploration

In a recent case study, T. Tilley, W. Hesse and R. Duke have considered the example of a mass transit ticketing system and have shown how the use case description can be refined and completed through several iteration cycles [35]. In this paper, the FCA-based class structure is compared to an alternative version based on Object-Z and the resulting similarities and divergences are discussed. While the similarities confirm the plausibility of the whole approach, it also has its limitations:

Firstly, how do we know when to stop iterating? The iterations during analysis result from questions about the expected position of objects and attributes in the line diagram. As a consequence of discussion between the analysts and the system owners and/or users the formal context is modified and the new line diagram examined. This could be seen as an ad hoc and informal form of *attribute exploration* applied to the objects [32]. An alternative would be to formally apply *object exploration* interactively until all the valid object implications in the context have been explored. This represents a continuation of the implication approach described above. However, such an approach might affect the readability of growing line diagrams and might become unmanageable for large examples.

4   Analysis and Structuring of Class Hierarchies

The sub/superconcept ordering inherent in a concept lattice means that class hierarchy analysis and (re-)structuring are obvious applications for FCA. This domain has been investigated by several authors, and in particular by the groups of R. Godin and G. Snelting (cf. e.g. [11, 21, 30]). A number of the approaches are outlined below.

Godin et al. consider a context where the formal objects are messages (methods in the Smalltalk programming language) and the formal attributes are classes [10, 13]. Concepts that have an empty attribute contingent, i.e. those not labeled by a class, are considered as new class candidates. In later work Godin et al. further incorporate static call graph information into the concept lattice where they consider the classes as formal objects and attributes and operations as the formal attributes [11].

Snelting and Tip also present a mechanism to reorganise class hierarchies based upon FCA. In their analysis of programs written in the C++ programming language, variables are taken as formal objects [28, 30]. The methods and fields of the objects to which the variables refer are then taken as formal attributes. If a variable is used to access a method or field then they are also associated in the formal context. Rules are employed to account for issues like assignment between variables and also to conservatively account for dynamic dispatch. The focus of the analysis is the objects that are created during an actual program run and how these are accessed via program variables.

In the work of Schupp et al. class hierarchies in the C++ standard template library (STL) are analysed using FCA [27]. They consider classes as formal objects and properties of the classes listed in the library documentation as formal attributes. To describe class libraries they introduce the notions of *"well abstracting"*, *"lacking orthogonality"* and *"lacking refinement"*. These notions are used to categorise libraries based upon the overall structure of the lattice. However, rather than inspecting various aspects of the structure using subcontexts or conceptual scaling, they attempt to construct and render the whole concept lattice from which they then draw their conclusions.

In addition to the applications outlined above there is also a related body of literature describing the application of FCA to the creation and maintenance of class hierarchies in databases and knowledge bases, for example, the work of Yahia et al. [38] and Godin, Mineau and Missaoui [12]. While databases form the backbone of many CASE tools these papers are beyond the scope of this chapter and are not discussed here. More recently, however, Godin and Valtchev have presented an overview of approaches for class hierarchy design in object-oriented software development [13]. Their contribution appears as a separate chapter in this volume.

Beyond the identification, analysis and structuring of class hierarchies, FCA has also found other applications in SE and the surveys by Snelting [28] and Tilley et al. [34] in this volume provide more general overviews. Across the literature describing the application of FCA to SE there is a wide variety of choices for both formal objects and formal attributes. Table 1 presents a summary of the approach described in this chapter along with a number of selected approaches. In addition to describing the choice of objects and attributes, the aim of the approach and an interpretation of what the ordering and concepts represent within the approach are also presented.

5   Conclusions and Outlook

The preceding section and the summarising table have shown that FCA has received increasing attention in the SE field during the last decade. While program analysis and searching software libraries belong to the "classical" applications of conceptual analysis in this field, there are new and evolving application areas in the fields of Requirements Engineering and of Object Oriented Analysis and Design (OOAD).

One of the principal ideas of OOAD is to design software systems along the concepts of their application domain and maintain, refine and extend these concepts through the whole development process. In this sense, OO software development is a particular form of *concept-based software development* (CBSD) which might be characterised by the following steps:

– During the requirements analysis and elicitation stage key concepts of the application domain are identified and formally defined.

Table 1. Table of selected approaches

Goals and aims	Formal objects	Formal attributes	Formal concepts	Meaning of order relation	References
Finding class candidates	"Things" relevant for use cases	Use cases	Class candidates	Specialisation of functionality	Düwel and Hesse [6, 7]
Analysis of class structure	Classes	Attributes and operations	Abstract classes	Generalisation	Godin et al. [11]
Analysis of class structure	Program variables	Attributes and operations	Classes	Generalisation	Snelting and Tip [30]
Merging use case descriptions	Sentences	Word phrases	Overlap in descriptions	Description merging	Richards et al. [26]
Analysis of Software Structure	Packages, classes, methods, or attributes	Packages, classes, methods, or attributes	Aspect similarity	Generalisation	Cole and Tilley [5]
Visualise Specification Structure	Specification Schemas	Markup elements or schemas	Similarity between schemas	Specialisation or schema composition	Tilley [33]
Debugging Temporal Specifications	Execution Traces	Finite Automaton Transitions	Trace Clusters	Trace similarity	Ammons et al. [1]
Modularisation of legacy systems	Program procedures	Program variables	Modules (of maximal cohesion)	Nesting of modules	Lindig and Snelting [21]
Configuration analysis	Code segments	Controlling expressions	Configurations	Specialisation of configurations	Lindig and Snelting [20, 29]
Searching software libraries	Software modules	Keywords	States during search	Specialisation of search results	Lindig [19]
Searching component libraries	Software components	Formal specifications	States during search	Specialisation of components	Fischer [9]
Project control	"Things" relevant for projects	Project activities	States of project progress	Grade of project progress	Vogt [37]

- These key concepts are discussed and re-worked together with systems owners and users, are related to existing ontologies (if any) and are checked for quality criteria such as completeness and consistency.

- The resulting concepts form the basis of OO (class and object) models and the evolving OO system design.
- The concept base is further enhanced by more system-oriented concepts which eventually form the basis for the system implementation. Even tests and integration policies can be defined and built around concepts.

It is quite obvious that FCA has to play a central role in such a CBSD process. Particular analyses may relate requirements to their ingredients, use cases to their involved things, classes to their methods and fields, programs or modules to their functions, procedures and variables, test cases to their data stores and accessing functions, components and subsystems to their contained modules and functionality etc. One or several FCA tool(s) might accompany the developer through the whole process and support him or her through visualisations, search facilities, questioning, and answer evaluation procedures etc.

Such an approach seems to be particularly promising if software development projects are not considered as isolated tasks but as a continuing engineering enterprise dealing with a certain application domain and its organisation and processes. Its concepts form an evolving knowledge base of the domain – its ontology [16]. FCA mechanisms and tools are used for a continuous enhancement and refinement of such a knowledge base. It is still too early to judge the viability of such a vision but the work done so far in the field of applying FCA to software analysis and modelling tasks is a good starting point for this approach.

References

1. G. Ammons, D. Mandelin, R. Bodik, and J.R. Larus. Debugging temporal specifications with concept analysis. In *Proceedings of the Conference on Programming Language Design and Implementation PLDI'03*. ACM, June 2003.
2. U. Andelfinger. *Diskursive Anforderungsanalyse. Ein Beitrag zum Reduktionsproblem bei Systementwicklungen in der Informatik*. Peter Lang, Frankfurt, 1997.
3. B.W. Boehm. Software engineering. *IEEE Transactions on Computers*, C-25(12):1216–1241, 1976.
4. G. Booch. *Object-Oriented Analysis and Design with Applications*. Benjamin/Cummings, 1994.
5. R. Cole and T. Tilley. Conceptual analysis of software structure. In *Proceedings of Fifteenth International Conference on Software Engineering and Knowledge Engineering, SEKE'03*, pages 726–733, USA, June 2003. Knowledge Systems Institute.
6. S. Düwel. *BASE - ein begriffsbasiertes Analyseverfahren für die Software-Entwicklung*. PhD thesis, Philipps-Universität, Marburg, 2000.
7. S. Düwel and W. Hesse. Bridging the gap between use case analysis and class structure design by formal concept analysis. In J. Ebert and U. Frank, Hrsg., *Modelle und Modellierungssprachen in Informatik und Wirtschaftsinformatik. Proceedings "Modellierung 2000"*, pages 27–40, Koblenz, 2000. Fölbach-Verlag.
8. S. Düwel and W. Hesse. BASE - ein begriffsbasiertes Analyseverfahren für die Software-Entwicklung. In K. Lengnink et al., Hrsg., *Mathematik für Menschen, Festschrift für R. Wille*, TU Darmstadt, 2003.
9. B. Fischer. Specification-based browsing of software component libraries. In *Automated Software Engineering*, pages 74–83, 1998.

302   Wolfgang Hesse and Thomas Tilley

10. R. Godin and H. Mili. Building and maintaining analysis-level class hierarchies using galois lattices. In *Proceedings of the OOPSLA'93 Conference on Object-oriented Programming Systems, Languages and Applications*, pages 394–410, 1993.
11. R. Godin, H. Mili, G. W. Mineau, R. Missaoui, A. Arfi, and T.-T. Chau. Design of class hierarchies based on concept (galois) lattices. *Theory and Application of Object Systems (TAPOS)*, 4(2):117–134, 1998.
12. R. Godin, G. Mineau, and R. Missaoui. Incremental structuring of knowledge bases. In *Proceedings of the International Knowledge Retrieval, Use, and Storage for Efficiency Symposium (KRUSE'95)*, LNAI, pages 179–198. Springer-Verlag, 1995.
13. R. Godin and P. Valtchev. Formal Concept Analysis-based class hierarchy design in object-oriented software development. In this volume.
14. T. Gruber. A translation approach to portable ontologies. *Knowledge Acquisition*, 5(2):199–220, 1993.
15. W. Hesse. Theory and practice of the software process - a field study and its implications for project management. In C. Montangero, editor, *Software Process Technology, 5th European Workshop, EWSPT 96*, LNCS 1149, pages 241–256. Springer, 1996.
16. W. Hesse. Ontologie(n). Das aktuelle Schlagwort. *Informatik-Spektrum*, 25(6):477–480, 2002.
17. W. Hesse and H.V. Braun. Wo kommen die Objekte her? Ontologisch-erkenntnistheoretische Zugänge zum Objektbegriff. In K. Bauknecht et al., Hrsg., *Informatik 2001 - Tagungsband der GI/OCG-Jahrestagung*, Bd. II, pages 776–781. Computer-Gesellschaft, books_372ocg.at; Bd. 157, Österr, 2001.
18. I. Jacobson. *Object-Oriented Software Engineering - A Use Case Driven Approach*. Addison-Wesley, revised printing edition, 1993.
19. C. Lindig. Komponentensuche mit Begriffen. In S. Braunschweig, Hrsg., *Proceedings Software-Technik '95*, pages 67–75, Oktober 1995.
20. C. Lindig. Analyse von Softwarevarianten. Technical Report Informatik-Bericht 98-04, Technische Universität Braunschweig, Januar 1998.
21. C. Lindig and G. Snelting. Assessing modular structure of legacy code based on mathematical concept analysis. In *Proceedings of the International Conference on Software Engineering (ICSE 97)*, pages 349–359, Boston, 1997.
22. J. Ludewig. Models in software engineering - an introduction. *Software and Systems Modelling*, 2(1), March 2003.
23. J. Martin and J. Odell. *Object-Oriented Analysis and Design*. Prentice Hall, 1992.
24. H.C. Mayr and Ch. Kop. A user centered approach to requirements modeling. In M. Glinz and G. Müller-Luschnat, Hrsg., *Modellierung 2002 - Model-lierung in der Praxis - Modellierung für die Praxis*, LNI P-12, pages 75–86. Springer, 2003.
25. B. Meyer. *Object oriented software construction*. Prentice Hall, 1988.
26. D. Richards, K. Boettger, and O. Aguilera. A controlled language to assist conversion of use case descriptions into concept lattices. In *Proceedings of 15th Australian Joint Conference on Artificial Intelligence*, 2002.
27. S. Schupp, M. Krishnamoorthy, M. Zalewski, and J. Kilbride. The "right" level of abstraction - assessing reusable software with formal concept analysis. In G. Angelova, D. Corbett, and U. Priss, editors, *Foundations and Applications of Conceptual Structures - Contributions to ICCS 2002*, pages 74–91. Bulgarian Academy of Sciences, 2002.
28. G. Snelting. Concept lattices in software analysis. In this volume.
29. G. Snelting. Reengineering of configurations based on mathematical concept analysis. *ACM Transactions on Software Engineering and Methodology*, 5(2):146–189, April 1996.

30. G. Snelting and F. Tip. Understanding class hierarchies using concept analysis. *ACM Transactions on Programming Languages and Systems*, pages 540–582, May 2000.
31. H. Stachowiak. *Allgemeine Modelltheorie*. Springer, Wien, 1973.
32. G. Stumme. Attribute exploration with background implications and exceptions. In H.H. Bock and W. Polasek, editors, *Data Analysis and Information Systems: Statistical and Conceptual approaches, Proceedings of GfKl'95. Studies in Classification, Data Analysis, and Knowledge Organization 7*, pages 457–469, Heidelberg, 1996. Springer.
33. T. Tilley. Towards an FCA based tool for visualising formal specifications. In B. Ganter and A. de Moor, editors, *Using Conceptual Structures: Contributions to ICCS 2003*, pages 227–240. Shaker Verlag, 2003.
34. T. Tilley, R. Cole, P. Becker, and P. Eklund. A survey of formal concept analysis support for software engineering activities. In this volume.
35. T. Tilley, W. Hesse, and R. Duke. A software modelling exercise using FCA. In B. Ganter and A. de Moor, editors, *Using Conceptual Structures: Contributions to ICCS 2003*, pages 213–226. Shaker Verlag, 2003.
36. Unified Modeling Language (UML) 1.5 documentation. OMG documentformal/03-03-01, 2003. as of 18th Aug.
37. F. Vogt. Supporting communication in software engineering: An approach based on formal concept analysis. Technical Report Preprint Nr. 1926, Technische Universität Darmstadt, Fachbereich Mathematik, 1997.
38. A. Yahia, L. Lakhal, J. P. Bordat, and R. Cicchetti. An algorithmic method for building inheritance graphs in object database design. In B. Thalheim, editor, *Proceedings of the 15th International Conference on Conceptual Modeling, ER'96*, volume 1157 of *Lecture Notes in Computer Science*, pages 422–437, Cottbus, Germany, October 1996. Springer.

Formal Concept Analysis-Based Class Hierarchy Design in Object-Oriented Software Development

Robert Godin[1] and Petko Valtchev[2]

[1] Département d'informatique, UQAM, C.P. 8888, succ. "Centre Ville"
Montréal (Qc), Canada, H3C 3P8
[2] DIRO, Université de Montréal, C.P. 6128, Succ. "Centre-Ville"
Montréal, Québec, Canada, H3C 3J7

Abstract. The class hierarchy is an important aspect of object-oriented software development. Design and maintenance of such a hierarchy is a difficult task that is often accomplished without any clear guidance or tool support. Formal concept analysis provides a natural theoretical framework for this problem because it can guarantee maximal factorization while preserving specialization relationships. The framework can be useful for several software development scenarios within the class hierarchy life-cycle such as design from scratch using a set of class specifications, or a set of object examples, refactoring/reengineering from existing object code or from the observation of the actual use of the classes in applications and hierarchy evolution by incrementally adding new classes. The framework can take into account different levels of specification details and suggests a number of well-defined alternative designs. These alternatives can be viewed as normal forms for class hierarchies where each normal form addresses particular design goals. An overview of work in the area is presented by highlighting the formal concept analysis notions that are involved. One particularly difficult problem arises when taking associations between classes into account. Basic scaling has to be extended because the scales used for building the concept lattice are dependent on it. An approach is needed to treat this circularity in a well-defined manner. Possible solutions are discussed.

1 Introduction

An important part of object software development is the class hierarchy. The design and maintenance of such a hierarchy has been recognized as a difficult problem [1, 29]. The difficulty increases with the size of the hierarchy and the possible evolution of the software requirements that may require the incorporation of modifications in the hierarchy.

A large body of work has focused on problems related to hierarchy construction and reconstruction. Various development scenarios have been addressed (see [11]), such as:

- Building the hierarchy from scratch using:
  - objects [24],
  - class specifications [8, 12],
- Evolution of the class hierarchy to accommodate new requirements:
  - unconstrained class addition [8, 12],
  - addition constrained by backward compatibility with a previous hierarchy [28] or existing objects [16],

B. Ganter et al. (Eds.): Formal Concept Analysis, LNAI 3626, pp. 304–323, 2005.

- Reengineering of an existing class hierarchy:
  - from the relationship between classes and their attributes/methods [2, 4],
  - using code analysis tools [7, 15],
  - by applying refactorings [9, 27],
  - from UML models including associations [19],
  - from access patterns in applications [32],
  - prompted by detecting defects using software metrics [30],
- Reengineering procedural code into an object environment [31, 34],
- Merging existing hierarchies [33].

In many cases, the proposed approaches rely on algorithms that are not grounded on well-established theoretical results. Thus, the corresponding methods may yield unpredictable results. In some cases, the exact form of hierarchies depend on adjustable parameters of the procedures. In contrast, Formal Concept Analysis (FCA) provides a natural theoretical framework for class hierarchy design and maintenance and several researchers have adopted this framework ([7, 12, 17, 27, 32, 36]). Hierarchies produced within this framework have a well-defined semantics that remains independent from the concrete algorithms used. In addition, the produced hierarchies tend to conform to general quality criteria such as *simplicity, comprehensibility, reusability, extensibility* and *maintainability*.

These high-level criteria represent desirable features of the final result that very much depend on its usage during further stages of the software process. However, these high-level criteria are knowingly favored by two more concrete quality criteria that may be measured directly on the target software artifacts:

1. *Minimizing redundancy.* Having each artefact defined in one single place in the code/specifications is a well-known software design principle that a class hierarchy should promote [9, 20, 21]. In contrast, keeping several definitions of the same artefact at possibly different locations may lead to inconsistencies between copies. Moreover, redundancy increases the complexity of the resulting software and, more dramatically, speaks about possible flaws in the design since repeating code/specification chunks is a hint that these have not given rise to the appropriate abstractions that help embed them into a single software unit. Besides, lessons from building large class libraries [26] show that it is hard to identify good abstractions a priori and it is often necessary to reorganize a library to reflect the undetected commonalities.

2. *Subclasses as specializations.* Inheritance hierarchies are sometimes created for code reuse purposes, especially those in code libraries. Thus, the inheritance between classes in the hierarchy my not correspond to any particular reality in the corresponding domain but rather help optimize code sharing in the hierarchy. However, as observed by [5], in the long run such a designing free of semantic concerns may produce libraries that are difficult to understand and hence to reuse. Therefore, many authors have advocated the enforcement of consistency with specialization in inheritance hierarchies ([2–4, 20, 22, 25]) in particular, in order to achieve better comprehensibility and reusability.

Hierarchies produced by methods based on FCA are guaranteed to meet these criteria. Depending on the design goals and available specifications, several alternative hierarchy types may be considered within this framework. These hierarchies can be viewed as ideal structures similar to relational database normal forms, with each normal form addressing a particular design goal.

In the following, a set of normal forms for class hierarchy design is described, all of them based on the FCA framework. These normal forms synthesize the previous propositions in a unified framework. Section 2 introduces the basic idea by defining the attribute factored lattice form and relating it to a concept lattice. Section 3 introduces the more compact attribute factored subhierarchy form which is based on the set of object and attribute concepts of the concept lattice. Section 4 proposes normal forms for factoring methods taking into account the distinction between signature and body and the possibility of method redefinitions. Section 5 discusses the factoring of associations and the complications introduced by circular dependencies. Available software tools are listed in Section 6 whereas Section 7 provides an overview of some on-going industrial projects involving FCA and lattices.

2   Attribute Factored Lattice Form

Fundamental constructs of object software are the notions of *object* and *class*. A class is an abstraction for a set of objects that share the same characteristics. In programming languages, these characteristics, also called *members* of the class, are *attributes* and *methods*. In modelling languages, *associations* are also used to relate classes. An *attribute*, also called instance variable or data member, contains data used to model the state of an object. This section is concerned with the attributes of the classes. Methods and associations will be examined in the following sections.

When using formal concept analysis for class hierarchy design, the set of formal objects G is a set of software artefacts, i.e., classes, objects or program variables, which are used as a starting point in the search for a suitable class hierarchy. The set of formal attributes M corresponds to properties of the classes or objects. Relevant properties include attributes (instance variables), methods (body and/or signature of the method) or associations (in the case of classes). Further information may be available such as values of the variables in objects or links to specific objects for associations. In this paper, we only consider the case where the starting point is a set of class specifications, i.e., G is a set of classes. Nevertheless, the principles are directly transposable to the case of example objects or program variables.

First, we consider the case of factoring out attributes of the classes. Let's take a simple example to illustrate the basic idea. Suppose that we have a specification of the attributes for a set of four concrete classes as illustrated in Figure 1. The specification could be interpreted as the exact set of concrete classes that the hierarchy must contain, i.e., these classes will be the only ones to "produce" objects in an application. Other classes of the hierarchy can be used to factor common specifications.

This input specification may be produced in several ways depending on the development scenario. For example, with a forward engineering process, classes and their attributes are first specified in the analysis phase of the process. Thus, they are produced by the analyst and typically expressed by means of a modeling language such as UML (as in Figure 1 on the left). Within a re-engineering process, the classes are already organized in a possibly larger hierarchy, with their respective specifications spread over the entire set of classes in the hierarchy. In this case, the attribute set of each concrete class is compiled from all its super-classes in the hierarchy. The goal of the corresponding reengineering scenario is now to refactor an existing hierarchy, i.e., to suggest a different organization of specifications within a new set of classes while preserving the semantics of the initial hierarchy. The semantics here is limited to the behavior of objects from all concrete classes, an approach that allows the modifications in the already existing source code that uses services from the initial hierarchy to be kept to a minimum. The incidence relation I of the formal context $\mathbb{K}$ representing the set of four classes and their instance variables is shown in Figure 1 on the right. The context is drawn as a cross table with classes identified by integers and the variables by letters.

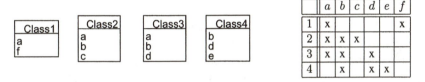

Fig. 1. Left: Example specification; **Right**: Corresponding context.

As the problem is to organize these classes in a hierarchy, a concept lattice is used as a guideline for the design of such a hierarchy (in some sense, it provides an ideal design). To that end, each formal concept is interpreted as a class of the hierarchy. Moreover, the sub-concept relation links are seen as specializations between classes. Figure 1 (on the left) shows the line diagram of the concept lattice with a *reduced labeling* of concepts. The labels assigned to the concepts indicate where, i.e., in which class, a particular attribute should be declared. For example, the attributes a and b will have to be placed at two general classes that are located immediately below the root class of the hierarchy. It is noteworthy that for class hierarchies, the bottom concept is dropped since it is of no use.

Figure 2 shows the *attribute factored lattice form* hierarchy that corresponds to this interpretation of the concept lattice. The four initial classes remain in the hierarchy but there are fewer declared attributes in these classes because of the factoring produced by the concept lattice. New classes (classes 5 through 9) are added that factor out common attributes. These are abstract classes because instances are created only for the four initial classes. The nature of the reduced labeling of the concept lattice guarantees that each attribute appears exactly once in the hierarchy. Object attributes in the initial concrete classes remain unchanged. However, part of them are now inherited from some new classes. Globally, all subclasses are specializations since they inherit the attributes of parent classes with no exception. There are no cancellations.

From a client point of view, using this hierarchy will produce the same effect as using the initial four classes. Therefore the generated hierarchy can be interpreted as a refactoring of the initial four class specifications.

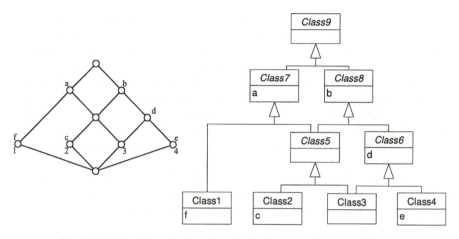

Fig. 2. Attribute factored lattice form for the input specification of Figure 1.

There is a large number of possible designs that can minimize redundancy. The concept lattice attains this goal minimizing the number of classes and the amount of multiple inheritance, which is often considered as undesirable since more complex. This is achieved by grouping attributes in classes whenever possible, as illustrated by the following example. Figure 3 shows two input classes. The attribute factored lattice form that appears in Figure 4 on the left factors out the common attributes a and b in the new Class3. The design presented in Figure 4 on the right also factors out the common attributes but is unnecessarily complex since it contains two classes, one for each attribute, thus capturing classes 1 and 2 in a multiple inheritance pattern. In contrast, the design in Figure 4 on the left is simpler while still providing the same quality criteria of redundancy avoidance and conformance to specialization.

Class1	Class2
b	a
a	b
c	d

Fig. 3. Input specification.

An important hypothesis underlying the approach is that identical attribute names identify properties that can be matched from a semantic point of view. If the matching is based on attribute names, care should be taken to identify semantic commonalities and rename attributes as necessary.

3 Attribute Factored Subhierarchy Form

The concept lattice is an exhaustive representation of commonalities among a set of concrete classes. As its size could grow rapidly, one may think of skipping some of its

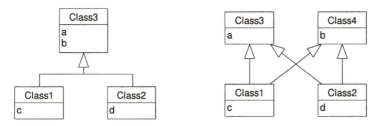

Fig. 4. Attribute factored lattice form for the input specification of Figure 3 and alternate factoring.

nodes to keep the whole structure manageable. Thus, a first idea could be to remove abstract classes that declare no properties. These classes, often called *empty classes*, can be removed without violating the formal quality criteria, i.e., no redundancy and specialization[1]. In the example in Figure 2 on the right, the empty classes, Class5 and Class9, could be omitted (see Figure 2). Even though Class3 declares no attributes, it has to be kept because it is not abstract.

The structure that occurs after the removal of all empty classes, called Galois subhierarchy in [7], corresponds to the set of all *attribute* and *object concepts* of the concept lattice. We recall that given an attribute, its *attribute concept* is the maximal concept in whose intent the attribute appears. Intuitively, the attribute concept of an attribute a is labeled by "a" in the diagram with reduced labeling. The notion of *object concept* is dual, i.e., object concepts have at least one object label. When re-engineering a class hierarchy within the FCA framework, the set of attribute and object concepts constitutes the minimal part of the concept lattice that should be preserved in order to satisfy both concrete formal quality criteria while respecting the initial class specification. In fact, object concepts have to be kept because they correspond to the concrete classes that are used by client code, in particular because they are the classes that can be instantiated[2]. Attribute concepts are in turn necessary because they correspond to classes that declare attributes which are further inherited by their subclasses.

The class hierarchy produced from the Galois subhierarchy constitutes what we call the *attribute factored subhierarchy form*. The class hierarchy for our example is shown in Figure 5 on the right.

Compared to the lattice, the resulting structure has fewer classes while still preserving the quality criteria. Between the lattice form and subhierarchy form, many alternative designs can be produced by selectively keeping subsets of the empty classes. These intermediate designs also preserve the quality criteria. There are no clear cut obvious formal criteria for assessing the usefulness of the empty classes. However, there are some cases where they clearly have value. For example, when a large set of classes inherit from another set as in the subhierarchy form in Figure 6 on the right, the intermediary empty class of the lattice form simplifies the design in the sense that it reduces the

---

[1] Here "formal" is used in the sense of measurable, as opposed to "informal" quality criteria, e.g., comprehensibility, which are hard to measure

[2] For environments that support multi-instanciation, the concrete classes could also be omitted

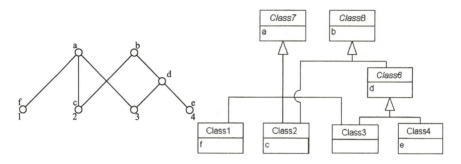

Fig. 5. Galois subhierarchy for the context in Figure 1 and corresponding attribute factored subhierarchy form.

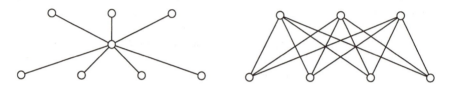

Fig. 6. Lattice and subhierarchy forms.

number of multiple inheritance situations. Indeed, multiple inheritance is knowingly more complex and is not necessarily allowed by all object programming languages. Therefore, it may sometimes be undesirable in the class hierarchy design. To return to our example in Figure 6, in the worst case, the removal of an empty class results in all possible direct links between parent and child classes. Therefore, $n + m$ specialization relationships in the lattice, with n and m being the number of parent/child classes, respectively, might be replaced by nm such relationships in the subhierarchy.

The above situation is an extreme case and need not to occur each time. When n and m are small, the value of the empty class is less obvious. Moreover, the value nm is an upper bound for the effective number of links that need to be created. In many cases a pair of parent and child classes will not create a new link since both classes are already linked through an alternative path of links in the hierarchy. For example, the removal of the empty Class5 in the lattice form of Figure 2 does not require a new link between its parent Class3 and its child Class8 in the subhierarchy form of Figure 2 because attribute b of Class3 is also inherited from the path going through Class6. Some work has been done on guiding the choice of empty classes using class hierarchy metrics [14].

An important feature of the above hierarchical normal forms is that usually they produce multiple inheritance. However, this does not automatically mean that the approach is useless for single inheritance environments. In fact, the normal forms represent an ideal structure that can be used as a starting point from which a good single inheritance hierarchy can be extracted. For example, to reduce multiple inheritance to a single one, a common practice is to choose a single parent class to keep in the hierarchy while replacing the remaining links by delegation references. A less elegant, but sometimes unavoidable, strategy for eliminating multiple parents is to duplicate locally the information that is inherited from the disconnected parents.

4   Method Factored Forms

Another important part of class hierarchies in OO development are the behavioral specifications incorporated in the class descriptions in the form of methods. Method specifications may be divided into two parts. Method signatures specify the way a method is invoked (name, parameters, return type), while the actual processing carried out by a method is specified by its body. The entire set of method signatures for a given class, also called its *interface*, is usually considered as its contract, i.e., the set of services the class must offer. If we want simply to factor out the method bodies, we can use the same approach as for attributes.

Figure 7 shows on the left an example of input specifications for five classes and the methods they support. Objects of Class3 need to "respond" to calls invoking methods b1() and c1(). Here, methods whose names begin with the same letter, i.e., a, b, or c, share the same signature while implementing it in different bodies (but we shall ignore this for now and come back to it later in the text). Figure 7 shows the corresponding context on the right.

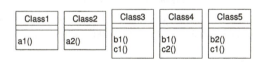

		$a1$	$a2$	$b1$	$b2$	$c1$	$c2$
1		x					
2			x				
3				x		x	
4			x				x
5					x	x	

Fig. 7. Example specification and corresponding context.

As for the attributes case, the concept lattice of this context reveals an organization of the class hierarchy that guarantees no redundancy and conformance to specialization. The reduced labeling in Figure 8 on the left indicates the location in the hierarchy where each method body should be specified. Shown in Figure 8 on the right is the corresponding class hierarchy called *method body factored lattice* form where each concept is interpreted as a class. Here again, when omitting the empty classes, we obtain the *method body factored subhierarchy* form. In our example, the difference with the lattice form is made up of the abstract Class8 which is skipped in the subhierarchy.

With method factoring that distinguishes method signatures from their bodies, the computation gets more complex. In our example, methods sharing the same letter in their names represent different method bodies, i.e., implementations, of the same method signature. For instance, methods a1() and a2() represent two different implementations of the a() signature. Such a distinction is sensible here since in many object-oriented development environments and languages, the signature and body of a

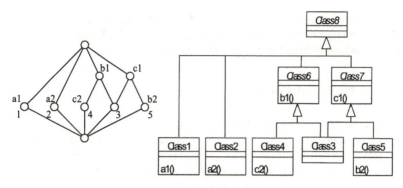

Fig. 8. Left: Reduced labeling of the concept lattice for the context in Figure 7; **Right:** Method body factored lattice form of the hierarchy.

method can be declared separately and there may be more than one body for the same signature. For example, in Java, a method can be declared as a mere signature (an *abstract* method) while leaving one or more implementations as a responsibility for the subclasses that not only inherit the signature but also need to effectively carry-out the specified work. Under this circumstance, it becomes necessary to determine the class where each aspect (signature and body) will be declared in the hierarchy.

The appropriate FCA constructs that help formalize the factoring of methods including their signatures are many-valued contexts and conceptual scaling. Thus, for each method signature m, we define a many-valued attribute m. In our example, there are three many-valued attributes for the three signatures a, b and c (see Figure 9 on the left). The values of a many-valued attribute are the method body names. The values a1 and a2 represent method bodies for a.

	a	b	c
1	$a1$		
2	$a2$		
3		$b1$	$c1$
4		$b1$	$c2$
5		$b2$	$c1$

$\mathbb{S}_a$	a	$a1$	$a2$
a	x		
$a1$	x	x	
$a2$	x		x

Fig. 9. Left: Many-valued context representing the shared method signatures; **Right:** Scale $\mathbb{S}_a$ for the a method signature.

Figure 9 on the right, shows the scale $\mathbb{S}_a$ for the many-valued attribute a. The corresponding concept lattices for each scale are illustrated in Figure 10. The special value a in the scale for the multi-valued attribute of the same name represents the declaration of the method signature. The values for the different bodies are children nodes of this special value in the scale.

The general case of method m ( ) with bodies m1 ( ), m2 ( ), . . ., mn ( ) is illustrated in Figure 11. The scale $\mathbb{S}_m$ is built by adding attribute m to a nominal scale for scale

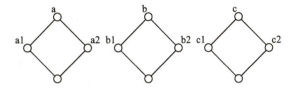

Fig. 10. Concept lattices of the scales for the three method signatures, a, b and c.

$\mathbb{S}_m$	m	$m1$	$m2$	...	mn
m	x			...	
$m1$	x	x		...	
$m2$	x		x	...	
...	...	...	...	...	...
mn	x			...	x

Fig. 11. Left: Scale $\mathbb{S}_m$ for the general case; **Right:** Concept lattice for the scale $\mathbb{S}_m$.

	$\mathbb{S}_a$			$\mathbb{S}_b$			$\mathbb{S}_c$		
	a	$a1$	$a2$	b	$b1$	$b2$	c	$c1$	$c2$
1	x	x							
2	x		x						
3				x	x		x	x	
4				x	x		x		x
5				x		x	x	x	

Fig. 12. One-valued context derived from the many-valued context in Figure 9 after scaling.

objects m1, m2, . . . , mn. Attribute m is assigned to all scale objects in order to represent the fact that every method body mi () implements signature m (). Later, we will show how more general scales are used when taking method redefinitions into account.

The one-valued context derived from the scaling for method signatures with our example appears in Figure 12.

The reduced labeling of the concept lattice (see Figure 9 on the left) produced from the derived one-valued context shows where each method signature and body parts should be declared. The corresponding class hierarchy called *method signature/body factored lattice* form is illustrated in Figure 14 on the right. The notation m () in the UML diagram represents the declaration of the signature while mn () represents the declaration of a method body corresponding to the m () signature. Here again, the class hierarchy is guaranteed to have no redundancy because each method signature and body is declared exactly once. Once again, the resulting class hierarchy conforms to specialization. As previously, in the method signature/body factored subhierarchy form, all empty classes are dropped out.

Finally, we consider the method factoring problem in its most general settings, i.e., when overriding, or redefinition of inherited methods, is allowed. For example, in most

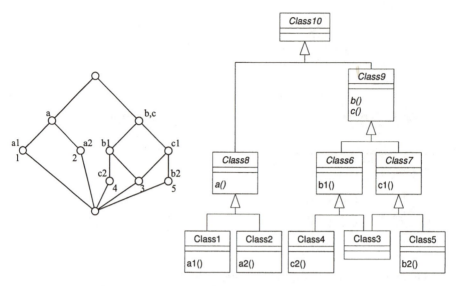

Fig. 13. Concept lattice of the derived one-valued context of Figure 12 and its method signature/body factored lattice form.

object languages, a class `Class1` which inherits a method `a2()` with a signature `a()` from a class `Class2` could still declare its own method `a1()` of the same signature `a()`. In this case, `a1()` overrides `a2()` in the sense that only `a1()` is directly available for objects from `Class1`. More precisely, the invocation of the signature `a()` on an object of `Class1` will in fact call method body `a1()` while `a2()` remains hidden for objects of `Class1`. Such redefinitions should conform to specialization. More sophisticated redefinitions can also be supported where the signatures are not identical. Notice that our framework is orthogonal to the type of redefinitions that are supported (e.g. *covariant* or *contravariant*).

As an illustration, suppose that in the previous example, signature b is a specialization of signature a and method body b2 is a specialization of b1. If the development environment supports redefinitions, this knowledge can be used to produce a hierarchy with finer factoring. For this purpose, we simply use an enhanced scale taking into account the relationships induced by the specialization order among method signatures and bodies. For our example, the relationships between the methods for the a and b signatures are represented in the scale of Figure 14.

In addition, the scale of Figure 15 reflects the fact that method `c2()` is a specialization of method `c1()`. The graphical representation of the concept lattices of these scales is given on the right part of both Figures 14 and 15.

The concept lattice produced by applying these scales and the corresponding method redefinition lattice form class hierarchy are shown in Figure 16. The resulting hierarchy reveals where each method signature and body should be declared without redundancy and in conformance with specialization. Again, one could consider omitting empty classes.

$\mathbb{S}_{ab}$	a	$a1$	$a2$	b	$b1$	$b2$
a	X					
$a1$	X	X				
$a2$	X		X			
b	X			X		
$b1$	X			X	X	
$b2$	X			X	X	X

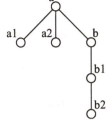

Fig. 14. Scaling and graph, both representing the specialization relationships for a and b.

$\mathbb{S}_c$	c	$c1$	$c2$
c	X		
$c1$	X	X	
$c2$	X	X	X

Fig. 15. Scaling and graph, both representing the specialization relationships for c.

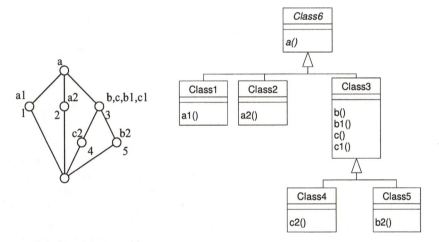

Fig. 16. Concept lattice after applying the scaling of Figures 14 and 15, and the corresponding method redefinition factored lattice form.

When taking specialization relationships between redefined properties into account, absence of redundancy is formalized by the notion of maximal factorization [7].

A class hierarchy is *maximally factorized* iff whenever two properties p_1 and p_2, which are declared by the classes C_1 and C_2, respectively, have a least upper bound in the specialization hierarchy of the properties, say p_3, then there is a common superclass C_3 of C_1 and C_2 declaring p_3. The special case of $p_1 = p_2 = p$, implies that p is declared in C_3.

A hierarchy produced from scaling based on the order relationships between properties is guaranteed to be maximally factorized [7]. For example, observe that a1() and b1() have a() as an upper bound in the scale. In the class hierarchy, the signature a() is declared in Class6 which is a superclass of the classes where a1() and b1() are defined, i.e., Class1 and Class3.

5   Factoring Associations

Many environments for object-oriented analysis and design admit explicit representations of inter-class *associations* and these are an important part of the UML description arsenal. Associations are to be seen as a generic expression of the links that connect individual objects, e.g., kinships, spatial and time relations, part-of relations, etc. Most of the time, they correspond to a classical binary relation between the objects in the extensions of the related classes. For example, the UML model of Figure 17 shows an association between class C1 and class C3 and another one between class C2 and C4. For the purpose of illustration, the classes also have some attributes.

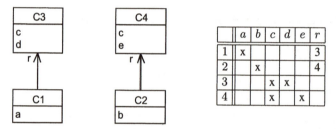

	a	b	c	d	e	r
1	x					3
2		x				4
3			x	x		
4			x		x	

Fig. 17. Example UML model with associations and corresponding context with a formal attribute for the UML association role.

Factoring associations can also be considered in the design process as inheritance spreads over associations too [19]. In fact, the specialization among associations appears naturally as most of the time associations models admit specialization links among associations. Moreover, in UML, associations can have their own properties. However, for the present discussion, we take a simplified model of an association: associations are directed and their only descriptor is a name. Thus, our model corresponds to what is called an *association role* in UML, which allows them to be further assimilated to object attributes whose values are other objects. This is similar in spirit to representing associations by object-valued attributes, or references, in object languages. Thus, a UML association role may be represented in a context by introducing a many-valued attribute named by the corresponding role. Such a transformation yields the context of Figure 17 with the many-valued attribute r. It is noteworthy that the classical object attributes are represented as formal attributes of the context, as before. In contrast, the values of the many-valued attribute r are actually the identifiers of the classes that are pointed to by the association. Thus, the numbers 3 and 4 stand for the classes C3 and C4, respectively. In terms of FCA, this means that the values of the many-valued attribute correspond to formal objects.

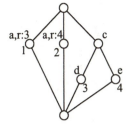

	a	b	c	d	e	$S_r = \mathfrak{J}$ 3	4
1	x				x		
2		x				x	
3			x	x			
4			x	x			

Fig. 18. One-valued context derived by a nominal scale for the association role r and its concept lattice $\mathfrak{B}^0$.

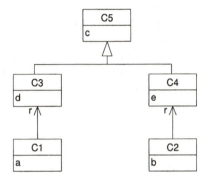

Fig. 19. Class hierarchy from $\mathfrak{B}^0$ after first iteration.

This introduces a circular dependency of the context on itself since the only way of constructing a lattice out of it is to scale the attribute r whose domain is (a part of) the context. Logically, such a scale would require a conceptual structure to be built on top of the context in order for meaningful abstractions to be made available as scaling targets. However, the construction of a meaningful structure is exactly what is the global analysis process is aimed at. In summary, to construct the conceptual hierarchy of the context, there must be another hierarchy on the same context to play the role of a scale and any consistent processing would reasonably require both hierarchies to be identical. As indicated in [35], the resulting apparent deadlock could be successfully resolved by a simple bootstrapping strategy. More precisely speaking, the proposed approach applies an iterative procedure that alternates lattice constructions and scaling. At each iteration, the lattice resulting from the previous iteration is used as a scale that helps enrich the current context and therefore leads to a more precise lattice at the next step. The process halts in a finite number of steps with a lattice which remains stable along two consecutive steps. The iterative approach is illustrated in the following.

In a first iteration, we consider a nominal scale $\mathfrak{J}$ for the r association role. The resulting one-valued context and the concept lattice noted $\mathbb{B}^0$ appear in Figure 18.

Interpreting the concept lattice as a class hierarchy produces the design in Figure 19. The first iteration generates a new superclass C5 for C3 and C4 based on the recognition of their common attribute c. This new class would have been produced by factoring

	a	b	c	d	e	$\mathbb{S}_r$ 3	4	5
1	x					x		x
2		x					x	x
3			x	x				
4			x		x			

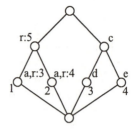

Fig. 20. Second iteration context enriched with the scale derived from the first iteration concept lattice $\mathbb{B}^0$ and its concept lattice $\mathbb{B}^1$.

attributes alone and therefore does not bring value to the classical FCA-based factoring. However, by taking associations into account, we can go further by factoring the association role. Thus, given the r role from C1 to C3, we can infer that there is also an association role r from C1 to C5 because C3 is now a subclass of C5. The same is true for the r role from C2 to C4. This can be taken into account by enriching the first context with a scale that incorporates the superclass relationships discovered in the concept lattice $\mathbb{B}^0$ of the first iteration context. The result is the second context which yields the concept lattice $\mathbb{B}^1$, both shown in Figure 20.

This leads to the discovery of a common association role abstracted in a new concept labeled r:5. This new concept produces the new class C6 in the new hierarchy of Figure 21, which factors the common association role referencing C5. Again, the newly discovered class is used to enrich the second iteration context by incorporating it in the scale. In this way, the third iteration context arises. In our example, the resulting hierarchy is isomorphic to the previous one, thus yielding a fixed point of our iterative process. This constitutes the final design whereby the resulting fixed point is called the *the association factored lattice form*.

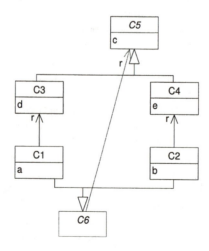

Fig. 21. Second iteration hierarchy.

The above simplified procedure has been applied to the re-engineering of UML analysis models, i.e., UML class diagrams with a rich set of descriptors. Taking into account all those descriptors that translate relevant aspects of both classes and associations in a UML class diagram, e.g., association multiplicity factors, property visibility, etc., requires a full-scale translation of the software object landscape elements into FCA. As a suitable representation, the notion of *Relational Context Family* has been proposed which may be thought of as a cluster of formal contexts whose formal objects are linked by a set of binary relations. A detailed presentation of the relational framework in FCA may be found in [18]. The application of the framework to the analysis of UML class diagrams and the underlying Iterative Cross-Generalization (ICG) method is described in [6].

6   Tools

The main algorithmic challenge of the FCA-based hierarchy design is the construction of the Galois subhierarchy. A number of methods for this task have been designed, starting with the work of Godin and Mili [13], followed by the publication of the algorithms ARES [7], AISGOOD [10], and *Ceres* [23]. It is noteworthy that most of the published methods are incremental procedures, i.e., they construct the hierarchy by acquiring the input data, e.g., the classes, one-by-one and integrating each newly inserted class into the current structure. A summary of the methods for class restructuring that do not rely on a FCA results may be found in [17].

Most of the methods for Galois subhierarchy manipulation have been designed to work on software-related datasets. The authors have provided generic implementations, however there is no code repository of all the original implementations. Instead, recently, the implementation of a generic platform for FCA and further lattice manipulations, called GALICIA[3], has been launched.

GALICIA is intended as an integrated software platform including components for the key operations on lattices and related partially ordered structures such as the Galois subhierarchy that might be required in practical applications or in more theoretically-oriented studies. It was designed to cover the whole range of basic tasks that make up the complete life-cycle of a lattice/subhierarchy: data input, construction and visualization. The platform is implemented in Java. On the algorithmic side, GALICIA includes conform implementations of the major Galois subhierarchy methods that are often accompanied by a set of experimental versions. Moreover, an entire component of the platform is dedicated to the ICG framework that produces several subhierarchies on a set of mutually related formal concepts representing a UML class diagram.

7   Recent Applications

One of the recent and promising application of the FCA-based methods for class design has been carried out within the MACAO[4] project. MACAO is a joint project of France

---

[3] See the website at: http://www.iro.umontreal.ca/~galicia

[4] http://www.lirmm.fr/~macao

Télécom, LIRMM, and SOFTEAM[5], a French software company specialized in CASE tool development. It is aimed at enhancing the Objecteering[6] CASE tool, an "all-in-one" environment that combines the Eclipse[7] development environment with model support (via full UML compliance) and code generation.

As part of the project goals, the ICG procedure within GALICIA has been connected to Objecteering. ICG thus provides to Objecteering users, i.e., software developers, the possibility to analyze the UML class models they have created within the CASE tool and to receive valuable suggestions as to possible improvements in these models. Operationally speaking, the UML model from the main tool is exported as a RCF and loaded into GALICIA. The result of ICG running on the RCF, once translated back into UML is fed into Objecteering. The initial and the re-engineered diagram can then be compared and the differences are evaluated.

The Objecteering - ICG tandem has been experimentally applied to a set of existing models of France Télécom, including medium-sized (e.g., a common user data model for several telecom services) and large-sized ones (e.g., a design model of an information system). The user feed-back about the relevance of the suggested new classes and associations was positive, as the ICG tool has discovered many abstractions that would be difficult to extract manually.

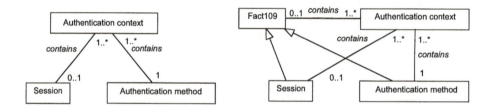

Fig. 22. Example of the creation of a new association (adapted from [6]).

Figure 22 depicts a part of the class diagram in one of the projects that was included in the study. The left hand side shows the initial diagram, whereas the right hand side represents the ICG suggestion. The remarkable element is the abstraction of a new association out of the initial associations named contains. The creation of the new association has further led to the discovery of a new class, Fact109. For a more detailed description of the experimental results for ICG, the reader is referred to [6].

8   Conclusion

We have presented a set of normal forms for the class hierarchy design problem in object oriented development. Each normal form addresses the factoring of different aspects of class properties based on the FCA framework. Although the factoring of attributes, methods and associations was presented separately, they could evidently be

---

[5] http://www.softeam.fr/

[6] http://www.objecteering.com/

[7] http://www.eclipse.org/

combined. The ultimate normal form, called fully factored lattice/subhierarchy form, consists in factoring out every aspect of the class specifications: attributes, method signatures/bodies with redefinition and associations.

These normal forms can be used as guides for the design of class hierarchies within several development scenarios. They could be incorporated in integrated development environment tools by automating the generation of the normal forms. A large body of algorithmic procedures are available to produce the underlying concept lattices efficiently. In practice, as is the case of normal forms for database design, the class hierarchy normal forms should be considered as ideal structures from which some deviations might be considered based on considerations that are not taken into account in the normalization process. Within tool support, we consider that the design process should not be seen as completely automated and some form of user interaction should be provided to produce the final hierarchy possibly by taking a normal form as a starting point or by contrasting some existing design with a normal form, thus revealing potential design anomalies.

Acknowledgment

The most recent research described in the paper was made possible by the respective Canadian NSERC (National Science and Engineering Research Council) grants hold by the authors as well as the team grant from FQRNT (Fonds de Recherche sur la Nature et les Technologies) of Quebec. The authors would like to thank as well their colleagues Rokia Missaoui and Marianne Huchard for the fruitful discussions and critical remarks. Thanks go to all the members of the software engineering team behind the Galicia platform.

References

1. G. Booch. *Object Oriented Analysis and Design with Applications, Second Edition*. Addison-Wesley, 1994.
2. E. Casais. *Managing Evolution in Object Oriented Environments : An Algorithmic Approach*. PhD thesis, Université de Genève, 1991.
3. D. Coleman, P. Arnold, S. Bodoff, C. Dollin, H. Gilchrist, and P. Jeremaes. *O-O Development — The FUSION Method*. Prentice Hall, 1993 1993.
4. W.R. Cook. Interfaces and Specifications for the Smalltalk-80 Collection Classes. In A. Paepcke, editor, *Proceedings of the Xth OOPSLA*, pages 1–15. ACM Press, 1992.
5. B.J. Cox. Planning the Software Revolution. *IEEE Software*, 7(6):25–33, November 1990.
6. M. Dao, M. Huchard, M. Rouane Hacene, C. Roume, and P. Valtchev. Improving Generalization Level in UML Models: Iterative Cross Generalization in Practice. In H. Delugach K. E. Wolff, H. Pfeiffer, editor, *Proceedings of the 12th Intl. Conference on Conceptual Structures (ICCS'04)*, volume 3127 of *Lecture Notes in Computer Science*, pages 346–360. Springer Verlag, 2004.
7. H. Dicky, C. Dony, M. Huchard, and T. Libourel. On Automatic Class Insertion with Overloading. In *Special issue of Sigplan Notice – Proceedings of ACM OOPSLA'96*, pages 251–267, 1996.
8. J. Dvorak. Conceptual Entropy and Its Effect on Class Hierarchies. *IEEE Computer*, 27(6):59–63, 1994.

9. M. Fowler. *Refactoring: Improving the Design of Existing Code*. Addison-Wesley, Reading, MA, 2002.

10. R. Godin and T.T. Chau. Comparaison d'algorithmes de construction de hiérarchies de classes. *L'Objet*, 5(3):321–338, 2000.

11. R. Godin, M. Huchard, C. Roume, and P. Valtchev. Inheritance And Automation: Where Are We Now? In *Object-Oriented Technology ECOOP Workshop Reader*, 2002.

12. R. Godin and H. Mili. Building and maintaining analysis-level class hierarchies using Galois lattices. In *Proceedings of OOPSLA'93, Washington (DC), USA*, pages 394–410, 1993.

13. R. Godin and H. Mili. Building and maintaining analysis-level class hierarchies using Galois lattices. In *Proceedings of OOPSLA'93, Washington (DC), USA*, special issue of ACM SIGPLAN Notices, 28(10), pages 394–410, 1993.

14. R. Godin, H. Mili, A. Arfi, G. W. Mineau, and R. Missaoui. A Tool for Building and Evaluating Class Hierarchies Based on a Concept Formation Approach. In *Proceedings of the OOPSLA 94 Workshop on Artificial Intelligence for Object-Oriented Software Engineering*, Portland, Oregon, 1994.

15. R. Godin, H. Mili, G. Mineau, R. Missaoui, A. Arfi, and T.T. Chau. Design of Class Hierarchies Based on Concept (Galois) Lattices. *Theory and Practice of Object Systems*, 4(2), 1998.

16. M. Huchard. Classification de classes contre classification d'instances. Evolution incrémentale dans les systèmes à objets basés sur des treillis de Galois. In *Actes de LMO'99: Langages et Modèles à Objets*, pages 179–196. Hermés, 1999.

17. M. Huchard, H. Dicky, and H. Leblanc. Galois lattice as a framework to specify algorithms building class hierarchies. *Theoretical Informatics and Applications*, 34:521–548, January 2000.

18. M. Huchard, M. Rouane Hacene, C. Roume, and P. Valtchev. Relational concept discovery in structured datasets. *Discrete Applied Mathematics*, submitted, 2004.

19. M. Huchard, C. Roume, and P. Valtchev. When concepts point at other concepts: the case of UML diagram reconstruction. In *Proceedings of the 2nd Workshop on Advances in Formal Concept Analysis for Knowledge Discovery in Databases (FCAKDD)*, pages 32–43, 2002.

20. R. Johnson and B. Foote. Designing reusable classes. *Journal of Object-Oriented Programming*, pages 22–35, June/July 1988.

21. T. Korson and J. D. McGregor. Technical Criteria for the Specification and Evaluation of Object-Oriented Libraries. *Software Engineering Journal*, 1992.

22. W.R. Lalonde. Designing families of data types using examplars. *ACM Transactions on Programming Languages and Systems*, 11(2):212–248, 1989.

23. H. Leblanc. *Sous-hiérarchies de Galois : un modèle pour la construction et l' evolution des hiérarchies d'objets (Galois sub-hierarchies: a model for construction and evolution of object hierarchies)*. PhD thesis, Université Montpellier 2, 2000.

24. K.J. Lieberherr, P. Bergstein, and I. Silva-Lepe. From Objects to Classes: Algorithms for Optimal Object-Oriented Design. *Journal of Software Engineering*, 6(4):205–228, 1991.

25. B. Liskov. Data abstraction and hierarchy. *ACM SIGPLAN Notices*, 23(5):17–34, May 1988.

26. B. Meyer. *Conception et programmation par objets pour du logiciel de qualité*. Intereditions, Paris, 1990.

27. I. Moore. Automatic Inheritance Hierarchy Restructuring and Method Refactoring. In *Proceedings of OOPSLA'96, San Jose (CA), USA*, pages 235–250, 1996.

28. P. Rapicault and A. Napoli. Evolution d'une hiérarchie de classes par interclassement. *L'Objet*, 7(1-2), 2001.

29. J. Rumbaugh, M. Blaha, W. Premerlani, F. Eddy, and W. Lorensen. *Object Oriented Modeling and Design*. Prentice Hall, 1991.

30. H. Sahraoui, R. Godin, and T. Miceli. Can Metrics Help to Bridge the Gap Between the Improvement of OO Design Quality and its Automations? In *Proceedings of the International Conference on Software Maintenance*, pages 154–162, 2000.
31. H.A. Sahraoui, H. Lounis, W. Melo, and H. Mili. A Concept Formation Based Approach to Object Identification in Procedural Code. *Automated Software Engineering*, 6:387–410, 1999.
32. G. Snelting and F. Tip. Understanding class hierarchies using concept analysis. *ACM Transactions on Programming Languages and Systems*, 22(3):540–582, May 2000.
33. G. Snelting and F. Tip. Semantics-based composition of class hierarchies. In *Proceedings of the 16th European Conference on Object-Oriented Programming (ECOOP 2002)*, Malaga, Spain, June 2002.
34. P. Tonella. Concept analysis for module restructuring. *IEEE Transactions on Software Engineering*, 27(4):351–363, 2001.
35. P. Valtchev, M. Hacene Rouane, M. Huchard, and C. Roume. Extracting Formal Concepts out of Relational Data. In E. SanJuan, A. Berry, A. Sigayret, and A. Napoli, editors, *Proceedings of the 4th Intl. Conference Journées de l'Informatique Messine (JIM'03): Knowledge Discovery and Discrete Mathematics, Metz (FR), 3-6 September*, pages 37–49. INRIA, 2003.
36. A. Yahia, L. Lakhal, R. Cicchetti, and J.P. Bordat. iO2 – An Algorithmic Method for Building Inheritance Graphs in Object Database Design. In *Proceedings of the 15th International Conference on Conceptual Modeling ER'96*, volume 1157, pages 422–437, 1996.

The ToscanaJ Suite for Implementing Conceptual Information Systems

Peter Becker[1] and Joachim Hereth Correia[2]

[1] School of Information Technology and Electrical Engineering (ITEE)
The University of Queensland
QLD 4072, Australia
peter@peterbecker.de
[2] Institut für Algebra
Dresden University of Technology
D-01062 Dresden
heco@math.tu-dresden.de

Abstract. For over a decade, work on Formal Concept Analysis has been accompanied by the development of the TOSCANA software. TOSCANA was implemented to realize the idea of *Conceptual Information Systems* which allow the analysis of data using concept-oriented methods. Over the years, many ideas from Formal Concept Analysis have been tested in TOSCANA systems while the real-world problems encountered led to new theoretical research. After ten years of development, the TOSCANAJ project was initiated to solve some outstanding problems of the older TOSCANA versions. The TOSCANAJ suite provides programs for creating and using Conceptual Information Systems. The experience with older TOSCANA implementations has been applied to the design of the programs. A workflow that developed through many TOSCANA projects has now been integrated into the tools to make them easier to use. Implemented as an Open-Source project and embedded into the larger Tockit project, TOSCANAJ is also a starting point for creating a common base for software development for Formal Concept Analysis. In this paper, we present the features of the TOSCANAJ suite and how they can be used to implement Conceptual Information Systems.

1 Introduction

Formal Concept Analysis is able to reveal and visualize conceptual structures inherent in data while neither adding nor removing information from the underlying data. However, line diagrams displaying concept lattices tend to activate more background knowledge than other representations, even though the information carried is the same as in the original data. This effect is attracting researchers and practitioners from many different domains to apply its methods to analyze their data.

Some of the examples where Formal Concept Analysis has been used for data analysis or information retrieval are:

B. Ganter et al. (Eds.): Formal Concept Analysis, LNAI 3626, pp. 324–348, 2005.

- analysis of medical data about children with diabetes (see [23])
- exploration of laws and regulations in civil engineering (see [19]))
- book retrieval in a library (see [18])
- assisting engineers to design pipings (see [33])
- analysis of verb paradigms in linguistics (see [9])
- code refactoring and class hierarchy design in software engineering (see [4, 20])
- investigation of international cooperation in political sciences (see [17])
- On-Line Analytical Processing (OLAP) and Knowledge Discovery in Databases (KDD) in the area of Data Warehousing (see [13, 14, 28])

Since its beginnings in the early 1980s, the research group on Formal Concept Analysis at Darmstadt University of Technology has collected experience with more than 200 research projects where these methods have been applied.

In many of those projects the many-valued data had to be mapped to formal contexts. In theory, methods to do this had been developed very early, but it was a major effort to make the necessary calculations by hand. When Rudolf Wille coworked with Beate Kohler-Koch on one of the first major projects applying Formal Concept Analysis to real-world data in the late 1980s (see [16]), the need for software support became apparent. Consequently, Rudolf Wille developed the idea of a TOSCANA system, which transferred the tedious tasks to the computer. This allowed the creation of interactive systems applying the methods of Formal Concept Analysis to be used by users without mathematical background.

However, after ten years of development it became hard to maintain and extend the source code of older TOSCANA programs and to adapt them to new requirements and newer versions of the Windows operating system. Out of the resulting limitations the idea was born to create a new TOSCANA. The TOSCANAJ project was thus started with the objective of creating a more extensible program addressing the issues of the older versions.

The TOSCANAJ project is not only concerned with reimplementing the older TOSCANA programs. It rather aims to provide a complete suite of tools for creating and using Conceptual Information Systems in many different ways, learning from the experiences with the various projects in the past. Additionally, it is also considered to be the starting point of a community effort to develop new software for Formal Concept Analysis. As an Open Source project it provides interested users and developers with source code for using, extending, and developing applications of Formal Concept Analysis.

In this paper, we will present the current state of the TOSCANAJ project. Section 2 describes the basic notions of Conceptual Information Systems in general and the development and architecture of TOSCANA and TOSCANAJ information systems, which are a special case. In Section 3, the history of TOSCANA and TOSCANAJ is detailed, together with a reflection on some of the fundamental decisions made at the very beginning of the project, including licensing, programming language, and project scope. After this, we present in Section 4 the components of the TOSCANAJ suite, highlighting some of the new capabilities.

Section 5 describes briefly the steps involved in the creation of a Conceptual Information System with TOSCANAJ. Much of the effort in the TOSCANAJ project has been spent on making the creation of these systems easier and less error-prone by using a workflow based on the experience with the older TOSCA-NA programs. After a short comparision with related projects in Section 6, the paper is concluded by an outlook on the next goals of the project in Section 7.

2   Conceptual Information Systems and Toscana

We understand Conceptual Information Systems to be systems that store, process, and present information using concept-oriented representations supporting tasks like data analysis, information retrieval, or theory building in a human-centered way. The scenario for Conceptual Information Systems typically starts with a data set which a user wants to explore using concept-oriented methods. In Fig. 1 the principal components of a Conceptual Information System and related roles are shown (see also [35, 37]).

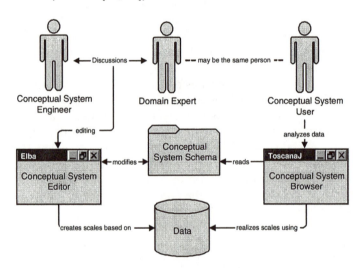

Fig. 1. Components and Workflow of a Conceptual Information System

The system is created by a *conceptual system engineer* in cooperation with a *domain expert*. They combine the engineer's knowledge about tools and theory with the expert's knowledge about the domain. Together they define the conceptual structures used to access the information in the system. These structures may be understood as parts of the experts knowledge being made explicit and therefore available to all users of the system. Using the *conceptual system editor*, the engineer stores the information about the conceptual structures and other information into a central repository, called a *conceptual system schema*. The schema is read by the *conceptual system browser* which allows the *user* to interact with the information using the conceptual structures.

When implementing a Conceptual Information System using methods of Formal Concept Analysis, the data is modeled mathematically by a *many-valued context* and is transformed via *conceptual scaling* [7]. This means that a formal

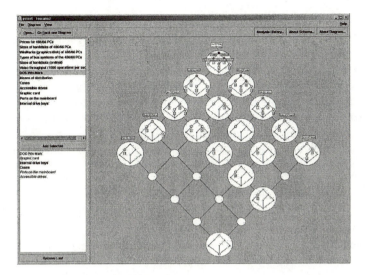

Fig. 2. TOSCANAJ displaying a nested diagram

context called *conceptual scale* is defined for each of the many-valued attributes which has the values of the attribute as objects. Structurally, there is no distinction between conceptual scales and formal contexts, but the notion of a conceptual scale does imply that the context is used to interpret an aspect of the many-valued context.

If a many-valued context and a conceptual scale are given, we can derive the *realized scale* – a formal context which has the objects of the many-valued context as objects and the attributes of the scale as attributes. In the realized scale, an object has an attribute if the value assigned to the object in the many-valued context has the attribute in the conceptual scale. Using the semi-product on contexts, we can combine simple scales into more complex ones. An important result for TOSCANA systems is that the concept lattice of the realization of the semi-product of conceptual scales can be embedded preserving joins into the direct product of the concept lattices of the conceptual scales. Using *nested diagrams* [39, 40] for displaying the direct product of the concept lattices of simple conceptual scales, we have a way to avoid the otherwise difficult problem of automatic drawing of concept lattices. If line diagrams for each simple conceptual scale are available, line diagrams for all created complex conceptual scales can be derived by nesting. Fig. 2 shows a nested diagram displayed by TOSCANAJ. This idea has been first elaborated in [23, 35, 37] and led to the development of Conceptual Information Systems with TOSCANA: the systems allow users to compose views using diagrams of the conceptual scales.

2.1 The Toscana Information System Life-Cycle

As the conceptual scales and the line diagrams have to be defined beforehand, the life-cycle of a TOSCANA information system can be separated into two phases, namely the *creation* and the *usage phase*. In the creation phase, the conceptual system engineer creates the set of scales and diagrams in cooperation with the domain expert. This information is stored together along with information for accessing the database and optionally with additional visualization information in the conceptual system schema. Secondly, in the usage phase, users are able to analyze and explore the data by re-using those diagrams and choosing views from the manifold of possible composite conceptual scales (see [24]).

TOSCANA information systems aim to facilitate the second phase. To allow the user to concentrate on the data, the user-interface is restricted to three simple operations, which nevertheless allow to explore the data in many different ways. The user can:

1. *select a scale* to see the objects mapped to it,
2. *nest or unnest scales*, creating new views combining the aspects of the respective simple scales,
3. concentrate on a subset of objects by *filtering* (sometimes called *"zooming"*) into the extent of a chosen concept.

Every alteration of the system's schema is done in the first phase, in the second phase the user is restricted to exploring the data. For this reason, the first phase is extremely important for the success of the final TOSCANA information system. The development of the conceptual scales together with the domain expert is a highly iterative process. First of all, it has to be clarified what data is available. Then a first set of conceptual scales is discussed and implemented by the engineer. Even a first, very small system, helps the domain expert to get a feeling for the way TOSCANA works and therefore helps to decide what kind of scales should go into the system. Another reason for the necessity of the iterative process is that those first prototypical systems may already reveal inconsistencies or gaps in the data which have to be solved. The accompanying data cleaning is a welcome side-effect of creating TOSCANA information systems.

When creating Conceptual Information Systems with TOSCANA 2, the TOS-CANA version most often deployed in the past, this step was the most difficult one. The program used to create the conceptual system schema was ANACONDA, basically a front-end to the C++ library for Formal Concept Analysis [34], both developed by Frank Vogt. ANACONDA is a very flexible program, allowing the definition and manipulation of contexts and line diagrams with various graphical enhancements. Additionally, the user can derive a line diagram of the concept lattice from a given context and – vice-versa – derive the context matching a line diagram. ANACONDA also supports the more complex aspects of the language CONSCRIPT, which was used for encoding the schemas of these systems. This includes *abstract, concrete,* and *realized scales* (cf. [19]) and extra information needed to activate some special features of TOSCANA 2. Being a powerful multi-purpose tool, ANACONDA was not guiding the user to follow a certain workflow and therefore caused the process of creating the conceptual system schema to be error-prone.

TOSCANA 3 liberated the engineer from knowing the distinctions of abstract, concrete, and realized scales. Only when looking directly at the file storing the schema, these concepts had to be understood. In most cases this was not needed, as TOSCANA 3 asserted the correct creation of all derived scales. Addressing more issues of the creation phase by implementing a particular workflow represented an important aspect of the development for the TOSCANAJ suite.

3   Origin and Design of ToscanaJ

TOSCANA's history starts in the early 90's with a first prototypical version. The first official version, TOSCANA 2, was implemented by Martin Skorsky based on the Formal Concept Analysis Library written by Frank Vogt [34]. Together with the Conceptual System Editor ANACONDA, developed by Frank Vogt, it was the first widely distributed set of components for Conceptual Information Systems. They were first presented at the computer fairs CeBit'93 and '94 and the Graph Drawing conference 1994 in Princeton (see [36]). In 1994 the consulting and software development company NaviCon GmbH[1] was founded by former members of the Formal Concept Analysis research group at Darmstadt University of Technology. Since then TOSCANA has also been applied in commercial projects and is now part of the NaviCon Decision Suite.

In 2000, Bernd Groh developed the last version of TOSCANA based on the original source code: TOSCANA 3. By abandoning some of the legacy code and functions, he was able to solve some of the major problems. Additionally, he introduced some features that accelerated the development of Conceptual Information Systems.

Around that time it was decided to start a new project instead of trying to completely refactor the existing code. An Open Source project was started to provide a tool that is well adapted to the workflow of TOSCANA systems on the user interface level, and at the same time to provide source code that is easily extensible for interested programmers.

In mid 2001, the KVO workgroup[2] started working on this project. Two KVO members had previously worked on TOSCANA: Peter Becker on the commercial version at NaviCon, while Bernd Groh had turned the academic version into version 3.0. Both were well aware of the problems with this program, many of which were due to the fact that TOSCANA was still based on the Borland library OWL, which was not supported anymore and caused problems on more modern Windows platforms.

One of the ideas created in the early stages of this project is to create a large, flexible framework for conceptual knowledge processing. This idea has by now been established as a separate project called Tockit[3]. In the long run, TOSCANAJ is expected to be just one of many Tockit applications. What we try to achieve is to move as much code as possible into a Tockit library once certain parts are stabilized.

The user interface and workflow design of TOSCANAJ itself is kept close to the original TOSCANA programs. This is done for a number of different reasons. First

---

[1] http://www.navicon.de
[2] http://www.kvocentral.org
[3] http://www.tockit.org

of all it is supposed to replace TOSCANA in installed systems, therefore a similar workflow has to be supported and a similar user interface eases the transition process. Secondly, reimplementing a well-known interface helped focusing on the technical aspects of the task, while a new user interface would most likely have caused constant changes to the underlying codebase due to requirement changes.

On the other hand, the TOSCANAJ project had aims to change other aspects of the program. Right from the start of the project, one of the major aims was to create a user interface for TOSCANAJ that was simpler than the existing one. In many ways small enhancements to the user interface have been made to remove clutter and to make important features more easily accessible. This is of course not as efficient as redesigning the workflow, but still the interface of TOSCANAJ is simpler without posing restrictions on the user.

The TOSCANAJ project also aims at interoperability, reuse and flexibility. This was the reason to choose Java as the programming language. As it was implemented with the current Java development kit, it is possible to run TOSCANAJ not only on Windows, but also on UNIX/Linux and Mac OS operating systems for which the Java runtime environment 1.4 is available. Instead of implementing everything ourselves or becoming dependent upon proprietary libraries, we integrated many Open Source libraries and tools. Moreover, several plugin interfaces are offered to be extensible without the need to change the code of TOSCANAJ itself.

In many ways the TOSCANAJ project is considered a bootstrapping project. The different aims for the project are:

- to create reusable code for Tockit;
- to revive software development for FCA in general;
- to create a testing ground for defining a suitable file format for Conceptual Schemas;
- to develop the technical skills of the people involved;
- to create a user base to get more feedback from real-world applications.

For the latter, having a proper deployment process, a website and documentation is about as important as is the quality of the software itself.

Another area where ToscanaJ is treading new ground is in the development process. It is developed as Open Source project on Sourceforge[4] and KVO involvement is not restricted to staff members, but also PhD and Honours students contribute to the project with their research work. Members of the Formal Concept Analysis research group from Darmstadt University of Technology contribute with their expertise and mathematical background and use the latest versions in their research projects, which creates valuable feedback in form of quality assurance and requirements analysis.

Being Open Source means that the code is accessible for everyone, and furthermore everyone is invited to join the development effort in multiple ways. The mailing lists, the bug tracker, the feature request tracker and all other management tools used are accessible online and people are welcome to contribute in form of code, comments, documentation, translations and other means.

---

[4] http://sourceforge.net/projects/toscanaj

What we are trying to achieve with the TOSCANAJ project is the creation of a professional tool, coming out of a research environment and still supporting research. This bridge between academia and industrial application is a challenge, and we consider it to be a research experiment of its own.

4   The ToscanaJ Program Suite

This section describes the features of TOSCANAJ and relates them to similar features in TOSCANA if possible. It is written for an audience without prior knowledge about TOSCANA, but tries to emphasize the development that has been made and how this can be used in new TOSCANA systems using TOSCA-NAJ.

The TOSCANAJ workflow for implementing a Conceptual Information System is quite similar to the workflow with older TOSCANA versions, even though many details have been changed. The description of the workflow given here is informal. For a formal introduction to the TOSCANA workflow, we refer to [15], which gives a mathematical description of TOSCANA systems.

4.1   The Core Workflow

As described in Section 2, the conceptual system schema of a TOSCANA system defines a set of diagrams, and the Conceptual System User can choose diagrams from this set to be displayed on screen. In TOSCANAJ, these diagrams can be selected in a list presented at the left hand of the screen (see Fig. 2). This part can be hidden by moving or clicking the separating bar, but normally the list remains visible. With modern screen resolutions this is a reasonable thing to do and makes the diagrams more directly accessible compared to the dialog used in the older TOSCANA versions.

Selected diagrams get added to a list of diagrams to be viewed. The old TOSCANA workflow of using preselected diagrams is kept, although there is one major difference: diagrams can be selected multiple times, so they can be revisited after the set of available objects has been changed.

The set of objects is changed by double-clicking on a node in a displayed diagram. This will filter the objects to the ones in the extent or the object contingent[5] of the concept denoted by this node. The default setting is to filter into the extent, although the user can change this setting in a menu. This can be done even in retrospective – whenever the user changes the filter option, the diagram contents will be recalculated.

The diagram layout is kept fixed to avoid complexity and broken layouts. The user can move the labels though, and in a typical TOSCANA information system he will be able to select from a number of different contents of the object labels. The default options allow displaying for the extents or object contingents either:

---

[5] The object contingent of a concept is the set of all objects in the extent of this concept but of none of its subconcepts.

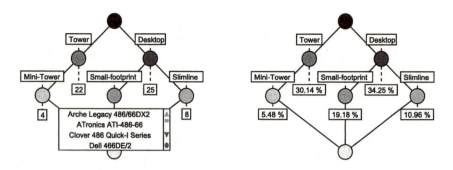

Fig. 3. Different label types in TOSCANA and TOSCANAJ

- **Count:** the number of objects in the set;
- **List:** a textual representation (usually an identifier) of the items in the set;
- **Distribution of Objects:** the percentual distribution with respect to the current object set in the diagram.

Fig. 3 shows the same diagram twice: on the left with the object contingent counts displayed (the default setting) and one label changed to display the list of objects, and on the right with the percentual distribution displayed.

If the data used is stored directly in the TOSCANAJ information system itself, these three options are the only ones available. The ability to store the whole context in the conceptual system schema is a reintroduced feature in TOSCANAJ, the recent versions of TOSCANA 2/3 required a connection to a relational database system (early versions of TOSCANA had allowed internal data storage too). If TOSCANAJ runs connected to a relational database, the conceptual system engineer can customize label contents by giving SQL expressions in a specific XML syntax. The additional features for database connected systems will be explained in further detail in Section 4.4.

TOSCANAJ also creates nested diagrams, although only one level of nesting is produced. This is done for simplicity and due to the fact that the authors are not aware of any application of deeper nestings. An additional feature is that an outer node will not be expanded if the whole inner diagram is not realized. This is similar to the behavior of TOSCANA 3 and can be considered as an implementation of a small aspect of *local scaling* as described in [26]. An example of this can be seen in Fig. 2.

With both simple and nested diagrams a highlighting function is available. Whenever the user clicks on a node, its filter and ideal are highlighted with stronger colors, while the rest of the diagram is slightly faded. All labels that belong to the highlighted part are raised, while the others are faded. This way the structure of the diagram can be more easily understood, which is especially useful for beginners. An example of highlighting is shown in Fig. 6 (a).

4.2   Additional Layout Options

Similar to the late commercial TOSCANA 2 versions and CERNATO (another commercial FCA tool by NaviCon), TOSCANAJ displays the extent size of the concepts underlying a diagram as color gradient on the nodes. This allows for easier recognition of the object distribution and quick detection of extreme values as shown in the left diagram of Fig. 4. This can be changed to display the object contingent sizes instead, although these features are turned off by default to give a simpler and more consistent user interface.

Another feature turned off by default is to change the node size with the size of the extent or object contingent of the concept it represents. This is disabled by default to avoid giving too much information at once, but can easily be enabled in any TOSCANAJ installation. The right diagram in Fig. 4 shows an example of mapping the object contingent size to the node radius.

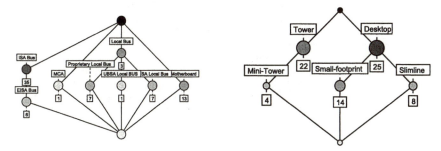

Fig. 4. Displaying information using color and node size

Table 1 shows all different dimensions of layout options in a diagram. In addition to this, the conceptual system engineer can add separate views on the data as described in Section 4.4.

4.3   Printing and Graphic Export

Printing in TOSCANAJ has been simplified. TOSCANAJ always scales the diagram such that it fits the page layout as set up for the printer. In the older TOSCANA versions the user has to set up the diagram size manually which affects the display on the screen, while TOSCANAJ always reproduces the diagram as shown on screen with appropriate scaling. This is a limitation since the user has no control on the size of the printout, but experience has shown that this is by far the most common case and that other printing options are used typically in the context of other documents, which is handled by the different graphic export options.

Since TOSCANAJ exports into a number of different graphic formats, including two vector-based ones, a workflow to create good printouts of any size does exist. These graphic exports are handled using a plugin interface, which makes it easy to add support for new graphic formats.

Table 1. Different options to change the diagram layout in TOSCANAJ

Object Label Queries	Default options are object count (absolute and relative) and the list of items. This can be extended or changed in the conceptual system schema, if connected to a database.
Set in Object Label	Can be either extent or object contingent.
Node Color	Can represent the extent size or the object contingent size. Extent size is displayed by default, but it can be changed to contingent size in the program configuration, alternatively menu options can be turned on. The colors of the gradient can be changed in the configuration file.
Node Size	Can represent the extent size or object contingent size or can be fixed. Fixed by default, it can be changed in the program configuration, or the menu options can be turned on.
Label Styles	Colors and fonts of the object and attribute labels can be defined in the conceptual system schema, independently for each label.
Other Colors	Other colors like the background color, line colors, colors used for the highlighting feature etc. can be set in the program configuration.

At the time of writing, two bitmap formats (PNG and JPEG) and two vector formats (SVG[6] and PDF) are supported. The four formats together cover all applications of graphic export encountered so far, typically one might find the following usages:

- PNG is used for classic websites, based on HTML;
- SVG is used for modern, XML-based websites and for documents in office applications;
- PDF is used for pdfLATEX and professional printing purposes.

The two vector-based formats allow a great deal of flexibility, since they are scalable and in the case of SVG even editable formats. PDF exports are used throughout this article, all figures showing lattice diagrams are graphic exports from TOSCANAJ, without any further editing.

4.4   Additional Features for Database Connected Systems

While TOSCANAJ does support memory-mapped systems, its full potential can only be used in combination with a relational database system. To connect to

---

[6] http://www.w3.org/Graphics/SVG/Overview.htm8

such a database system, JDBC (see [38]) is used, which allows connecting to most common RDBMSs directly. Even more can be accessed using the JDBC-ODBC bridge. To allow for easy deployment of smaller databases, TOSCANAJ also comes with an embedded database engine[7]. This way a database engine is available in each TOSCANAJ installation, which avoids the need for setting up a database engine or being bound to Windows and the Jet Engine (the database engine behind MS Access) with all its limitations. The engine embedded in TOSCANAJ does not need any setup at all, TOSCANAJ will just read an SQL script defining the database with `CREATE TABLE` and `INSERT INTO` statements and execute it on an internal database system.

Compared to older TOSCANA versions, the features for displaying database contents have been greatly enhanced in TOSCANAJ. They can be grouped in two parts: firstly, TOSCANAJ is able to display information based on SQL queries in the object labels. This gives the conceptual system engineer options to adjust the object labels to the needs of a particular TOSCANA system. Secondly, TOSCANAJ offers a plugin API for database viewers, with some implementations provided in the standard distribution. These two feature sets will be discussed in further detail in the next two subsections.

SQL Definitions for Label Content. As discussed above, older TOSCANA versions offer three types of content for the object labels: the number of objects in an extent or contingent, a list of their names (typically the keys in the system) and a percentage showing the relation of the size of the set to the corresponding size for the top concept.

The general idea of displaying some information about the object set determined by filtering/nesting and the view settings is kept in TOSCANAJ and the default options for the content types are the same as in older versions. The difference is that in TOSCANAJ the conceptual system engineer can either add new options or replace the old ones with new versions. This is done using SQL expressions to define different queries.

TOSCANA and TOSCANAJ find the object set when realizing scales against a database by creating an SQL query of the form:

```
SELECT [querypart] FROM [table] WHERE [whereclauses];
```

The `table` is determined by the database setup, while the `whereclauses` are generated from the current context of a concept in the diagram and the settings the user made. The relevant aspects for the `whereclause` generation are:

- which concept in the lattice is displayed;
- if its extent or object contingent is displayed;
- which concepts of other diagrams have been selected for filtering;
- if filtering is done by extents or object contingents;
- if nesting is used (which can add another filter expression).

---

[7] http://hsqldb.sourceforge.net/

The `querypart` can be set independently from this. The three default queries query either the `COUNT` of the object set, the object key values or the `COUNT` of the set divided by the `COUNT` of the top concepts extent. The object key is set up together with the database table in the definition of the database connection in the Conceptual Schema.

```
<aggregateQuery name="Average Price (relative)">
   <queryField format="$ 0.00" separator=" (">
      AVG(price)
   </queryField>
   <queryField format="0.00 %" relative="true" separator=")">
      AVG(price)
   </queryField>
</aggregateQuery>
```

Fig. 5. A system-specific query definition

A definition for alternative label contents is shown in Fig. 5. Here an aggregate querying the average price of the object set is defined. This is done twice, first to be displayed as an absolute number formatted in an appropriate format, then relative to the same value of the top concept, displayed in a standard percentage format. Additional attributes define how the different results are formatted and combined into the label contents as shown in Fig. 6 (b). The different label columns can use the same database column as shown, but they can also refer to different columns. It is also possible to use different SQL aggregates at once, for example to display the price range by querying for the `MIN` and the `MAX` of the price column and format the results as a range.

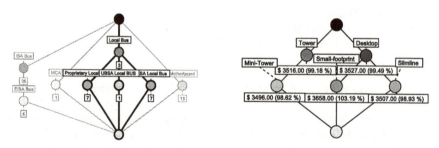

Fig. 6. (a) Highlighting a concept in a diagram. (b) The result of the query in Fig. 5. Hundred percent equals the average price of all objects in the diagram

Similar capabilities are available to display lists. List results can be either treated as lists in the way they get provided by the database system, or the SQL keyword `DISTINCT` may be added to apply set semantics instead. As with the aggregate queries, multiple columns can be used with the same formatting options.

These features allow computing numerical aspects in the context of the Conceptual Information System as requested by Stumme and Wolff in [30, 31]. The possibility to apply statistical functions is especially important, as data analysis is usually associated with numerical analysis. Now, both conceptual and statistical methods can be applied together in TOSCANAJ.

Database Viewers. If more data is to be displayed than can be shown within an object label, a database viewer can be used. The notion of a database viewer generalizes the idea of opening forms and reports in MS Access, as it is done in classic TOSCANA systems, or starting external programs like a browser to show results from the database or a template filled with such results as done in TOSCANA 3.

The database viewers in TOSCANAJ are written using another plugin API. Implementing additional viewers is particularly easy since other parts in TOSCANAJ handle most aspects of the database connectivity as well as the configuration for the viewers. The viewers themselves just request the parameter values from the manager interface and then send statements to the database connection provided and interpret those into a view. For viewing a set of objects with a viewer for single objects a framework with paging buttons is provided.

At the time of writing, the TOSCANAJ distribution comes with four implementations of this plugin interface, two more are in planning. The first of the four database viewers provided is a simple, text-based one, where some column names are given with delimiters in a text file. For example, a string "$$name$$" would be replaced with the value of the name attribute for this object, assuming start and end delimiters are set to double dollar symbols.

Far more powerful is the HTML database viewer. Its template is basically XHTML, but extended with two additional elements: <field> and <repeat>. A <field> entry gets replaced with the contents of the column given in its definition, while the <repeat> element gets repeated for each object in an object set. This allows for defining database reports as the one shown in Fig. 7, with the result shown in the left screenshot of Fig. 8. The HTML viewer used in TOSCANAJ allows using most HTML commands plus basic CSS commands, thus offering a range of formatting options.

A third viewer does actually not show anything but calls an external program. Queries can be included in the same fashion as with the simple text viewer. This way one can for instance open a browser to display a URL retrieved from the database. In Fig. 8 in the right hand picture also the buttons of the paging framework are visible which allows to turn a view for a single object into a view for a set of objects.

The last viewer demonstrates the graphical possibilities. It displays bar charts as shown in the right screenshot of Fig. 8 and is configured by supplying the columns to be displayed, their on-screen names and the color ranges in XML.

In planning are the integration of the JFreeChart[8] library for a broad range of charting tools and a viewer for opening documents or URLs without speci-

---

[8] http://www.jfree.org/jfreechart/index.html

```
<h1>DB Report</h1>
<table border="1">
  <tr>
    <th>Name</th>
    <th>Price</th>
  </tr>
  <repeat>
      <tr>
        <td><field content="PCname"/></td>
        <td><field content="price"/></td>
      </tr>
  </repeat>
  <tr>
    <td><i><field content="COUNT(*)"/> PCs</i></td>
    <td><i>Av: <field content="AVG(price)"/></i></td>
  </tr>
</table>
```

Fig. 7. HTML-base reporting: a definition of a simple report

Fig. 8. Two different views on the data in the database

fying a particular program. These so-called "shell executes" start the program configured in the setup of the underlying desktop environment.

5   Creating Conceptual Information Systems with ToscanaJ

In developing TOSCANA information systems we can distinguish two cases of data origin. In one case the underlying data is already present and the system is built on top of an existing database. In the second case the data is not available as digital data source but has to be transformed first. For example, this might be the case if the user wants to evaluate the results of a questionnaire. In the times of TOSCANA 2, a database had to be created and the data to be entered there.

However, for small systems the necessary effort is out of proportion: to install a database management system first, to create a database and data table, and then to enter the data. With TOSCANA 3 the situation improved slightly: it is possible to use TOSCANA 3 without database if the data was entered directly in the context of the scales. However, the data was distributed on several contexts and thus hardly readable. Besides, only scaled data was saved, not the original values.

An important goal in the development of TOSCANAJ was to make the application of the tools easier to handle. The entry barrier for using TOSCANAJ tools should be as low as possible. For this reason the TOSCANAJ suite supports both cases – it can create and display systems with or without an external database. To avoid combining too many features into one tool, there are two editors: ELBA supports the creation of Conceptual Information Systems where a relational database is already present, while SIENA supports input and management of data, but no live connection to external data sources. Some import options are available to allow for porting existing data into a conceptual information system.

5.1 System Creation with Elba

For many legacy TOSCANA information systems, a relational database with the data existed at the time of their setup. As the TOSCANAJ development also aims at supporting those legacy systems, an editor which manages this case was the first to be developed: ELBA. ELBA supports the conceptual system engineer in creating TOSCANAJ systems based on data stored in a relational database. A screenshot of ELBA editing a line diagram can be seen in Fig. 9.

Connecting to a Database. When creating a new system, the user is presented with a dialog to choose the type of database to connect to (see Fig. 10 on the left). The TOSCANAJ tools support connections to the embedded database engine, which is able to read data from standard SQL scripts, to JDBC and ODBC connected databases or directly to MS Access databases as files. The latter option is important to port existing TOSCANA information systems to TOSCANAJ, as MS Access is a common database for many small-scale research projects. After choosing one type, the user enters the necessary information for connecting to the specific database. ELBA will then try to connect to the specified database.

If this is successful, ELBA retrieves information about the available data tables and the names of their columns and presents it in the last step of the dialog (see Fig. 10 on the right), in order to support the specification of a mapping from the data table into a many-valued context. The user decides which table will be used for the analysis, and which column is used to identify the objects. In smaller databases this is often a column with the object names.

Defining Conceptual Scales. After this step is completed, the user can start defining the conceptual scales. When creating a new diagram, a selection of

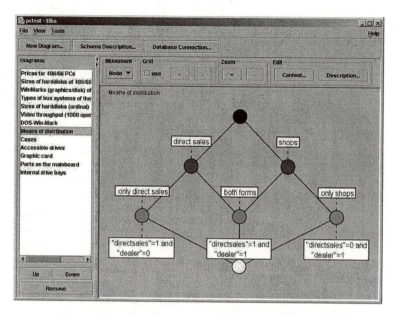

Fig. 9. ELBA's main window while a diagram is edited

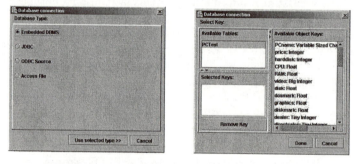

Fig. 10. Dialogs for defining the database connection in ELBA

methods to create scales appears. Selecting one opens the corresponding dialog (two are shown in Fig. 11). The available methods are:

- *Attribute List* is a first step towards the implementation of logical scaling [22]. Attributes are defined by SQL clauses and ELBA creates the corresponding lattice by either supposing that all combinations are possible (resulting in a boolean concept lattice) or by querying what combinations are matched by objects in the database.
- *Context Table* is the most flexible method: the user may enter arbitrary strings for objects and attributes and select the incidence relation as required. In the end, however, the object names must be valid WHERE clauses such that each object in the database is matched by exactly one of the clauses. On demand, ELBA verifies if this holds true.

Fig. 11. Two of the scale creation dialogs (Attribute List and Ordinal Scale)

- *Nominal Scale* enables the user to select single values from the set of all values as attributes of the scale. The user can also combine values via AND or OR.
- *Ordinal Scale* is used for ordered values represented by numbers. The resulting line diagram is a simple chain. The user simply enters the separating values, how they should be ordered and if an object with the exact value belongs to the upper or lower node. As a variation, interordinal scales can be created.
- *Grid Scale* allows the user to build the product of two ordinal scales in one diagram. If both scales refer to the same many-valued attribute, the resulting scale is the standard interordinal scale. If different attributes are chosen, the diagram visualizes the direct product of the two ordinal scales.

In all these scale generators, ELBA tries to support the user by supplying as much information about the database as possible. This includes database schema information such as column types as well as information from the tables themselves, for example the value ranges for nominal and ordinal scales.

After the creation of the scale, a diagram with the same name is created and the name appears on the left panel of the ELBA main window. For simple cases like ordinal, inter- or crossordinal scales, these diagrams are usually layed out well, in more complicated cases the user might be required to manually lay out the diagram. If the user chooses one of the entries in the list, the corresponding diagram is shown in the main area, where it can be changed.

Editing the Line Diagrams and Contexts. There are several options to help the user when interacting with the diagram: a number of movement manipulators can be chosen, for instance the user can drag a node either by itself or together with all nodes above or below (in the order theoretical sense). Another manipulator is called *NDim* and refers to a mapping of the concept lattice into a n-dimensional structure. Describing the exact algorithm would exceed the extent of this paper (see [1] for an early implementation of this idea). This manipulator allows for dragging any node in the diagram with the only exception of the top, while it is ensured that the underlying semantic structure gets kept as far as it can be identified by the program.

Another editing help in combination with the manipulators is an optional grid with changeable gridsize, which causes nodes to snap to the intersections while they are moved. This makes aligning nodes easier. Two more buttons allow for *zooming* in or out of the diagram to change the relative sizes of the nodes and labels compared to the distance between the nodes. If there is an actual error in the scale, the user can fix it by opening the scale in the Context Table Editor, where all object names (SQL clauses) and attribute names can be changed.

The *description editor* allows the Conceptual System Engineer to give more detailed information about a conceptual scale that is accessible to the TOSCA- NAJ user later on. This way a description of the scale can be given, that can help the TOSCANAJ user in understanding its meaning in the context of the whole system. Such a description feature is also available for the system as a whole and for the attributes.

Additional Features. The `Tools` menu contains two additional functions. The *Export Realized Scales* feature allows to dump a snapshot of the scales realized with the current data. This may be used to look for interesting scales in the sense of knowledge discovery in databases (cf. [12, 14]). The second feature, *Export Database as SQL* transforms the database into an SQL script which then may be imported into most existing relational database systems, since these SQL scripts use a common subset of SQL. With the embedded database engine delivered with the TOSCANAJ suite, this tool allows creating Conceptual Information Systems that can be run on any platform supported by TOSCANAJ and that are not bound to a specific operating system or database management system – the SQL script can be deployed together with the Conceptual Schema File. Together with ELBAs ability to import the Conscript file format of TOSCANA 2/3 systems (CSC), an easy way to port old examples into TOSCANAJ systems is thus offered.

5.2   System Creation with Siena

While ELBA is replacing ANACONDA in the classical development of TOSCANA systems as tool for the conceptual system engineer, the goal of SIENA is slightly different: It aims at users who have only basic knowledge of Formal Concept Analysis and want to set up a small TOSCANA system themselves. Therefore, it has to be easier to use. The main technical difference is that no database connection is needed; instead the data is entered using SIENA and stored in the system's schema together with the data about conceptual scales.

SIENA is able to edit a many-valued context as spreadsheet view and it can import several formats: like ELBA it can import `csc` files created by ANACONDA. Additionally, it reads the Burmeister `cxt`-format which is used by many other tools due to its simplicity. And finally, SIENA is able to import the XML output from CERNATO.

Once the data is entered, either by import or by entering it manually, the many-valued attributes can be scaled step by step to create lattices and thus diagrams. The user chooses a many-valued attribute and a subset of values from its value range, which defines a single-valued attribute. From a mathematical point of view, this is equivalent to defining an attribute extent in a scaling context, although this is not made explicit to avoid requiring the user to understand the mathematical notion of scaling. If the data contains single-valued attributes by itself, these are treated as a many-valued attribute with only one value, which seamlessly embeds the single-valued case into the more flexible framework of a many-valued context.

6   Related Work

The development of TOSCANAJ is currently concentrated on establishing the features needed for the implementation of Conceptual Information System. More than the older versions of TOSCANA it is open for extensions to support more advanced methods for conceptual analysis and retrieval of data. Its development profits from related projects working with methods on formal concept analysis. The project SCOLA by Andreas Plüschke (see [21]) has been developed in cooperation with the TOSCANAJ-team. SCOLA represents a prototypic implementation of extensible data type components which are actually basic components for formal concept analysis. The tools CONEXP and QUDA (see [8, 41]) by Serhiy Yevtushenko have a focus on data analysis and support experts analyzing dependencies or interesting facts in data. They include not only conceptual but also numerous statistical methods for the analysis. However, they do not target the workflow of Conceptual Information Systems as presented in this paper. Some of the exploration methods for knowledge acquisition in CONEXP have originally been introduced in the tool CONIMP by Peter Burmeister which currently provides the most advanced method for attribute exploration with incomplete knowledge (see [2]). The tool GALICIA developed at the University of Montréal aims to become an alround tool to support both application-oriented and theory-developing tasks in the domain of formal concept analysis. Early implementations are available at http://www.iro.uomtreal.ca/galicia.

7   Outlook

With the current state of TOSCANAJ we have for the first time a large FCA application available as Open Source. It is implemented as Java application and thus runs on all common operating systems. This means that the features – exceeding the ones from any TOSCANA program before – are available to a broad audience and the source code can be extended or re-used by anyone interested. Many aspects of TOSCANAJ have been designed for ease of extensibility, to allow practitioners to adjust the feature set towards particular needs and researchers to test new ideas within an existing framework.

The idea of having a framework for research and new applications has been implemented as the Tockit project. The stable and re-usable parts of TOSCANAJ will be moved into the Tockit framework, which is meant to be a universal project for Open Source development for Conceptual Knowledge Processing

(CKP). Tockit is not restricted to Java, not even to programming activities – it is supposed to be a platform for communication of people interested in applying and developing CKP systems by providing mailing lists and web-content.

The TOSCANAJ project is an ongoing effort and many additional features are planned. This includes more features for database connectivity, more complete editors, new editing workflows and better interoperability with other FCA tools (e. g. ConExp[9] [41] or Galicia[10] [32]) as well as standard data formats, such as CSV and similar formats or generic XML.

Furthermore we plan to create new tools directly in the TOSCANAJ context. One of the most relevant ideas is a tool which implements SQL expressions as data structures with a partial order modeling the implications. This can then be used to enhance diagram layouting and user interaction. Another planned project is a workbench program allowing the user to manipulate the relevant data structures without a fixed workflow. A first approach to implement scriptable components for Conceptual Knowledge Processing has been made by Andreas Plüschke in his diploma project SCOLA [21].

In the long run, more complicated feature requests are considered for inclusion in the project. For example, more structured views on the set of scales will be offered. If the set of conceptual scales in a TOSCANAJ system is relatively large, the availability of too many scales may confuse the user. For this reason, a structured view on the scales is needed. A simple possibility are tree views as implemented in TOSCANA 3, or maybe a meta-system like a Conceptual Information System on the scales [27], or a graphical map, representing the scales in a geographical fashion, so-called *information maps* [5, 10]. These proposals will be realized in TOSCANAJ over time.

Other users requested more flexibility with regard to the underlying database structure. Typically, only a single data table is analyzed in a TOSCANA system, although several tables are in the database. In [11, 12] the idea to *pivot* the many-valued contexts is presented, where it is possible to change the object set from one database entity to another. In TOSCANA 2/3 this was only possible by using multiple instances of the program. TOSCANAJ should support this idea, so system integrating multiple views can be created more easily.

Conceptual Information Systems aim at supporting a wider range of tasks than only data analysis or document retrieval. More complex tasks like theory building (see [25]) are not fully implementable as computer programs but have to be seen as a communication-centered process where software tools are only a small part. Still TOSCANAJ should provide tools implementing this part as far as possible using mixed-initiative workflows.

Finally, since TOSCANAJ is still a very young tool, many of the existing features still have to be applied in the context of real-world projects. TOSCANAJ offers a large number of flexible features; it still has to be seen how exactly they can and will be used. This will be valuable feedback to enhance the program, in addition to the planned features and the ongoing work on performance and flexibility.

---

[9] http://sourceforge.net/projects/conexp

[10] http://www.iro.umontreal.ca/~valtchev/galicia/

Acknowledgements

The authors wish to thank the Distributed System Technology Centre, Darmstadt University of Technology, the University of Queensland and Griffith University for supplying resources which made the project TOSCANAJ possible. Furthermore we wish to thank all authors of former TOSCANA programs and systems, as well as Peter Eklund, Richard Cole, Bastian Wormuth, Sergey Yevtushenko, Thomas Tilley and Tim Kaiser for their valuable input. Karl Erich Wolff sponsored the development of a software part for Temporal Concept Analysis, which has not been described due to space restrictions.

The development of TOSCANAJ was substantially aided by a research project between the Formal Concept Analysis research group at Darmstadt University of Technology and the KVO Laboratories at the University of Queensland, jointly funded by DFG – Deutsche Forschungsgemeinschaft and ARC – Australian Research Council.

The paper also benefitted from the kind hospitality and support of the Centro de Ciências Matemáticas at the University of Madeira where one of the authors was a guest while working on this article.

References

1. Peter Becker. Multi-dimensional representations of conceptual hierarchies. In Guy Mineau, editor, *Conceptual Structures: Extracting and Representing Semantics*, Supplementary Proceedings ICCS, pages 33–46, Stanford University, California, USA, July 30th to August 3rd 2001. Department of Computer Science, University Laval. Contributions to 9th International Conference on Conceptual Structures.

2. Peter Burmeister. ConImp - ein programm zur formalen begriffsanalyse. In Stumme and Wille [29], pages 25–56.

3. Harry S. Delugach and Gerd Stumme, editors. *Conceptual Structures: Broadening the Base. 9th International Conference on Conceptual Structures, ICCS 2001*, number 2120 in LNAI, Stanford, USA, July 2001. Springer, Berlin – Heidelberg – New York.

4. Stephan Düwel and Wolfgang Hesse. Identifying candidate objects during system analysis. In Keng Siau, editor, *Third CAISE'98/IFIP 8.1 International Workshop on Evaluation of Modeling Methods in System Analysis and Design*, Pisa, June 8–9 1998.

5. Peter Eklund, Bernd Groh, Gerd Stumme, and Rudolf Wille. A contextual-logic extension of toscana. In Ganter and Mineau [6], pages 453–467.

6. Bernhard Ganter and Guy W. Mineau, editors. *Conceptual Structures: Logical, Linguistic and Computational Issues. 8th International Conference on Conceptual Structures, ICCS 2000*, number 1867 in LNAI, Darmstadt, Germany, 2000. Springer, Berlin – Heidelberg – New York.

7. Bernhard Ganter and Rudolf Wille. Conceptual scaling. In Frank Roberts, editor, *Applications of combinatorics and graph theory to the biological and social sciences*, pages 139–167. Springer, Berlin – Heidelberg – New York, 1989.

8. Peter Grigoriev and Serhiy Yevtushenko. QuDA: Applying formal concept analysis in a data mining environment. In Peter Eklund, editor, *Concept Lattices. 2nd Intl. Conference on Formal Concept Analysis, ICFCA 2004 Proceedings*, number 2961 in LNAI, pages 386–393, Sydney, Australia, February 23–26, 2004. Springer, Berlin – Heidelberg – New York.

9. Anja Großkopf. Formal concept analysis of verb paradigms in linguistics. In E. Diday, Y. Lechevallier, and O. Opitz, editors, *Ordinal and symbolic data analysis*, number 8 in Studies in classification, data analysis, and knowledge organization, pages 70–79, Berlin–Heidelberg, 1996. Springer–Verlag.

10. Markus Helmerich. Begriffliche Informationskarten – Orientierungs- und Navigationshilfen für Lernumgebungen mit kontextuell-logischer Grundlage. Diploma thesis, Darmstadt University of Technology, 2002.

11. Joachim Hereth. Formale Begriffsanalyse im Data Warehousing. Diploma thesis, Darmstadt University of Technology, 2000.

12. Joachim Hereth and Gerd Stumme. Reverse pivoting in conceptual information systems. In Delugach and Stumme [3], pages 202–215.

13. Joachim Hereth, Gerd Stumme, Rudolf Wille, and Uta Wille. Conceptual knowledge discovery in data analysis. In Ganter and Mineau [6], pages 421–437.

14. Joachim Hereth Correia, Gerd Stumme, Uta Wille, and Rudolf Wille. Conceptual knowledge processing - a human-centered approach. *Journal on Applied Artificial Intelligence, Special Issue on Concept Lattices for Knowledge Discovery in Databases, Taylor&Francis*, 17(3):281–302, March 2003.

15. Tim Kaiser. Conceptual data systems – providing a mathematical basis for TOSCANA-systems. Diploma thesis, Darmstadt University of Technology, August 2002.

16. Beate Kohler-Koch. Zur Empirie und Theorie internationaler Regime. In Beate Kohler-Koch, editor, *Regime in den internationalen Beziehungen*, pages 17–85. Nomos, Baden-Baden, 1989.

17. Beate Kohler-Koch and Frank Vogt. Normen- und regelgeleitete internationale Kooperationen. In Stumme and Wille [29], pages 325–340.

18. Wolfgang Kollewe, Christine Sander, Rudi Schmiede, and Rudolf Wille. TOSCANA als Instrument der bibliothekarischen Sacherschließung. In H. Havekost and H.-J. Wätjen, editors, *Aufbau und Erschließung begrifflicher Datenbanken*, pages 95–114, Oldenburg, 1995. (BIS)–Verlag.

19. Wolfgang Kollewe, Martin Skorsky, Frank Vogt, and Rudolf Wille. TOSCANA – ein Werkzeug zur begrifflichen Analyse und Erkundung von Daten. In Rudolf Wille and Monika Zickwolff, editors, *Begriffliche Wissensverarbeitung – Grundfragen und Aufgaben*, pages 267–288, Mannheim, 1994. B.I.–Wissenschaftsverlag.

20. Christian Lindig and Gregor Snelting. Assessing modular structure of legacy code based on mathematical concept analysis. In *Proceedings of the 19th international conference on Software engineering*, pages 349–359. ACM Press, 1997.

21. Andreas Plüschke. Design of a component based framework for conceptual knowledge processing. Diploma thesis, Darmstadt University of Technology, October 2002.
http://www.st.informatik.tu-darmstadt.de/public/Thesis.jsp?id=5.

22. Susanne Prediger. Logical scaling in formal concept analysis. In D. Lukose, H. Delugach, M. Keeler, L. Searle, and J. F. Sowa, editors, *Conceptual structures: Fulfilling Peirce's dream*, number 1257 in Lecture Notes in Artificial Intelligence, pages 332–341, Berlin–Heidelberg–New York, 1997. Springer–Verlag.

23. Patrick Scheich, Martin Skorsky, Frank Vogt, Cornelia Wachter, and Rudolf Wille. Conceptual data systems. In O. Opitz, B. Lausen, and R. Klar, editors, *Information and classification*, pages 72–84. Springer, Berlin – Heidelberg – New York, 1993.

24. Martin Skorsky, Gerd Stumme, Rudolf Wille, and Uta Wille. Reuse in the development process of TOSCANA systems. In F. Puppe, D. Fensel, J. Köhler, R. Studer, and Th. Wetter, editors, *Proc. Workshop on Knowledge Management, Organizational Memory and Reuse, 5th German Conf. on Knowledge-Based Systems*, Würzburg, Germany, March 3.–5. 1999.

25. Selma Strahringer, Rudolf Wille, and Uta Wille. Mathematical support for empirical theory building. In Delugach and Stumme [3], pages 169–177.

26. Gerd Stumme. Local scaling in conceptual data systems. In P. W. Eklund, G. Ellis, and G. Mann, editors, *Conceptual structures: Knowledge representation as interlingua*, number 1115 in Lecture Notes in Artificial Intelligence, pages 121–131, Berlin–Heidelberg, 1996. Springer–Verlag.

27. Gerd Stumme. Hierarchies of conceptual scales. In T. B. Gaines, R. Kremer, and M. Musen, editors, *Proc. Workshop on Knowledge Acquisition, Modeling and Management (KAW'99)*, volume 2, pages 78–95, Banff, October 16-22 1999.

28. Gerd Stumme. Conceptual on-line analytical processing. In K. Tanaka, S. Ghandeharizadeh, and Y. Kambayashi, editors, *Information Organization and Databases*, pages 191–203. Kluwer, Boston–Dordrecht–London, 2002.

29. Gerd Stumme and Rudolf Wille, editors. *Begriffliche Wissensverarbeitung – Methoden und Anwendungen*. Springer, Berlin – Heidelberg – New York, 2000.

30. Gerd Stumme and Karl Erich Wolff. Computing in conceptual data systems with relational structures. In *Proceedings of the International Symposium on Knowledge Representation, Use and Storage Efficiency*, pages 206–219, Vancouver, 1997.

31. Gerd Stumme and Karl Erich Wolff. Numerical aspects in the data model of conceptual information systems. In Y. Kambayashi, Dik Kun Lee, Ee-Peng Lim, M. K. Mohania, and Y. Masunaga, editors, *Advances in Database Technologies*, number 1552 in LNCS, pages 117–128. Springer, Berlin – Heidelberg – New York, 1999. Proc. Intl. Workshop on Data Warehousing and Data Mining.

32. P. Valtchev, D. Grosser, C. Roume, and M. Rouane Hacene. Galicia: an open platform for lattices. In Aldo de Moor, Wilfried Lex, and Bernhard Ganter, editors, *Using Conceptual Structures: Contributions to the 11th Conference on Conceptual Structures*, pages 241–254. Verlag Shaker, Aachen, 2003.

33. Niko Vogel. Ein begriffliches Erkundungssystem für Rohrleitungen. Diploma thesis, TH Darmstadt, 1995.

34. Frank Vogt. *Formale Begriffsanalyse mit C++: Datenstrukturen und Algorithmen*. Springer–Verlag, Berlin–Heidelberg–New York, 1996.

35. Frank Vogt, Cornelia Wachter, and Rudolf Wille. Data analysis based on a conceptual file. In Hans-Herrmann Bock and P. Ihm, editors, *Classification, data analysis, and knowledge organization*, pages 131–140, Berlin–Heidelberg, 1991. Springer–Verlag.

36. Frank Vogt and Rudolf Wille. TOSCANA – a graphical tool for analyzing and exploring data. In Roberto Tamassia and Ioannis G. Tollis, editors, *Graph Drawing*, pages 226–233. Springer, Berlin – Heidelberg – New York, 1995.

37. Cornelia Wachter and Rudolf Wille. Formale Begriffsanalyse von Literaturdaten. In DGD, editor, *Deutscher Dokumentartag 1991 - Information und Dokumentation in den 90er Jahren: Neue Herausforderung, neue Technologien*, pages 203–224, Frankfurt, 1992.

38. Seth White, Maydene Fisher, Rick Cattell, Graham Hamilton, and Mark Hapner. *JDBC(TM) API Tutorial and Reference: Universal Data Access for the Java(TM) 2 Platform.* Addison-Wesley, Reading, MA, 2nd edition, June 1999.
39. Rudolf Wille. Liniendiagramme hierarchischer Begriffssysteme. In Hans-Herrmann Bock, editor, *Anwendungen der Klassifikation: Datenanalyse und numerische Klassifikation*, pages 32–51. Indeks–Verlag, Frankfurt, 1984. Line diagrams of hierarchical concept systems (engl. translation). *Int. Classif.* **11** (1984), 77–86.
40. Rudolf Wille. Lattices in data analysis: how to draw them with a computer. In Ivan Rival, editor, *Algorithms and order*, pages 33–58, Dordrecht–Boston, 1989. Kluwer.
41. Serhiy Yevtushenko. System of data analysis "Concept Explorer". In *Proceedings of the 7th national conference on Artificial Intelligence KII-2000*, pages 127–134, Russia, 2000. In Russian.

Author Index